Social Interactive Television:
Immersive Shared Experiences and Perspectives

Pablo Cesar
CWI, The Netherlands

David Geerts
IBBT / K.U.Leuven, Belgium

Konstantinos Chorianopoulos
Ionian University, Greece

Information Science REFERENCE

INFORMATION SCIENCE REFERENCE
Hershey · New York

Director of Editorial Content: Kristin Klinger
Senior Managing Editor: Jamie Snavely
Managing Editor: Jeff Ash
Assistant Managing Editor: Carole Coulson
Typesetter: Lindsay Bergman
Cover Design: Lisa Tosheff
Printed at: Yurchak Printing Inc.

Published in the United States of America by
Information Science Reference (an imprint of IGI Global)
701 E. Chocolate Avenue, Suite 200
Hershey PA 17033
Tel: 717-533-8845
Fax: 717-533-8661
E-mail: cust@igi-global.com
Web site: http://www.igi-global.com/reference

and in the United Kingdom by
Information Science Reference (an imprint of IGI Global)
3 Henrietta Street
Covent Garden
London WC2E 8LU
Tel: 44 20 7240 0856
Fax: 44 20 7379 0609
Web site: http://www.eurospanbookstore.com

Library of Congress Cataloging-in-Publication Data

Social interactive television : immersive shared experiences and perspectives / Pablo Cesar, David Geerts, and Konstantinos Chorianopoulos, editors.

p. cm.

Includes bibliographical references and index.

Summary: "This book discusses the advent of social interactive television with its complex opportunities and challenges for media researchers and others today"--Provided by publisher.

ISBN 978-1-60566-656-3 (hardcover) -- ISBN 978-1-60566-657-0 (ebook) 1. Interactive television. I. Cesar, Pablo. II. Geerts, David, 1973- III. Chorianopoulos, Konstantinos, 1975-
HE8700.95.S63 2009
384.55--dc22

2008052441

British Cataloguing in Publication Data
A Cataloguing in Publication record for this book is available from the British Library.

Editorial Advisory Board

Table of Contents

Detailed Table of Contents

This chapter provides an introduction to and overview of social television, in an attempt to find the real meaning of the term. It explores the history and current state of social television, looks at a number of examples of Social TV systems and their features, compares different definitions of the term, and outlines dimensions of design that have been used to organize the topic. The author argues that historically the notion of social television is intimately bound up with television itself, and that the two remain difficult to separate even today. The convergence of content and communication to create social media is turning Social TV into a reality and in the process turning television into what it was originally intended to be.

Conventional broadcasting has the impressive power to create shared experiences over huge audiences or even entire populations. The sharing of such experiences deepens our sense of connectedness with others, which in turn arguably leads to positive effects in society as a whole. Interactive television and related technologies have the potential to further collapse distance and broaden these positive effects

studies with such systems. Finally, each heuristic is presented and explained in more detail, along with some more specific guidelines. Geerts hopes these heuristics will enable designers and evaluators of social interactive television systems to make sure they support the social uses of television and create sociable systems.

This chapter seeks to evaluate the attitudes and practices of media participation amongst young Portuguese aged between 12-18 years, with a particular focus on content creation and sharing through media and information and communication technologies (ICT). Audience participation in television and internet will be addressed, having as basis the results and findings of three empirical studies integrated in the PhD research project of the author, namely: an ethnographical study about the usage of media and ICT usage by 10 families conducted at their own domestic contexts, a quantitative survey about the usage of media and ICT by young people aged 12-18 with a total of 962 respondents and, finally, an evaluation study of a participatory media format which was tested and evaluated by 77 teenagers from three different schools. The main objective is to better understand the attitudes and behaviors of young people in Portugal towards the practices of creation and sharing digital content through media and ICT, providing empirical data about the range and frequency of experiences of content creation by this specific population, as well as their interest and adherence to participatory media formats.

Applications for interactive TV (iTV) addressing social aspects will only be successful, if the intended user and user community is taken into account during the development process. Existing methods for evaluating usability and user experience aspects of social interactive TV applications are not really enough to address and focus on the broad range of factors related to social user experience. This chapter presents various forms of user-centered methods enhancing usability and user experience aspects of interactive TV applications in general. Giving an overview on currently used user-centered methods to enhance and evaluate usability and user experience, some selected methods are presented for early and late development stages, reflecting the peculiarities stemming from the non-traditional environment the applications are used in—the home. The chapter will show how the methods can be adopted to focus more precisely on social aspects, especially social user experience. It enables the reader to get an overview on currently used user-centered methods in interactive TV and to learn about benefits and shortcomings of these methods as well as how to choose appropriate methods for their own purposes.

Television and the Internet have proven to be a popular combination for both broadcasters and viewers. Because of this popularity they are increasingly facing the consequences of central bottlenecks, which could be overcome by taking a different approach: Peer-to-Peer (P2P) technology. However, P2P systems can only be successful with as much cooperation among as many users as possible. This chapter explains how this cooperation is hard to enforce, and how inducing it might be more successful. Relevant psychological theories are listed that can be used to induce this user cooperation, along with possible applications of cooperation inducing mechanisms for Peer-to-Peer Television (P2P-TV) systems. The authors aim to provide practical criteria along which these mechanisms can be evaluated on their contribution to social activity in P2P-TV systems.

This chapter presents results from an ongoing social television project, in the context of other research in the field. The authors give a detailed description of the STV prototype used in their research, and summarize their studies, which provide the findings explained in the rest of the chapter. Three major research focuses are identified, namely evaluation and validation of Social TV systems, communication modality comparison, and detailed observation of user behaviors. Based on the findings in these areas, the authors list three major open questions and challenges for the field: multi-user support, new equipment requirements, and the creation of distinct and unique social television experiences. Finally, the chapter suggests that the emphasis within social television may be moving from research to design, implementation and deployment.

The combination of content and communication has proven to be a powerful and successful concept. Many online services not only allow for consumption of content, but also give their users the possibility to exchange views about the content among each other. YouTube for instance, not only allows its visitors to watch movie clips, but also to discuss them, to review them, and to send recommendations about them. Electronic Program Guides on the web more and more offer social functionality in addition to listing television programs; thus integrating content and communication. Users can discuss their favorite TV shows in forums that are organized around TV programs. ConnecTV is a social interactive TV service that combines communication with watching television. It makes watching TV a social activity, and aims to give its users the feeling of 'watching together'. ConnecTV was developed in the B@Home research project (B@Home). Among the project's goals was an investigation of the type of services that will become feasible when 'fiber to the home' is widely introduced, giving households broadband internet connections with significantly more capacity than today. Although the functional design of ConnecTV was carried out in close cooperation with experts from the media industry, the real test for end-user acceptance would be a field trial with the service. In addition, a trial would reveal the effects on the users' viewing behavior, and would give insight in how to make a positive business case around ConnecTV. In 2007 ConnecTV was implemented and a field trial was held in about 50 households in the town of Enschede, in The Netherlands. In literature, many systems have been proposed that combine social networks with consuming content in general, or with watching TV specifically. Examples are AmigoTV (Coppens, Trappeniers, & Godon, 2006), 2BeOn (Abreu, Almeida, & Branco, 2001), SocialTV (Harboe, Massey, & Metcalf, 2006), ChaT.V. (Fink, Covell, & Baluja, 2006) and CollaboraTV (Harrison & Amento, 2007). Although some of ConnecTV's functionality can be found in these services as well, other functions (like following a buddy, or switching to the most popular channel) are new. The main contribution of the research described in this chapter though lays in the field trial of ConnecTV. With some of the above-mentioned social TV services field trials have been performed (e.g. (Harboe, Massey, & Metcalf, 2006)), but they were limited in size (typically two or three groups), and did not use extensive logging of user activities. To our knowledge, a field trial of a social TV service at the scale of the ConnecTV field trial with such extensive logging, surveys and experience sampling has not been performed before. This chapter starts with explaining the functionality of ConnecTV and its implementation. The main focus of this chapter is the field trial: the set-up, the research questions, the research methodology and the results of the field trial are discussed extensively. The chapter concludes with a description of the most viable approaches for a positive business case, and gives an outline for future work.

With the advent of digital video recorders and video-on-demand services, the way in which we consume media is undergoing a fundamental change. People today are less likely to watch shows at the same time, let alone the same place. As a result, television viewing which was once a social activity has been reduced to a passive, isolated experience. CollaboraTV was designed to address this new mode of television viewing by directly supporting asynchronous communication. We demonstrated its ability to support this communal viewing experience through a lab study and a month-long field study. Our studies show that users understand and appreciate the utility of asynchronous interaction, are enthusiastic about CollaboraTV's engaging social communication primitives and value implicit show recommendations from friends. Our results both provide a compelling demonstration of a social television system and raise new challenges for social television communication modalities.

Chapter XIII
From 2BeOn Results to New Media Challenges for Social (i)TV 222
Jorge Ferraz Abreu, University of Aveiro, Portugal
Pedro Almeida, University of Aveiro, Portugal

This chapter focuses on traditional and emergent challenges for the Social (i)TV area focusing on explaining the development and evaluation of one of the first Social iTV prototypes and looking at the challenges new media is introducing to this research field. The authors begin by explaining the conceptualization, development and evaluation process of the 2BeOn system and continue with the most important results from it's evaluation with a particular focus on the results that can be important when developing any Social iTV platform. In the last part of the chapter recent developments in the broadcast of TV and Audiovisual content, namely considering the Internet as a medium, are addressed. In this scope authors propose a categorization of emergent online distribution platforms along with a set of social activities users perform on those platforms. Taking in consideration some of the challenges surrounding the presented scenario the chapter ends with the conceptualization of UMCA, a system that could increase social interaction activities performed during the consumption of online AV/TV content.

Section IV
Thinking Out-of-the-Box: Social TV on Large Screens, Mobile Devices and the Web

Chapter XIV
Examining the Roles of Mobility in Social TV 244
Konstantinos Chorianopoulos, Ionian University, Greece

Mobile TVs have been available for many years, without ever becoming very popular. Moreover, the first wave of research has been mostly concerned with technology and standards, which are necessary to ensure interoperability and market acceptance. Although, there has been a significant body of computer supported co-operative work (CSCW) and mobile human-computer interaction (HCI) research findings, there is limited investigation in the context of leisure activities, such as TV. In this chapter, the author proposes three concepts that drive the main paths for research and practice in mobile and social TV: (1) Mobile TV as a content format, (2) Mobile TV as user behavior and (3) Mobile TV as interaction

Foreword

Unconstrained interactivity of all kinds—with people, technology, environments—has reshaped life in the modern world. Concomitantly, the social imagination has expanded to incredible proportions with tremendous cultural and political consequences. Television looms at the center of all this activity.

Television has performed brilliantly as a socially-interactive medium from the time it first appeared in the mid 20th Century. No other communications medium has ever influenced the structural and relational properties and processes of everyday life the way television does. Anyone who has lived through any stage of television's short history intuitively understands the intrinsically social nature of the visual medium just by reflecting for a moment on his or her own viewing experiences.

Even more than newspapers, magazines, film, and radio before it, television's endless stream of images and ideas inevitably prompts discussions about content. From simple conversations to complicated discourses about politics and culture, television gives everyone plenty to talk about. Because television uniquely sends images and ideas directly into the most intimate venues for media consumption—domestic living spaces of all sorts—the medium immediately became more than a technology that conveniently delivers news and entertainment. The television set, programs, and viewing situations (form, content, and context) all became focal points of family life that changed the way people live. Withstanding and sometimes absorbing the challenge posed by other information and communications technologies, including the ubiquitous Internet, television has yet to be replaced as the most useful and dynamic medium of social exchange ever invented.

Although television viewing in the developed world has often become a decentralized, even isolated activity; watching TV will never be primarily a private experience. Even when individuals watch alone, the viewing experience doesn't end when the set is turned off. Its social significance often only begins then. Viewers routinely process program content—drama, news, sports, documentaries, specials, and countless niche genres—with other people inside and outside the home during and following consumption of content. Television sets off incessant waves of social interaction by the very nature of its constant and provocative content (of widely varying quality, of course, but that's another matter) in contexts that range from private chats and commentaries to global discussions and debates.

The vectors of television's social interactivity rotate and expand in ways that reflect prevailing trends of technological and cultural development. New forms of human expression and communication are created in the process. Social interactive television—the subject of this book—represents just such a change. Social interactive TV turns "viewers" and "audience members" into "interactants," "users," "participants," even dialogical "partners." Social communication about television and para-social interaction with television today are being joined, not replaced, by communication through television via interactive content and novel forms of mediated social connectivity. Moreover, television's convergence with other communications media—especially the Internet and mobile phone—are proving to be inherently social phenomena, not just technological adaptations.

Even within the boundaries of their quasi-social relationship with television, audiences have always been emotionally involved with the visual medium. But so far the TV-viewer communication system has operated largely out of balance; programming flows from the TV set to viewers who cannot directly respond. The digitalization of video and audio has further enhanced the emotional impact of television signals on their viewers. Still, the primordial impulse to interact with television—to express oneself back to the message source and to help influence the course of events—has always been there. Simply talking back to the box—even hurling insults and other abusive comments—has long been noted by ethnographic researchers and casual observers.

Now the "box" is listening and rushing to accommodate users' wants and needs.

Given the surge in the new varieties of mediated social interactivity, the very term "television" is becoming out of date. The idea of television as the transmission of video signals to receivers across distance doesn't capture the dynamic nature of what's happening at the frontier of the medium's development. The more comprehensive idea of "telecommunications," which used to seem too abstract, technical, or theoretically inappropriate to describe television transmission and reception now seems to be a more accurate and relevant descriptor. The "natural capabilities of television" that the editors of this book promote have begun to correspond in practice with the principles of any healthy form of human communication—real two-way or multi-way interaction.

Viewer ratings don't come close to revealing the kinds of social experiences TV routinely facilitates. The intricacies and subtleties of the connection between television and its viewers inherently resist measurement, especially quantitative assessments. Ethnographic research and the other forms of qualitative investigation that first emerged in academic circles in the 1970s were proving to be better at penetrating the deeper levels of television's social significance. Those more naturalistic techniques were later appropriated by commercial researchers and advertisers when market differentiation and the need for a greater understanding of how viewers actually experience (and relate to) television promised increased revenues. The advent of social interactive television poses complex opportunities and challenges for academic media researchers and others today. The theoretical and methodological approaches represented in the following pages will certainly help propel the next generation of television research in exciting—indeed necessary—new directions that are fitting for the times.

James Lull
San José State University, California, USA

James Lull *is Professor Emeritus of Communication Studies at San Jose State University, California, specializing in media and cultural studies. Dr. Lull has also taught in the United States at the University of California-Santa Barbara, the University of California-Berkeley, the University of San Francisco, and the University of Pittsburgh's Semester at Sea Program. He received a Doctoral degree at the University of Wisconsin-Madison and a Masters degree (with Honors) at the University of Oregon. James also holds an Honorary Doctorate in Social Sciences from the University of Helsinki, Finland and an Honorary Professorship at Aalborg University, Denmark. Professor Lull has been granted two Fulbright Senior Scholar Awards (Brazil and Mexico) and a Leverhulme Trust Fellowship (England). He has taught courses at universities in the People's Republic of China, Brazil, Venezuela, Finland, Denmark, England, Mexico, Chile, Argentina, and Sweden. Dr. Lull is author or editor of a dozen books in English. His books have been translated into Spanish, Italian, Chinese, Portuguese, Indonesian, Japanese, Turkish, Romanian, and Ukrainian. Fluent in Spanish, James has published many articles in that language, and regularly teaches and conducts research in Mexico, South America, and Spain.*

Preface

SOCIAL INTERACTIVE TELEVISION RESEARCH

Television since its invention has been considered a social link between people. People chat and discuss television programs, recommend shows to be watched, and send SMS messages when there is something of interest on. Like previous technologies, the digitalization of the medium imposed a number of technological priorities such as transmission mechanisms or middleware standardization. Once the technology is stable, the second phase of research focused on the immediate interactivity options, such as content navigation and interactive content. The next frontier of research focuses on the actual natural capabilities of the medium: Social Interactive Television. Social Interactive Television: Immersive Shared Experiences and Perspectives aims to set the basic research lines of this exciting topic that is social interactive television. The final intention of the book is to provide the reader an easy-to-access compendium of the research around this emergent topic.

STRUCTURE OF THE BOOK

It is our belief that the future of interactive television, and of interactive digital media for that matter, is to provide experiences and not only services. Television experience is a shared one, thus this book concentrates on systems, interaction design, and evaluation methods that focus on social experiences around digital media both in the leaving room and beyond.

In order to introduce in detail the topics of the book, the first chapter "Social Television: From Beginning to End" by Gunnar Harboe, surveys social interactive television research with a historic perspective, indicating the challenges ahead. This chapter contextualizes the rest of the contributions of this book and provides the reader with the necessary tools to explore a world yet to be discovered. The book, then, is divided into four sections:

- The Broad Picture. Frameworks and Applications
- Who, What, Why and How? Methodology and Audience Studies
- Making It Work. Social Television Systems
- Thinking Out-of-the-Box. Social TV on large screens, mobile devices and the web

The first section, The Broad Picture, includes contributions on basic human and technological aspects. Stefan Agamonolis gives an overview of the many projects he was involved with during his time at MIT Media Lab Europe and Distance Lab, related to social interactive television. Even though most projects can not be classified as social television as such, each project includes social and or audiovisual aspects and teaches valuable lessons that can benefit the design of social television systems. Cristian Hesselman and colleagues describe a technological framework for social interactive television, promoting an open infrastructure that includes applications from third-party developers. Tom Gross, Thilo Paul-Stueve, and Mirko Fetter look at the related field of Computer Supported Cooperative Work (CSCW) and encourage us to use the concepts from this research area when creating social interactive television. Finally, Dick Bulterman, Pablo Cesar, Rodrigo Guimarães, and Jack Jansen report on the Ambulant Annotator. The Ambulant Annotator is an extension to Personal Video Recorders that allows the user to enrich and share television content with selected members of his peer group. The chapter explores asynchronous communication mechanisms that enhance the connectedness between people.

As any new topic, there is a need for new methodologies and methods for measuring social systems. The next section, Who, What, Why and How?, provides an insight on such new methods. David Geerts presents a list of 12 sociability heuristics that can be used to design and evaluate social interactive television, especially focussed on the social interaction enabled by these systems. Célia Quico describes her detailed studies of younger Portuguese and their families when using cross-media applications on television and the Web, and how that influences audience participation. Regina Bernhaupt, Marianna Obrist, and Manfred Tscheligi explain the different ways that users can be involved in designing and evaluating social interactive television applications, stressing the importance of user involvement for creating usable and engaging designs. They especially focus on the home environment, as this new context has specific characteristics that have to be adequately addressed. Finally, Jenneke Fokker, Huib de Ridder, Piet Westendorp, and Johan Pouwelse detail several psychological theories that explain how users can be motivated to share content with other users.

The following section, Making It Work, provides an overview of a number of relevant and innovative systems. Gunnar Harboe, Elaine Huang, Noel Massey, Crysta Metcalf, Ashley Novak, Guy Romano, and Joe Tullio explain in detail the several iterations of the Social TV system by Motorola Labs, including voice and text chat, a buddy list and ambient awareness devices. They present the results of several user studies in the lab and in the field they have conducted, and that have guided the design of their system. Erik Boertjes, Jente Klok, Omar Niamut and Martijn Staal describe the field study they have performed with the ConnecTV system by TNO, and the lessons they learned from that trial. They also draw attention to the business models that could support the introduction of social interactive television systems on the market, and make it a success. Brian Amento, Christopher Harrison, Mukesh Nathan and Loren Terveen introduce the CollaboraTV prototype, specifically focussing on the asynchronous interaction this system provides, and also present results from an extensive field study with users. Finally, Jorge Abreu and Pedro Almeida describe how they designed one of the first social interactive television systems, 2BeOn, and how they consider AV-content on the Internet as an important platform for stimulating social interactions while watching television.

The final section of the book, Thinking Out-of-the-Box, studies social television systems out of the set-top box, including social WebTV systems, mobile TV, and big public screens. Konstantinos Chorianopoulos introduces us to mobile social television, and discusses which aspects are specific to the mobile context. Kenton O'Hara and Maxine Glancy take us outside, literally, and explain how people interact with television content on large screens placed in public spaces. David Shamma and Yiming Liu return to more familiar ground, the Web, and link instant messaging with television and video content, showing how this can enhance the chat experience. Finally, Justin Weisz looks at how watching television online while chatting can create a social experience, and at the same time influences distraction, entertainment and sociability. He draws several conclusions that can help designing more sociable online television applications and systems.

The book tries to cover all the relevant aspects of this emergent topic that is here to improve the lives of the viewers; from technology to business models and without forgetting that the user is the central piece of any computing system.

Pablo Cesar
CWI, The Netherlands

David Geerts
IBBT / K.U.Leuven, Belgium

Konstantinos Chorianopoulos
Ionian University, Greece

Acknowledgment

This book builds upon the collective effort of many researchers in the area of social interactive television. We are grateful to all those who have participated in previous conferences, workshops, panels, and informal discussions. We are especially grateful to the program committee members and participants of the series of workshops on social interactive television that were organized in conjunction with the EuroITV conference series. Finally, we wish to express our gratitude to the advisory board of this book, which was composed of the following eminent researchers:

- Anxo Cereijo Roibás – Vodafone, UK
- Nicolas Ducheneaut - Palo Alto Research Center, USA
- Ian Kegel - BT, UK
- Newton Lee - ACM, USA
- James Lull - San Jose State University, USA
- Kris Luyten – EDM, Belgium
- Daniel Salber – Joost, The Netherlands

Chapter I
In Search of Social Television

Gunnar Harboe
Motorola, USA

ABSTRACT

This chapter provides an introduction to and overview of social television, in an attempt to find the real meaning of the term. It explores the history and current state of social television, looks at a number of examples of Social TV systems and their features, compares different definitions of the term, and outlines dimensions of design that have been used to organize the topic. The author argues that historically the notion of social television is intimately bound up with television itself, and that the two remain difficult to separate even today. The convergence of content and communication to create social media is turning Social TV into a reality and in the process turning television into what it was originally intended to be.

INTRODUCTION

The term "social television," or "Social TV," is not new, but over the last few years it has acquired a specific technical meaning. It is used to refer to a variety of experimental systems that claim to support social experiences for television viewers, and to the research into such experiences. This book testifies to the brisk activity going on in this area of study.

To understand what all the research is about, we should first understand more precisely what Social TV is. That is the question explored in this chapter. It is investigated from a number of perspectives. First, we look at the history of social experiences for the TV, a history that stretches back much further than usually acknowledged. We will see that it has its roots in the very earliest inspirations for television itself, and has been a persistent motif of science fiction and a recurring ambition of developers ever since. Then, examples of modern Social TV systems of different types are presented, drawing attention to the range of disparate experiences supported and the overlap with other topics. At the same time, a set of typical features is identified; features which include sharing of TV presence information, support for shared experiences, freeform communication, and the publication of TV viewing data.

With social television examples from the idea's inception to the present thus catalogued, we examine the various attempts to offer a definition of the field. These are shown to follow one of two forms, and the limitations of both are discussed. Finally, we look at theoretical models and taxonomies that lay down the important dimensions of the social television space. Using these dimensions, different Social TV experiences can be characterized.

Even after a comprehensive review of the topic from all these different angles, a precise and straightforward statement of what social television is remains difficult to articulate. However, the search shows that such uncertainties are inherent to the field, and by the end of this chapter the reasons for the vagueness should be clear. Furthermore, a consideration of these reasons suggests that the future adoption and assimilation of Social TV is likely.

HISTORY

It is difficult, if not impossible, to establish where the concept of social television first originated. In one form or another it predates television itself. In the late nineteenth century, long before video and communication technologies crystallized into their familiar modern forms, science fiction writers such as Albert Robida dreamt up extraordinary devices like the all-purpose *telephonoscope*; television, videoconference terminal and proto-Internet browser all in one screen. Truly a converged technology! Among its many functions, the telephonoscope allowed audiences to experience theatrical performances in their homes and share the experience with other viewers as if sitting in the same theater:

"So, one can applaud?" asked Barnabette.
"Of course!" Mr. Ponto replied. "Home viewers can also offer their own [applause]. Here, let me connect out to the theater. You may applaud, if you wish."
"So," a laughing Barbe inquired, "we could also send out boos if we wished to?"
"Definitely not!" answered Mr. Ponto. "That's forbidden! You understand that if expressions of disapproval were permitted, any practical joker could disturb the shows from their own armchairs."
"But then," Barbe continued, "home viewers who find the play boring are not entitled to let it be known? That's quite unpleasant; one must repress one's feelings and keep them to oneself."
"Absolutely not, you silly girl. The home viewer may boo to his heart's content whenever a play bores him, but he must make sure to shut off the connection to the theater. Thus he can express his bad mood without causing any disturbance to the theater." (Robida, 1882/2004, pp. 55–56)

Social television-like technologies, particularly variations on video telephony, went on to become a staple of science fiction. In *Nineteen Eighty-Four*, George Orwell provides an especially sinister example:

The voice came from an oblong metal plaque like a dulled mirror which formed part of the surface of the right-hand wall. The instrument (the telescreen, it was called) could be dimmed, but there was no way of shutting it off completely. [...] The telescreen received and transmitted simultaneously. Any sound that Winston made, above the level of a very low whisper, would be picked up by it, moreover, so long as he remained within the field of vision which the metal plaque commanded, he could be seen as well as heard. (Orwell, 1949/1990)

Although the function of the telescreen is primarily monitoring, in the novel it is also occasionally used for communication and interactive content. At the time of the book's writing in 1948, video links were already in use, and by 1956 AT&T had created the first experimental prototype of PicturePhone, their famous videophone system (AT&T, 2008). Although years of user tests were not encouraging, PicturePhone was finally commercialized in 1970. It was a market failure, the first of many to come for video telephony (Noll, 1992).

This form of video telephony, which did not involve television content, falls outside social television as it is now generally understood. Of course, it had always been possible to socialize around the TV set, or communicate about TV content via other means (such as by telephone). After the listener/viewer call-in program format made the jump from radio to television in the 1960s (Newcomb, 1997), a small number of viewers could even join the broadcast from their own homes, in an early form of interactive programming.

Prestel, a videotex system (a type of early computer network initially accessed on TV terminals) from the UK Post Office/British Telecommunications that was launched in 1979 and remained in operation until 1994, introduced a form of email between users. However, neither this nor later videotex features such as bulletin boards and multi-player games were integrated with viewing TV content, and the services were frequently accessed from dedicated terminals rather than on the TV.

The first system to offer something akin to a remote social television experience appears instead to have been Spacephone, launched in 1980 by the US TV-maker Zenith. Spacephone was a feature offered on a number of high-end TV models which allowed users to talk on the telephone through their TV set, using the TV speakers and a built-in microphone as a speakerphone ("Zenith Electronics," 2008). Users could answer the phone and place calls with their remote control, and watch television while talking. However, the TV audio was muted during calls. Nevertheless, two Spacephone users talking to each other over the system would have experienced a simple version of social television. Unfortunately, Spacephone was not a success, and a faltering Zenith discontinued the feature in the mid-eighties (Perry, 1988).

A few years later, in 1988, a comedy show that would exert considerable influence on social television premiered on a local cable station in Minnesota (Cornell & Henry, 2002). *Mystery Science Theater 3000* (MST3K) involved a janitor marooned in space, watching old movies with his robot friends, and ridiculing them in order to preserve his sanity. MST3K's instantly recognizable visual format, showing the cast in silhouette sitting in front of the movie screen, has inspired several Social TV systems.

Social television finally started to gain traction around the turn of the century. Over a period of about two years, starting in 2000, SMS TV Chat was launched in many countries in Europe, becoming a big (and very profitable) hit. SMS TV Chat is essentially a chat room where all the text is displayed on the television as part of the broadcast. Participants write to the chat room by sending text messages from their cell phones. There are a number of different program formats. Sometimes the chat takes up the whole screen, so that there is no other TV content. Sometimes an unrelated program runs in the background or in a part of the screen. And sometimes a fully interactive live program is built around the chat (Beyer, Enli, Maasø, & Ytreberg, 2007). Especially this third case undeniably creates a social experience around the television content. However, the fact that the conversation is displayed as part of the broadcast severely constrains the type of communication and contact that is possible. All utterances are public, and the volume of messages and the possibility of delay can restrict an individual's ability to get a word in, or for groups of people to meaningfully converse (Zelenkauskaite & Herring, 2008).

Also in 2000, America On-Line (AOL) launched AOLTV, a line of retail set-top boxes to compete with Microsoft's WebTV. Like WebTV, it allowed users to surf the web on their television, but more importantly to read email, instant message (IM) and chat in chat rooms while watching TV on the same screen (Hill, 2001). AOLTV did not integrate television context with the IM presence information or provide ways for users to send links to TV content, but otherwise the box offered a near fully-fledged social television experience. AOLTV went off the market in 2002, but the chat features remain part of many Interactive TV services (Hu, 2003).

Soon afterwards, the research community started exploring social television in earnest, starting with Abreu, Almeida and Branco's (2001) description of 2BeOn. This activity continues up to the present. However, while a number of corporate research labs (including groups at Alcatel-Lucent, Microsoft, Google, PARC, Motorola, AT&T and Siemens) have published about Social TV prototypes and concepts, or even announced systems, as of late 2008 none of these have actually been launched.

At the same time, social sharing of video is becoming a reality online. YouTube[1] and other video sites have long offered the ability to share links to videos and leave comments as a form of asynchronous communication. Netflix[2] shares viewing history, ratings, comments and reviews of DVDs and online videos with a user's contacts and the wider community through Netflix Buddies. Now synchronous viewing and communication (via text, voice or video chat) are part of sites such as YouTube Streams,[3] Lycos Cinema[4] and Wi-Fi TV.[5] BuddyTV[6] offers a "TVj" feature to arrange similar sessions around TV shows, but does not provide the program itself: participants have to access it separately. Asynchronous communication is also becoming more advanced, with the ability to attach comments to particular playback positions in the video stream, so-called *deep tagging*, widely available (Click TV,[7] VeoTag,[8] ReviewTube,[9] InnerToob,[10] etc.), and several sites offering media-rich annotations and enhancements such as speech bubbles, graphics and animations (BubblePly,[11] Vidavee Graffiti[12]).

While video sites are moving into live chat, IM clients are starting to offer co-viewing of video. Zync is being integrated into Yahoo! Messenger (Shamma, Bastéa-Forte, Joubert, & Liu, 2008), and Microsoft is adding Messenger TV to their MSN Messenger[13] client. Both provide similar functionality, namely the ability for people in a chat session to watch an online video clip in near-synchronization. Finally, the internet TV application Joost[14] includes IM and "Channel Chat" plugins, where the latter is a public chat room associated with each Joost TV channel. (Given the dot-com industry's notorious turbulence, it is very likely that other sites and services will have complemented or replaced many of the ones listed above by the time this sees print.)

In some ways, history is coming full circle. Early on, as Robida's ideas became reality, parts of the telephonoscope split and diverged into different devices and services. Over the next hundred years, efforts to bring them together led to many different innovations, but a genuinely social kind of television remained science fiction. Now the technologies that have been separate for so long are converging again, and we are getting close to realizing the original thought in whole. However, currently available solutions still fall short of a comprehensive and seamless experience. They are limited in the types of content and the types of communication they support. In some systems, users have to coordinate multiple separate devices, and many do not share the video context or TV presence with other users, so that even if the communication *is* integrated with the media playback in a single device or application, the two activities remain essentially separate. For a fully integrated social television experience, we have to look to prototype systems in the research community.

EXAMPLES

Although the idea has a long history, social television has only recently come to be recognized as a distinct research topic within the field of interactive television (iTV). The best way to convey what Social TV means today may be by example—or, rather, multiple examples. There are many to choose from in the literature. Those that have been described in some detail include 2BeOn, from the University of Aveiro (Abreu et al., 2001; Abreu & Almeida, 2008), AmigoTV, from Alcatel (now Alcatel-Lucent) (Coppens, Trappeniers, & Godon, 2004; Coppens, Vanparijs, & Handekyen, 2005); Media Center Buddies, from Microsoft Labs (Regan & Todd, 2004a; 2004b); and ConnecTV, from TNO (Boertjes, 2007; Boertjes, Klok, & Schultz, 2008).

However, there is one particular example that is especially close at hand: At Motorola we have over the last few years developed a series of social television prototypes, collectively known as STV, for use in lab and field trials (Harboe, Massey, Metcalf, Wheatley, & Romano, 2008a; Harboe et al., 2008b). The design goal throughout development has been to allow a small group of friends or relatives to share a feeling of contact or togetherness while watching TV. In the current version, STV3, the key elements are:

- **Television presence** information through a buddy list and ambient display.
- Support for setting up **shared viewing experiences** through program suggestions and the ability to join what someone is watching.
- **Freeform communication** through text chat and group voice calls.
- Awareness of **viewing habits** and preferences, provided by viewing history lists and integration of viewing information into the electronic programming guide (EPG) grid.

In the selection and type of features, STV3 closely resembles a number of other social television systems. 2BeOn, AmigoTV, Media Center Buddies and ConnecTV all incorporate more or less the same capabilities. (The exception is the ambient display, which is apparently unique to STV.) However, in other systems there are some variant solutions to the same problems, as well as additional features not included in this particular prototype.

While STV3 supports creating shared viewing sessions through the presence information, the ability to join what a buddy is watching, and suggestions, some other systems take a more direct approach. ConnecTV has a "follow-friend" mode that switches channels automatically whenever a buddy does. Goldenberg (2007) and the Living@Room team (Ghittino, Iatrino, Modeo, & Ricchiuti, 2007) both propose a "shared remote control" paradigm where users who are watching together remotely all can take control of the joint content stream, pausing or rewinding everybody's TVs at the same time. The IM video player Zync uses a similar shared playback control (Shamma, Bastea-Forte, Joubert, & Liu, 2008). Coates (2005) suggests a type of voting system or game to resolve disputes over what to watch.

One feature not included in STV3 is video communication, to allow participants who are watching together to actually see each other. Video is the basis for Reflexion (Agamanolis, 2008), and is included in AmigoTV (Coppens et al., 2005) and Web Zync (Shamma et al., 2008), as well as in many social television concepts (Ghittino et al., 2007; Coates, 2005; Gross, Fetter, & Paul-Stueve, 2008).

In Telebuddies, the social interaction takes place within a multiplayer quiz game that is part of an interactive TV show (Luyten, Thys, Huypens, & Coninx, 2006). Creating content-related activities like games, wagers or fantasy sports around communication opens up many new possibilities for television-based socializing.

The increasing prominence of non–live-broadcast content, such as programs recorded on Digital Video Recorders (DVRs) or streamed on-demand from the web, has led to many proposals for ways in which users can interact around content asynchronously. A widespread idea is to record a first user's interactions as a kind of "commentary track" (whether in the form of voice, text, or non-verbal annotations such as emoticons), and play it back later for other users, synchronized to the content (Fink, Covell, & Baluja, 2006; Ducheneaut, Moore, Oehlberg, Thornton, & Nickell, 2008). This concept is implemented in CollaboraTV (Harrison & Amento, 2007; Nathan et al., 2008).

When the recorded interactions and annotations become sufficiently complex and powerful, it makes sense to consider them as a form of end-user enrichment of the TV content (Cesar et al., 2008). Such enhancements may include virtual edits, links to related content, freehand drawings or "scribbles" on top of the video, metadata tags, as well as the basic voice, text and graphical comments. It may even take the form of user-supplied closed captions and translations.

From such end-user enrichment, it is only a short step to user-generated content (Alliez, 2008) and community television (Vasconcelos, Fava, Kampf, Schilling, & Furtado, 2007; Obrist, Beck, Kepplinger, Bernhaupt, & Tscheligi, 2008). This in turn leads naturally to integration with blogs and social networks (Mantzari & Vrechopoulos, 2007; Baca & Holtzman, 2008), and to open forums where strangers meet and interact (as opposed to the fairly closed buddy lists for connecting family and friends presupposed by many other Social TV systems), described by Hess (2007). Beyond all these examples, social television bleeds over into iTV in general, as well as into telepresence, video telephony, any kind of content sharing, multiplayer games, online social networks, and collaborative filtering, among other things.

The examples listed above, although they encompass systems that support quite different experiences, seem to have something in common, when taken as a whole. Exactly what that characteristic is is more difficult to answer. The great variety of social television features and the indistinct boundaries of Social TV as a research area raise the question: How can we tell what is and what isn't Social TV?

DEFINITIONS

There have been many attempts to define social television. On reviewing the literature, it turns out that these efforts can be grouped under two definitions, one narrow and one broad, with the same researchers often using one or the other depending on the context (Table 1).

The broad definition defines "social television" as *any* technology that supports social practices associated with TV. Current social practices include things like talking about upcoming TV shows, watching TV together (at home, or in public places like a bar), and talking about TV programs after the fact.

This classification is comprehensive, but it makes the potential scope of the topic staggeringly broad. Most things in life have a social dimension, and television-related information is more or less ubiquitous in the modern world. Is a tabloid magazine, filled with gossip about television stars and a source of much potential small talk, then a social television technology? It would seem so, especially if integrated somehow into an interactive TV system.

The other, narrow definition avoids this issue. Simply put, it states that "social television" describes systems that create an experience somehow "like" watching TV together, even though viewers are physically remote from each other. This is achieved by integrating communication technologies (such as voice communication or text chat) with the TV.

This categorization covers many aspects of the social television examples listed in the previ-

Table 1. The two types of definitions of social television found in the literature

Social Television Definitions

Narrow	Broad
Communication technologies that create a remote, shared experience of watching TV together	***Any technology that supports social practices associated with television***
Using communication technology to connect TV viewers, in order to create remotely shared experiences around TV content. (Harboe et al., 2008b)	*Interactive TV (iTV) systems that support the sociable aspects of TV viewing. This includes improvements to collocated interaction.* (Harboe et al., 2008a)
An audio-video system which allows distant viewers to communicate with each other using several interpersonal communication modalities, such as open audio channel, instant messaging, emoticons, etc. (Chorianopolous, 2007)	*Technological support for the social practices that surround TV viewing.* (Chorianopolous, 2007)
Integration of television and computer technology to support sociable, computer-mediated group viewing experiences. [...] Design for distributed, shared television viewing. (Ducheneaut et al., 2008)	*Leveraging [...] computing integration to remove the barriers to sociable interaction around video content.* (Ducheneaut et al., 2008)
Video services that integrate other communication services like voice, chat, context awareness, and peer ratings to support a shared TV experience with one's peer groups. (Klym & Montpetit, 2008)	*Provide groups of TV spectators with technical support for collocated and geographically distributed TV watching and social interaction.* (Gross et al., 2008)
Providing two or more remote TV consumers with a joint TV watching and communication experience. (Schatz, Wagner, Egger, & Jordan, 2007)	*Communication and social interactions—remote or co-located—in a TV-watching context, or related to a TV experience; and technology that supports these communications and interactions.* (Geerts, Harboe, & Massey, 2007)

ous section, but not all of them. For example, it would not seem to include the ability to send a recommendation of a program you *just watched* to a friend, yet that kind of functionality is part of several Social TV systems. More importantly, it leaves out whole categories of related systems and features that have been envisioned but not systematically explored, like the ability for two people sitting next to each other on the couch watching a quiz show to participate and compete against each other.

Another ambiguity, shared by both definitions, is what is meant by "television." In addition to live broadcasts, many other types of content are available or accessible on television sets, including video-on-demand (VOD), internet video, and other user-generated content. Some of these media types are clearly relevant to social television. Sharing a home movie between two TVs is arguably a social television experience. Sharing a photo album may also qualify. At the same time, it is becoming possible to watch TV on other devices, such as cell phones and mp3 players. So maybe sharing a home movie between two phones is another example of social television? But sharing photos between two cell phones can hardly qualify, or the term is diluted beyond reason!

Perhaps these difficulties have less to do with shortcomings of the definitions than they are simply reflections of a world in transition, where the boundaries between traditional media types and communication services are blurring. Social TV is just one *part* of the research on content plus communication, which also encompasses social music, social photos, and more. Ultimately, the most important dimension of a user's social media experience is probably not whether or not it is classified specifically as "social television".

DIMENSIONS

By now, it should be evident that the social television space encompasses many radically different experiences. In order to bring them into some kind of organization, several researchers have put forward dimensions of design for social television.

In the analysis of Coppens et al. (2004), a social television experience consists of a mix of three basic components: personal content (including personal favorites and personally recommended content), rich communication (including all the modalities already discussed: text, voice, video and graphical messages), and community support (presence, invitations, scheduling, and support for social networking). While this is a serviceable description of the AmigoTV system, the distinction between the categories is not always clear. Program suggestions, for example, could arguably fit in any of the three.

Ducheneaut et al. (2008) divide TV-related sociability into direct and indirect forms. Direct sociability is the socializing that takes place during or just before and after the television experience, while indirect sociability refers to socializing about the content *apart* from the actual experience of it (such as when talking about an upcoming program or something that has been watched previously). This is an important distinction, but remote communication, and especially asynchronous interaction, confuse the matter, since one of the people socializing may be in the middle of the television experience while the other is not.

Chorianopoulos (2007) uses the well-established CSCW time-space matrix (Baecker, Grudin, Buxton, & Greenberg, 1995) to describe television sociability in two dimensions: synchronous—asynchronous communication and collocated—distant presence. The resulting quadrants correspond to four scenarios that could benefit from social television systems. The most valuable contribution of this model is probably to identify interesting, non-obvious potential use-cases for Social TV, such as supporting people watching at the same place at different times. However, it should be recognized that in practice, both dimensions have more than two states. Text chat, for example, has aspects of both synchronous and asynchronous communication (Voida, Newstetter, & Mynatt, 2002), and the ability to record and play back interaction sequences synchronized to the content, found in systems like CollaboraTV, mimics certain characteristics of a synchronous experience in an asynchronous communication situation (Harrison & Amento, 2007). Along the space axis, social television has an important category of nearly-collocated communication: between people watching on different TVs in the same household.

Hess (2007) identifies seven aspects of community building for Social TV systems that bring people together. Of these, group size may be the most generally relevant, making a distinction between one-to-one communication and interaction in larger groups. Because Hess's focus is on a "Find-a-Friend" scenario, he assumes an open, public forum, where participants would mostly have weak-tie relationships, and does not consider more closed and close-knit groups of friends and family who have strong ties to each other.

The most extensive taxonomy of social television is due to Schatz et al. (2007). They define six dimensions of the design space for mobile social television, incorporating a number of the axes used by others: synchronous—asynchronous, verbal—non-verbal, integrated with—separate from the content, one-to-one—one-to-many, push—pull, and high bandwidth—low bandwidth. All of these dimensions are equally applicable to non-mobile social television. The degree to which the communication is integrated with the content (allowing people to communicate *via* the content, as when sending suggestions, sharing ratings, or adding tags, annotations or other enhancements) is an especially important factor not identified in the other models.

Gross et al. (2008) divide social television functionality into three components: group TV watching, presence info and communication, and recommendations. To this, they add social, spatial and temporal dimensions that generally correspond to those outlined above, and mention the possibility of group adaptive interfaces. On this basis they put forth a model of design dimensions.

In summary, theoretical models of social television have proposed a number of different dimensions of design. For the most part the models complement each other and can be used together. Many of the dimensions are simply aspects of communication technologies and online social networks in general. However, the integration with TV content does give rise to new dimensions that define the relationship between the communication and the content.

CONCLUSION

Like the topic itself, the search for social television remains a work in progress. Our ability to delineate it, either in terms of history, by example, through definition or by analysis of its characteristic dimensions, is limited by its variety, its interconnectedness with other social experiences, and its ongoing development. Yet if a definitive description is still elusive, the idea itself appears on the verge of finally being realized on a grand scale.

Whether or not any of the prototype systems mentioned above ever make it to market (and several of them are ready to take that step), it seems a near certainty that some sort of communication, probably text chat, will soon be easily available alongside the viewing of TV content. Online, a variety of social experiences around video and television content are emerging simply from the convergence of internet and video.

Ironically, just as convergence is turning Social TV into reality, it may also be rendering the notion obsolete. Without a clear distinction between a TV screen and other screens and between television programming and other content, and with communication and content sharing available ubiquitously, social television as a separate concept might no longer make sense. When first conceived of, Social TV was not thought of as distinct from the idea that would become television. As communication and video experiences are integrated in modern-day telephonoscopes, perhaps the ability to cheer along with an audience spread out all over the world will soon be commonplace—and disconnecting before booing just good manners.

ACKNOWLEDGMENT

I would like to thank the many people who have contributed to my thinking on social television, including fellow Social TV researchers, workshop participants, and Motorola coworkers. I also particularly wish to thank Elaine Huang, Joe Tullio, Ashley Novak and Noel Massey for their help revising the manuscript for this chapter.

REFERENCES

Abreu, J., Almeida, P., & Branco, V. (2001). 2BeOn: Interactive Television Supporting Interpersonal Communication. In J. A. Jorge, N. Correia, H. Jones, & M. B. Kamegai, (Eds.),

Proceedings of the Sixth Eurographics Workshop on Multimedia 2001 (pp. 199–208). New York: Springer Verlag.

Abreu, J. F., & Almeida, P. (2008). From Scratch to User Evaluation: Validating a Social iTV Platform. In A. Lugmayr, S. Kemper, M. Obrist, T. Mirlacher, & M. Tscheligi (Eds.), *Adjunct Proceedings of EuroITV 2008* (pp. 31–32). Tampere, Finland: Tampere University of Technology.

Agamanolis, S. (2008). At the Intersection of Broadband and Broadcasting: How Interactive TV Technologies can Support Human Connectedness. *International Journal of Human-Computer Interaction, 24*(2), 121–135.

Alliez, D. (2008). Adapt TV Paradigms to UGC by Importing Social Networks. In A. Lugmayr, S. Kemper, M. Obrist, T. Mirlacher, & M. Tscheligi (Eds.), *Adjunct Proceedings of EuroITV 2008* (pp. 29–30). Tampere, Finland: Tampere University of Technology.

AT&T. (2008). *Technology Timeline: Picturephone.* Retrieved September 5, 2008, from http://www.corp.att.com/attlabs/reputation/timeline/70picture.html

Baca, M., & Holtzman, H. (2008). Television meets Facebook: Social Networks through Consumer Electronics. In A. Lugmayr, S. Kemper, M. Obrist, T. Mirlacher, & M. Tscheligi (Eds.), *Adjunct Proceedings of EuroITV 2008* (pp. 35–36). Tampere, Finland: Tampere University of Technology.

Baecker, R. M., Grudin, J., Buxton, W. A. S., & Greenberg, S. (1995). *Readings in Human-Computer Interaction: Toward the Year 2000.* San Francisco, CA: Morgan Kaufmann.

Beyer, Y., Enli, G. S., Maasø, A. J., & Ytreberg, E. (2007). Small Talk Makes a Big Difference: Recent Developments in Interactive, SMS-Based Television. *Television & New Media, 8*(3), 213–234.

Boertjes, E. (2007). ConnecTV: Share the Experience. In A. Lugmayr & P. Gołębiowski (Eds.), *Adjunct Proceedings of EuroITV 2007* (pp. 139–140). Tampere, Finland: Tampere International Center for Signal Processing.

Boertjes, E., Klok, J., & Schultz, S. (2008). ConnecTV: Results of the Field Trial. In A. Lugmayr, S. Kemper, M. Obrist, T. Mirlacher, & M. Tscheligi (Eds.), *Adjunct Proceedings of EuroITV 2008* (pp. 21–22). Tampere, Finland: Tampere University of Technology.

Cesar, P., Bulterman, D. C. A., Geerts, D., Jansen, J., Knoche, H., & Seager, W. (2008). Enhancing Social Sharing of Videos: Fragment, Annotate, Enrich, and Share. In *Proceedings of the 16th international Conference on Multimedia*, in press. New York: ACM Press.

Chorianopoulos, K. (2007). Content-Enriched Communication: Supporting the Social Uses of TV. *The Journal of The Communications Network, 6*(1), 23–30.

Coates, T. (2005, March 23). *Social Software for Set-Top Boxes.* Retrieved September 5, 2008, from http://www.plasticbag.org/archives/2005/03/social_software_for_settop_boxes/

Coppens, T., Trappeniers, L., & Godon, M. (2004). AmigoTV: Towards a Social TV Experience. In *Proceedings of EuroITV 2004.*

Coppens, T., Vanparijs, F., & Handekyen, K. (2005). *AmigoTV: A Social TV Experience Through Triple-Play Convergence* (Technology White Paper). Retrieved September 5, 2008, from http://www1.alcatel-lucent.com/com/en/appcontent/apl/T0205-Amigo_TV-EN_tcm172-195461635.pdf

Cornell, C., & Henry, B. (2002). The Almost But Still Not Quite Complete History of MST3K. *Satellite News.* Retrieved September 5, 2008, from http://www.mst3kinfo.com/history/

Ducheneaut, N., Moore, R. J., Oehlberg, L., Thornton, J. D., & Nickell, E. (2008). Social TV: Designing for Distributed, Sociable Television

Viewing. *International Journal of Human-Computer Interaction, 24*(2), 136–154.

Fink, M, Covell, M., & Baluja, S. (2006). Social—and Interactive—Television: Applications Based on Real-Time Ambient Audio Identification. In G. Doukidis, K. Chorianopoulos, & G. Lekakos (Eds.), *Proceedings of the 4th European Conference, EuroITV 2006* (pp. 138–146).

Geerts, D., Harboe, G., & Massey, N. (2007). *Social Interactive Television Workshop* [Presentation slides]. Retrieved September 5, 2008, from http://soc.kuleuven.be/com/mediac/socialitv/results.htm

Ghittino, A., Iatrino, A., Modeo, S., & Ricchiuti, F. (2007). Living@Room: A TV-Based Social Experience for Multimedia Content Streaming. In A. Lugmayr & P. Gołębiowski (Eds.), *Adjunct Proceedings of EuroITV 2007* (pp. 207–212). Tampere, Finland: Tampere International Center for Signal Processing.

Goldenberg, S. (2007). Digital Video Recorders and Micro-Social Networking: Recreating the Shared Watching Experience of Television. In A. Lugmayr & P. Gołębiowski (Eds.), *Adjunct Proceedings of EuroITV 2007* (pp. 213–217). Tampere, Finland: Tampere International Center for Signal Processing.

Gross, T., Fetter, M., & Paul-Stueve, T (2008). Toward Advanced Social TV in a Cooperative Media Space. *International Journal of Human-Computer Interaction, 24*(2), 155–173.

Harboe, G., Massey, N., Metcalf, C., Wheatley, D., & Romano, G. (2008a, May). The Uses of Social Television. *Computers in Entertainment, 6*(1), 1–15.

Harboe, G., Metcalf, C. J., Bentley, F., Tullio, J., Massey, N., & Romano, G. (2008b). Ambient Social TV: Drawing People Into a Shared Experience. In *Proceeding of the Twenty-Sixth Annual SIGCHI Conference on Human Factors in Computing Systems* (pp. 1–10). New York: ACM Press.

Harrison, C., & Amento, A. (2007). CollaboraTV: Using Asynchronous Communication to Make TV Social Again. In A. Lugmayr & P. Gołębiowski (Eds.), *Adjunct Proceedings of EuroITV 2007* (pp. 218–222). Tampere, Finland: Tampere International Center for Signal Processing.

Hess, J. (2007). Supporting Community Building in iTV Environments. In A. Lugmayr & P. Gołębiowski (Eds.), *Adjunct Proceedings of EuroITV 2007* (pp. 129–130). Tampere, Finland: Tampere International Center for Signal Processing.

Hill, B. (2001). *AOLTV for Dummies.* New York: John Wiley & Sons.

Hu, J. (2003, February 18). America Online Confirms End of AOLTV. *CNET News.* Retrieved September 5, 2008, from http://news.cnet.com/2100-1023-984920.html

Klym, N., & Montpetit, M.J. (2008). *Innovation at the Edge: Social TV and Beyond* (VCDWG Working Paper). Cambridge, MA: MIT Communications Futures Program. Retrieved September 5, 2008, from http://cfp.mit.edu/publications/index.shtml

Luyten, K., Thys, K., Huypens, S., & Coninx, K. (2006). Telebuddies: Social Stitching with Interactive Television. In *CHI '06 Extended Abstracts on Human Factors in Computing Systems* (pp. 1049–1054). New York: ACM Press.

Mantzari, E., & Vrechopoulos, A. (2007). "My Social Tube": User Generated Content and Communication on Interactive Digital Television. In A. Lugmayr & P. Gołębiowski (Eds.), *Adjunct Proceedings of EuroITV 2007* (pp. 241–246). Tampere, Finland: Tampere International Center for Signal Processing.

Nathan, M, Terveen, L., Harrison, C., Yarosh, S., Stead, L., & Amento, B. (2008). CollaboraTV:

Making Television Viewing Social Again. In *Proceedings of UXTV 2008*, in press.

Newcomb, H. (Ed.). (1997). *Encyclopedia of Television*. (1st ed.). New York: Taylor & Francis. Retrieved September 5, 2008, from http://www.museum.tv/archives/etv/

Noll, A. M. (1992). Anatomy of a Failure: Picturephone Revisited. *Telecommunications Policy, 16*(4), 307–316.

Obrist, M., Beck, E., Kepplinger, S., Bernhaupt, R., & Tscheligi, M. (2008). Local Communities: Back to Life (Live) Through IPTV. In M. Tscheligi, M. Obrist, and A. Lugmayr (Eds.), *Proceedings of the 6th European Conference, EuroITV 2008* (pp. 148–157). Berlin: Springer Verlag.

Orwell, G. (1990). *Nineteen Eighty-Four*. London: Penguin Books.

Perry, T. S. (1988). The Longest Survivor Loses Its Grip. *Spectrum, 25*(8), 16–20.

Regan, T., & Todd, I. (2004a). *Media Center Buddies: Instant Messaging Around a Media Center* (Technical Report MSR-TR-2004-47). Redmond, WA: Microsoft Research. Retrieved September 5, 2008, from ftp://ftp.research.microsoft.com/pub/tr/TR-2004-47.pdf

Regan, T., & Todd, I. (2004b). Media Center Buddies: Instant Messaging Around a Media Center. In *Proceedings of the Third Nordic Conference on Human-Computer interaction* (pp. 141–144). New York: ACM Press.

Robida, A. (2004). *The Twentieth Century*. In P. Willems (Trans.), A. B. Evans, (Ed.). Middletown, CT: Wesleyan University Press. (Original work published 1882).

Schatz, R., Wagner, S., Egger, S., & Jordan, N. (2007). Mobile TV Becomes Social: Integrating Content with Communications. In V. Luzar-Stiffler & V .H. Dobric (Eds.), *Proceedings of the ITI 2007 29th International Conference on Information Technology Interfaces* (pp. 263–270). Zagreb, Croatia: SRCE University of Zagreb.

Shamma, D. A., Bastea-Forte, M., Joubert, N., & Liu, Y. (2008). Enhancing Online Personal Connections Through the Synchronized Sharing of Online Video. In *CHI '08 Extended Abstracts on Human Factors in Computing Systems* (pp. 2931–2936). New York: ACM Press.

Vasconcelos, P., Fava, F., Kampf, T., Schilling, A., & Furtado, E. (2007). Ethnographic Investigational Methodology and Evaluation on Local Television Channel Creation that Allows Interaction with the Community. In A. Lugmayr & P. Gołębiowski (Eds.), *Adjunct Proceedings of EuroITV 2007* (pp. 282–286). Tampere, Finland: Tampere International Center for Signal Processing.

Voida, A., Newstetter, W. C., & Mynatt, E. D. (2002). When Conventions Collide: The Tensions of Instant Messaging Attributed. In *Proceedings of the SIGCHI Conference on Human Factors in Computing Systems: Changing Our World, Changing Ourselves* (pp. 187–194). New York: ACM Press.

Zelenkauskaite, A., & Herring, S.C. (2008). Television-Mediated Conversation: Coherence in Italian iTV SMS Chat. In *Proceedings of the 41st Hawaii International Conference on System Sciences* (pp. 145–155). Los Alamitos, CA: IEEE Computer Society.

Zenith Electronics. (2008). In *Wikipedia, the free encyclopedia*. Retrieved September 5, 2008, from http://en.wikipedia.org/wiki/Zenith_Electronics

ENDNOTES

1. http://www.youtube.com/
2. http://www.netflix.com/
3. http://www.youtube.com/streams_main
4. http://cinema.lycos.com/

[5] http://www.wi-fitv.com/

[6] http://www.buddytv.com/

[7] http://www.click.tv/

[8] http://www.veotag.com/

[9] http://www.joeberkovitz.com/reviewtube/

[10] http://www.innertoob.com/

[11] http://www.bubbleply.com/

[12] http://graffiti.vidavee.com/

[13] http://messengertv.msn.com/

[14] http://www.joost.com/

Section I

The Broad Picture:

Frameworks and Applications

Chapter II
Broadening the Effects of Broadcasting:
How ITV can Collapse Distance and Transform Communication

Stefan Agamanolis
Distance Lab, UK

ABSTRACT

Conventional broadcasting has the impressive power to create shared experiences over huge audiences or even entire populations. The sharing of such experiences deepens our sense of connectedness with others, which in turn arguably leads to positive effects in society as a whole. Interactive television and related technologies have the potential to further collapse distance and broaden these positive effects of broadcasting—enabling new modes of communication, providing an enhanced sense of community, offering opportunities to meet new people, and allowing us to build relationships in new ways. This chapter surveys a number of research projects undertaken in the Human Connectedness group at Media Lab Europe and at Distance Lab that address these themes, as a way to suggest new trends at the intersection of television, networking, and computing.

INTRODUCTION

Many areas of the world are undergoing a transition from conventional analog terrestrial broadcasting to a digital television infrastructure. In addition to expensive new production equipment, this transition requires new television receivers or set-top boxes to be distributed throughout entire countries to ensure audiences can receive the new transmissions.

In building support for the transition to a digital television infrastructure within groups who will be bearing some of the costs, ITV advocates often highlight various governmental or commercial benefits: Advertisers will be able to target messages to specific households and individuals with pinpoint accuracy. Government departments will be able to offer a range of services directly in our living rooms. Businesses will be able to sell us products at the touch of a button on our remote controls. But does the consumer really care enough about these conveniences to spend hard-earned cash upgrading their television equipment, even if picture quality will be slightly improved?

This chapter argues that the greatest potential for new ITV technologies lies in the communicative domain. Conventional broadcasting has a special power of creating shared experiences over large groups of people or even entire populations. The feeling of sharing an experience with so many others arguably deepens our sense of connection to those around us, and this translates to positive benefits within our personal circles as well as society at large. Digital television systems being introduced, together with the broadband "back-channels" they will be connected to, have the potential to further collapse distance and broaden these positive effects of broadcasting—enabling new modes of communication, cultivating an enhanced sense of community, offering opportunities to meet new people, and allowing us to build and maintain relationships in new ways. These new capabilities and ways of "staying in touch", which are focused on the viewer/user, could serve a significant role in the swift acceptance and deployment of ITV technologies throughout the world.

The Human Connectedness research group at Media Lab Europe explored the theme of human relationships and how they are mediated by technology, with the mission of developing new technologies and experiences that allow people to build and maintain relationships in new ways. Several of the group's projects addressed the areas of radio, television, and broadcasting either directly or indirectly. This group's work is being continued at Distance Lab, a research institute that is bringing together technology, design and the arts to overcome the disadvantages of distance in learning, health care, relationships, culture and other domains. This article gives an overview of relevant projects, with the hope that they can reveal interesting trends and inspire further research in this area. Links to longer publications with additional information are included, which in turn reference related efforts in other research centers.

THE IMPACT OF TECHNOLOGY ON HUMAN RELATIONSHIPS

Human psychology is an enormously complex topic and our understanding of it is always changing, but a wide body of evidence and experimentation validates the basic claim that humans have a fundamental need for contact with other humans. Authors like Lewis, Amini & Landon (2000) detail the vital regulating effects that social contact and healthy relationships have on human mental and physical well-being. They also reveal the consequences that arise from a lack of these requirements, particularly during infancy and childhood.

The degree to which an individual is "self sufficient" or "independent", able to live on their own and not needing anything from anyone else, is often used as a measure of success in our modern society. However, humans, and indeed all mammals, are not in fact "self-regulating" creatures. Recent theories assert that human physiology is largely regulated and stabilized by others who are nearby. This stabilization happens via many different channels including facial expression, physical touch, hormonal signals, and so on.

The term "limbic regulation" refers to this system of mutual exchange and adaptation between mammals. Disruptions in this regulation can have serious consequences, particularly in the early stages of life. Infants who have not received sufficient physical and emotional contact with another human (such as a mother or another person) develop serious mental deficiencies and health problems that can be life-threatening depending on the severity of the isolation. However, this regulation is also critical throughout life:

"Adults remain social animals: they continue to require a source of stabilization outside themselves. That open-loop design means that in some important ways, people cannot be stable on their own—not should or shouldn't be, but can't be." (Lewis, Amini & Landon, 2000)

Other authors have reported on how disruptions in social contact compare to other typical health hazards:

"...social relationships, or the relative lack thereof, constitute a major risk factor for health—rivaling the effects of well-established health risk factors such as cigarette smoking, blood pressure, blood lipids, obesity, and physical activity." (House, Landis & Umberson, 1988)

In short, our interactions and relationships with other people form a network that supports us, makes our lives meaningful, and ultimately enables us to survive. Unfortunately, several factors might hinder us from forming and attaining balance in these relationships that we want and need to have with others. For example, someone who must live apart from family and friends as part of a job requirement (say in the oil industry, or in the military) may experience difficulties. Societal trends in social behavior are a subject of increasing study. In his book "Bowling Alone", Robert Putnam (1998) describes how people in American society increasingly lack social interactions and have fewer close friends, and he discusses how this deficit of social interconnectedness endangers the health of individuals and communities alike. His observations arguably also apply to societies in other parts of the world.

Technology clearly plays a significant role in the way we form and conduct relationships with others. For example, new forms of transportation enabled our circles of friends and family to become increasingly spread out over large physical areas. Meanwhile, various communication technologies ranging from the postal mail to the telegraph and telephone have allowed people to maintain close relationships that might not have otherwise survived those distances. Internet chat rooms and social networking sites have enabled users to meet people with similar tastes or interests that they might never have met otherwise. Flat-rate home broadband and PC-based video conferencing programs allow family members in distant locations to hold virtual gatherings and "visit" with each other for extended periods of time.

However, though they may introduce a time-saving convenience, new technologies are not always supportive of human contact. Automated bank machines, vending machines, self-service gasoline pumps, and automobiles themselves arguably reduce opportunities for casual contact with sales clerks, neighbors, and other passers-by that we might have had prior to the introduction of these perceived conveniences. On the other hand, the time saved by using such devices may allow an individual to invest more energy in an important relationship at a later time.

A study undertaken at Carnegie Mellon University has suggested the use of computers and the Internet may contribute to social isolation and higher stress levels (Kraut et al, 2002). For many people, the richness and instantaneity of Internet-mediated communication may be helpful in maintaining relationships over a distance, but for others these same features may emphasize the existence of that distance and place a strain on a relationship. Whether the benefits of using such technologies outweigh the costs is likely dependent on each individual's tendencies and tolerances as well as the character of the relationship in question.

THE IMPORTANCE OF DESIGN

The *design* and *context* of a communication device itself also clearly has an impact on the relationships they mediate. If a video conferencing system is difficult to use, the frustration evoked by dealing with it may transfer to the interaction it is supposed to facilitate. Certain designs may lend themselves more to business communications whereas others may be suited to informal or personal relations. Using a mobile phone while walking down a busy street may be just what's needed to complete a commercial transaction, but it's exactly not what's needed when you wish to have a relaxing and high-quality interaction with your distant partner. A professional video conference system situated in a formal conference room may be good for business meetings, but not at all good for informal interaction with friends or a catch-up with your kids in another location. Robustness of the design is also a factor. If a mobile phone system or Internet video chat program keeps freezing or disconnecting you repeatedly during important conversations, the aggravation may again transfer unexpectedly into your relationships with the people to whom you were talking.

This chapter asserts that, at this point in time, we generally know how to broadcast video, audio, and other data or to send it around the Internet. What we *don't* know (at least as much) is how to design devices and systems that are useable and meaningful to real people trying to form and maintain real relationships of many different kinds. This belief was partly the motivation behind the formation of the Human Connectedness research group at Media Lab Europe in 2001, as well as Distance Lab in 2006.

The Human Connectedness group aimed to invent new experiences and modes of communication that combat some of the negative effects mentioned above and enable people to form, maintain, and enhance relationships in new ways. In addition to applications that address individual relationships, the group worked on projects that involve, for example, supporting awareness and collaboration between distant groups of people, and on enabling new forms of cultural exchange.

After the closure of Media Lab Europe in 2005, Distance Lab was launched and has been extending and forming new branches of this work, providing an environment where technology, design and the arts are brought together with the mission of overcoming the disadvantages of distance in several domains including learning, health care, cultural applications, and relationships. Distance Lab places emphasis on building working prototypes and demonstrations and transferring these into new products and services in the global marketplace.

Like the Human Connectedness group, Distance Lab is interested in building a technological framework for new applications which includes taking advantage of the infinite bandwidth and processing-rich computing environments of the future and the opportunity to extend these networked media environments into our architectural surroundings as well as into interfaces that sit close to our bodies and are always with us. Consequently, the Lab's projects often reference the established research domains of ubiquitous computing and wearable computing. However, the Lab places special emphasis on forming a design framework that includes an understanding of sociological and psychological factors to help shape these systems in a fashion that reflects the needs and sensibilities of their intended users.

Some of the projects discussed in this chapter touch on the themes of television and broadcasting only indirectly. Nevertheless, many have argued that interesting insights are often found at the edges and intersections of different domains. It is hoped that, through the discussion surrounding these projects, interesting trends for digital and interactive television will be revealed that can inform and inspire future research efforts.

RESPONSIVE TELEVISION AND EQUALIZATION OF EXPERIENCE

Viper is a tool for creating television programs that can re-edit themselves based on different factors, such as viewer preferences or profiles or viewing conditions. Unlike traditional editing systems in which producers build a single linear video program, those using Viper create a database of annotated video and audio clips and other media objects. Together with this database, the producer develops a set of programmatic editing guidelines that describe which bits of this source material should be included and how they should be layered and assembled into complete high-quality programs for different viewing situations.

Viper consists of a graphical interface (Figure 1) for creating annotated media databases, and a framework of primitives, based in the Isis programming language, for expressing the editing guidelines. Unlike other systems that simply splice clips end to end, Viper's framework enables the use of complex editing constructs, such as inserts, AB rolls, graphics, and transitions of various sorts. This enables the creation of richly responsive television programs that have production values equivalent to what is expected in mainstream broadcasting.

Nevertheless, Viper aims to enable a new genre of video programming, distinct from traditional television, that offers new narrative possibilities and allows directors and producers to gain more

Figure 1. Viper interface for creating and annotating databases of media clips

control over how their programs are edited and exhibited in different viewing situations. Viper's playback engine supports client-side personalization, in which the final edit is performed on the viewer's receiving device, eliminating the need to transmit personal information to a distant and perhaps untrusted server.

When Viper was conceived in the Object-Based Media group at the MIT Media Lab, we were initially interested in the area of personalized advertising, in which a marketing message could be customized to aspects of a viewer's profile. As a somewhat sarcastic demonstration, we created a responsive political campaign ad that tailors its presentation to portray the candidate in the most persuasive way for each individual viewer (Agamanolis & Bove, 2003a). The mock candidate in our example is running for "student council" at the Media Lab, and the advertisement responds to factors like the viewer's job title (affects the way the candidate addresses the viewer), information about the viewer's concerns (affects the visual imagery and content of the candidate's message), and the viewer's favorite music genre (affects the background music). It is also possible to vary the pacing of the ad based on information about the viewer's attention span.

As the Viper project continued at the Human Connectedness group, we explored another important potential application area: *equalization* of experience. Normally personalized television implies a lack of shared experience, which is arguably the real power of broadcasting. The typical complaint is: If everyone watches their own personalized version of a TV program, then there is no longer a shared experience, and the discussions at work around the water cooler the next day would be very boring.

However, the same is true of traditional broadcasting—many viewers *see* the same thing but they may not *experience* the same thing because of varying educational backgrounds, skill levels, mental states, and so on. Tools like Viper could be used to create adaptive programs, such as educational documentaries or news shows, which aim to equalize experience across a population of viewers. This might involve introducing additional background material on subjects unfamiliar to a specific viewer, or adjusting pacing so that concepts that are harder to understand for certain viewers are not presented too quickly or are repeated more often.

ENHANCING SHARED EXPERIENCE WITH SHARED VIRTUAL SPACE

Anyone who uses video conferencing technologies on a regular basis is probably acutely aware of their limitations. Some systems use audio-based camera switching algorithms, which limits awareness of the inactive participants. Users of such systems often find themselves repeatedly asking each other "Are you still there?" Other multiparty systems display participants in separate windows, in scenes that often look like the 3x3 grid title sequence from the American TV show "The Brady Bunch". The visual or temporal separation characteristics of these designs could give rise to a negative confrontational dynamic. If graphical windowpanes or temporal cuts separate participants, the conference might feel less like a collaboration and more like a political talk show in which people should be arguing about something.

Reflexion is an interpersonal video communication system that operates like a "magic mirror" in which you see a reflection of yourself together with the reflections of other people in remote locations. All participants are layered together and can "touch" and interact with each other directly in the virtual video scene (Figure 2). This potentially creates a space with a more intimate social dynamic, one that may be more appropriate for many kinds of applications that are not well suited to conventional videoconferencing systems. (Cullinan & Agamanolis, 2002)

Figure 2. Screen shots of a 3-person Reflexion session

Figure 3. A Reflexion station in use at Media Lab Europe

A Reflexion station consists of a camera and video display connected to a computer (Figure 3). Each participant, of which there can be several, uses a separate Reflexion station. The computer extracts an image of the participant from his background and transmits a mirror image of it over the network to the other stations. The computer also receives extracted participant images from the other stations and combines them all together into a single video scene.

The system automatically monitors auditory cues and uses them to compose the scene in a way that enhances the interaction. For example, the system listens to who is speaking in order to judge who is the "center of attention". Active participants are rendered opaque and in the foreground to accentuate their visual presence, while other less-active participants appear slightly faded in the background in a manner that maintains awareness of their state without drawing undue attention. The system smoothly transitions the layering and appearance of the participants as their interactions progress. Every participant sees exactly the same composition, enhancing the sense of occupying a "shared space".

In our experiments with Reflexion, we found that one of the most satisfying ways to use it is to watch TV programs or movies together. Potentially, this could be done between people anywhere in the world. In this mode, each participant is miniaturized along the bottom of the screen so the video image is more visible while everyone can still see everyone else's facial expressions and gestures. The video playback or live television feed is synchronized at all locations so the viewers can discuss or laugh about the program together while it is showing.

Feedback from informal user tests suggests that Reflexion might be a useful device for family communication—for example, to connect the households of members of an extended family who want to gather and watch old home movies or perhaps a live soccer match on television. Other scenarios were also suggested, such as distance learning environments and multi-user remote interactive theater spaces. Digital interactive television platforms now being introduced in many parts of the world, combined with broadband upstream and downstream connectivity, could easily support these kinds of applications.

The Reflexion prototype uses a peer-to-peer networking strategy for audio and video transmission to achieve low latency. Multicasting is used if available to reduce the need for duplicate packet transmissions during multi-point conferences (we have tested up to four stations running simultaneously—limited by the amount of hardware we had available). A central server handles control messages that synchronize the screen compositions or movies at each station.

BACKGROUND AWARENESS AND AMBIENT MEDIA

TV is normally a foreground experience—it demands the attention of those nearby. But the same technology, coupled with a broadband back-channel, could be used in a more "ambient" mode in a variety of ways, for example to create a sense of community between remote households or work spaces. As we worked on this theme, we aimed to break out of the physical constraints of normal television screens and computer monitors and embed these kinds of systems in the surrounding architecture.

We created a system called **iCom** that connected several laboratory areas at Media Lab Europe and the MIT Media Lab 24 hours a day in order to support background awareness, informal chance encounters, and ad-hoc audiovisual meetings between remote research colleagues—things we felt were important to build a sense of family and togetherness between the two labs and at the same time were not well supported by other communication media like telephones and email. The goal was to make it feel like like we were working in one large lab rather than two different labs separated by thousands of miles. (Agamanolis, 2003b)

At each iCom location there was a large-screen projection and seating area integrated within a larger work space. (Figure 4) These screens displayed several live streams of video captured from at least two cameras in each location. The characteristics of each connected space were similar and the researchers located in these spaces generally knew each other. A sofa and coffee table in front of the screen emphasized use of the locations as informal socializing areas.

A trackball enabled users to rearrange the windows or turn on microphones for meetings or casual interaction with the remote sites. The design addressed some potential privacy issues by synchronizing the screen projections at each site: If you couldn't see yourself on the screen,

Figure 4. iCom stations at Media Lab Europe (top) and the MIT Media Lab (bottom).

then nobody else could see you—what you saw on the screen was what the other sites saw on their screens, and nothing was recorded or available to view in other places. The system conserved bandwidth by reducing frame rates where no activity was detected and by adjusting transmitted resolution to reflect the size of each video window. The use of connectionless networking protocols enabled the system to operate effectively on a commercial (and often choked) transatlantic Internet connection.

The iCom system operated more or less continuously for about four years. As such, it may be the longest running example of a *media space* connecting different locations on a continuous basis. It served as a lightweight communication tool for holding project meetings for cross-lab collaborations as well as for informal socializing between remote friends, including occasional flirting. An important goal of the iCom, like some media space efforts predating it (Bly, Harrison, & Irwin, 1993), was not to be "yet another video conferencing system", but rather to use audiovisual media in an architectural mode—to make it feel like the connected locations were *physically adjacent*, just like an conventional window provides a view between adjacent spaces. This desired architectural effect was often difficult to explain to visitors who spent only a few minutes in our space. Feedback from those working near the iCom for longer periods suggested that after an initial "novelty" period, those who frequented the iCom spaces began to think about the installations more in this spirit. Some reported feelings of isolation when their site was occasionally out of order due to a projector malfunction or network outage.

The iCom stations in our laboratories were designed to connect informal work/social areas inside open-plan office spaces. However, it is possible to imagine how the design could be altered for a home environment. Typical households have a "family room"—a room in which the family gathers for various activities, including watching television. An iCom-like system could enable a sort of "extended family room" that is connected in an ambient way to the households of distant family members or friends. This would make it possible, for example, to watch television with others who are far away in a shared space that, unlike Reflexion, extends into the physical environment.

Our group worked on other prototypes involving background awareness and ambient media. One project, the **Open Window**, involved creating an ambient audiovisual portal between hospital rooms and locations considered to be strengthening to a patient (Cullinan et al, 2004). **Habitat** explored using ordinary pieces of furniture as interfaces to convey information about the rhythms of connected households that could perhaps be used among members of a family living in different places (Patel et al, 2003). A new generation set-top box could easily serve as a communication hub for such ambient media interfaces. But, if it is not immediately practical to build such systems into a home's architecture or furnishings, traditional television hardware could still serve as a platform for these kinds of applications. For example, activity in remote households could be manifested as abstract graphics or overlays on traditional TV programmes, or live camera feeds between the connected locations could be offered on TV channels of their own.

MEETING NEW PEOPLE BY BECOMING A LOCAL BROADCASTER

Personal music players like the Walkman and the iPod allow users to insulate themselves from the outside world, which is often a positive thing given the level of stress and annoyance in modern city life. At the same time, sometimes people are curious about other people or may be interested in making new acquaintances, and this often proves difficult in an alienating urban environment where people are trained to keep to themselves. The **tunA** project addresses these desires, using the medium of music as a catalyst for breaking the ice between people who are nearby.

Developed on iPaqs and connected via ad-hoc Wi-Fi networks, tunA is a mobile wireless application that allows users to share their music locally through handheld devices. Users can listen to their own music just like a normal music player, or they can "tune in" and listen to what other people in physical proximity are listening to (Figure 5). When tuning in, tunA synchronizes any receiving devices and the source device using a patent-pending technique so that a group of users can gesture or dance together in synchrony, just as if they were listening to the same conventional radio station (Figure 6). This synchronization enables a sense of shared experience that isn't quite possible with the Internet audio streaming and podcasting paradigms in wide use today.

By default, every tunA user is their own mini radio broadcaster, potentially creating a new kind of music sharing experience that adds

Figure 5. Screen shots of the tunA prototype, which ran on iPaq handheld computers.

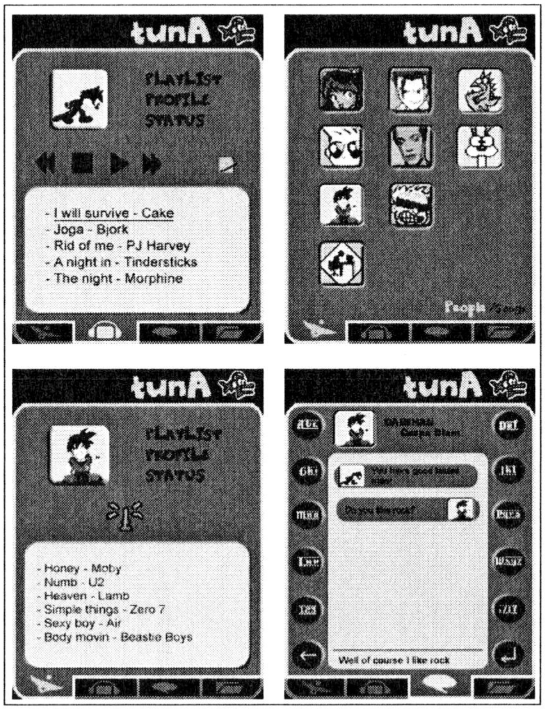

Figure 6. Depiction of the desired effect of tunA.

a social dimension to an activity that previously was usually done in individual isolation (Figure 5). tunA lets you see the playlists of other users in range and bookmark songs that others own so that you could buy them yourself later if you want. tunA lets you bookmark other users as well, so that you can be notified in the future if they come into range again.

Perhaps most importantly, tunA allows users to exchange short text messages, similar to SMS messages, without necessarily knowing anything about each other except what they are listening to. In initial user studies, this messaging feature was felt to be an essential ingredient to support the desired ice-breaking effect of the application. (Bassoli et al, 2006) If two tunA users who contact each other using this messaging feature don't end up becoming friends, the brief interaction may at least allow them to feel slightly less isolated from the communities they weave through every day.

With the advent of video iPods, 3G mobile phones, and other handheld video devices, the idea of tunA could easily be translated into the space of television. Instead of listening to someone else's music, you could watch the same TV program that they are watching. Just as in the music version of tunA, playback would be tightly synchronized, so you could have a shared viewing experience and laugh or react at the same time as other viewers around you. It is possible to imagine a fixed version of the application as well, perhaps one that allows other households in your local neighborhood to watch the same thing on their televisions that you are watching on yours, allowing you to become your own mini TV broadcaster.

SUPPORTING COMMUNICATION WITH NOVEL INTERFACES

Remote controls don't have to be the only way to interact with digital television applications, especially those that involve human communication and connectedness. In fact, one of the best activities in which people can engage to build a new relationship is sports. Traditional sports foster bonding and team spirit through the sharing of a physically strenuous competitive activity. Studies have shown that playing sports increases one's tendency to bond socially with team-mates and competitors (Zahariadis & Biddle, 2000).

Breakout for Two is a prototype "sports over a distance" installation that aims to build the same sense of community over a distance, not with email and instant messengers, but with real balls, sweat, and exertion. The system employs what we call an *exertion interface*—an interface that deliberately requires intense physical effort and can be expected to be physically exhausting when used for an extended period of time. In short, it gets your adrenaline moving and makes you sweat, just like any physical exercise or sport.

The Breakout for Two game is a combination of soccer, tennis, and the vintage video game *Breakout*. Players in remote locations must throw or kick a real soccer ball at a local physical wall to break through a projection of virtual "bricks" that overlay a live video image of the opposite player. The installation creates the effect of a "virtual game court" in which the competitors are separated by a transparent barrier through which they can speak to each other and interact (Figure 7). The bricks on each player's screen are synchronized—when one player breaks through a block, the same block disappears from the other player's screen. Players win the game by destroying the most bricks in a given time period. Games typically last several minutes and can incorporate varying levels of difficulty or two-on-two play (Figure 8).

A more recent "sports over a distance" prototype called **Remote Impact** is a variant of boxing. Unlike the Nintendo Wii and other console videogames, Remote Impact recognizes and registers intense brute force. A life-sized silhouette of the remote participant is projected on the interface, which resembles a mattress standing against a wall (Figure 9). A novel sensing system measures the location and intensity of each impact. Players can punch, kick, or throw their entire bodies against their projected opponent, and the system recognises when there has been a hit or a miss. Players can dodge hits by ducking or moving out of the way, just as in real boxing. Players can also talk to and hear each other through a voice connection between the locations.

Our theory is that augmenting an online communication environment with exertion will greatly enhance the potential for social bonding, just as playing an exhausting game of squash or tennis with a new acquaintance or co-worker helps to "break the ice" and build friendships. The heightened state of arousal induced by the exertion also potentially makes the interaction more memorable. We undertook an initial experiment to test these hypotheses and evaluate the effects of exertion interfaces, with encouraging results. For example, players of Breakout for Two said they got to know each other better, became better friends, felt the other player was more talkative, and were happier with the transmitted audio and video quality in comparison to a control group playing an analogous game using a traditional non-exertion keyboard interface. (Mueller et al, 2003) Further studies are planned as of this writing.

Breakout for Two and Remote Impact are not directly related to the realm of digital television beyond their use of video displays on a large scale. But certain lessons from these experiments are

Figure 7. A Breakout for two court installed at Media Lab Europe

Figure 8. Scenes from a match of breakout for two involving two players in each court

Figure 9. Remote impact uses a specially instrumented mattress as a projection surface and brute force game interface

potentially relevant to designers of digital TV experiences. Traditional TV viewing is often a passive experience for the audience, and the interfaces used in interactive television applications are typically hand-held remote controls with the obligatory confusing array of buttons. What if a television program required that you interact with it in a physical way, exerting yourself just like in a sport? Could this be a way of forging a richer relationship or sense of identification between a viewer and, say, the characters in the story? Video games like Dance Dance Revolution (Konami) have utilized physical exertion as an integral component of an experience, but these effects seem largely unexplored in the space of more traditional or narrative forms of television programming.

FINAL THOUGHTS

The intention of this chapter was to introduce the reader to a few projects from the Human Connectedness group and Distance Lab that may provide insight into potential future trends in digital and interactive television. At the root of all of these projects is an aim related to collapsing distance and transforming interpersonal communication and connectedness in some way, which this chapter contends is the most promising area in which to work in order to stimulate interest in and acceptance of the changes in progress in television systems in many parts of the world. Not all of the prototypes discussed relate directly to television or broadcasting, but hopefully the ideas underlying them are relevant and can inform and inspire potential future research and development efforts.

The Human Connectedness group operated under the direction of the author between 2001 and 2005 at Media Lab Europe, the Dublin-based European branch of the MIT Media Lab. Media Lab Europe closed in 2005 as a result of failed negotiations between MIT and the Irish government. However, many of the group's themes and projects are being continued within Distance Lab, a digital media research and commercialization center focused on overcoming the disadvantages of distance. Distance Lab, also under the direction of the author, was launched in 2006 in Scotland with the support of Highlands and Islands Enterprise, an economic and social development agency covering the north and west of Scotland. (http://www.distancelab.org)

ACKNOWLEDGMENT

I would like to thank all of the members of the Human Connectedness group and Distance Lab for their ideas and contributions to these projects. This research was supported by sponsors of Media Lab Europe and Distance Lab.

REFERENCES

Agamanolis, S., & Bove Jr., V. M. (2003a). Viper: A Framework for Responsive Television. *IEEE MultiMedia, 10*(1), 88-98.

Agamanolis, S. (2003b). Designing Displays for Human Connectedness. In K. O'Hara, M. Perry, E. Churchill, & D. Russell, (Eds.), *Public and Situated Displays: Social and interactional aspects of shared display technologies* (pp. 309-334), Kluwer.

Bassoli, A., Moore, J., & Agamanolis, S. (2006). tunA: Socialising Music Sharing on the Move. In K. O'Hara, & B. Brown (Eds.), *Consuming Music Together: Social and Collaborative Aspects of Music Consumption Technologies* (pp. 151-172), Springer.

Bly, S. A., Harrison, S. R., & Irwin, S. (1993). Media Spaces: Bringing People Together in a Video, Audio, and Computing Environment. *Communications of the ACM, 36*(1), 28-47.

Cullinan, C., & Agamanolis, S. (2002). Reflexion: a responsive virtual mirror. *Conference Companion, UIST 2002 Symposium on User Interface Software and Technology.*

Cullinan, C., Agamanolis, S., Roche, D., & Hegarty, F. (2004). Open Window: Reducing Isolation for Patients. *Adjunct Proceedings, UbiComp 2004.*

House, J. S., Landis, K. R., & Umberson, D. (1988). Social Relationships and Health. *Science, 241*(4865), 540-545.

Kraut, R., et al. (2002). Internet Paradox Revisited. *Journal of Social Issues, 58*(1), 49-74.

Lewis, T., Amini, F., & Landon, R. (2000). *A General Theory of Love*, Random House.

Mueller, F., Agamanolis, S., & Picard, R. (2003). Exertion Interfaces: Sports over a Distance for Social Bonding and Fun. *Proceedings of CHI 2003*, ACM Press.

Patel, D., & Agamanolis, S. (2003). Habitat: awareness of life rhythms over a distance using networked furniture. *Adjunct Proceedings, UbiComp 2003.*

Putnam. R. (2000). *Bowling Alone: The Collapse and Revival of American Community.* Simon & Schuster.

Zahariadis, P., & Biddle, S. (2000). Goal Orientations and Participation Motives in Physical Education and Sport: Their relationships in English schoolchildren. *Athletic Insight—The Online Journal of Sport Psychology, 2*(1).

Human Connectedness group web site: http://www.medialabeurope.org/hc

Distance Lab web site: http://www.distancelab.org

Chapter III
Interactive TV Together:
An Open Service Infrastructure for Enhancing Interactive TV Experiences

Cristian Hesselman
Telematica Instituut, The Netherlands

Joost Broekens
Telematica Instituut, The Netherlands

Mark Gülbahar
Institut für Rundfunktechnik, Germany

Florian Winkler
NEC Europe Ltd., Germany

Daniel Görgen
Philips Research, The Netherlands

Ferry de Jong
Logica, The Netherlands

John Vester
Logica, The Netherlands

Wijnand Derks
Telematica Instituut, The Netherlands

Robert de Groote
Telematica Instituut, The Netherlands

Andrew Tokmakoff
Philips Research, The Netherlands

Daniele Abbadessa
NEC Europe Ltd., Germany

João da Silva
NEC Europe Ltd., Germany

Dirk-Jan van Dijk
Telematica Instituut, The Netherlands

Xiaoming Zhou
Philips Research, The Netherlands

Remco Poortinga
Telematica Instituut, The Netherlands

Sander Smit
Logica, The Netherlands

Hugo ter Doest
Telematica Instituut, The Netherlands

Martin Snijders
Telematica Instituut, The Netherlands

Rogier van Laar
Logica, The Netherlands

ABSTRACT

One of the challenges in the world of interactive digital TV is to improve the user experience facilitated by these services. In this chapter, the authors discuss their approach towards reaching this goal, which is to integrate community and interactivity services ("Web 2.0"-style) of third-party providers from out-

side the world of IPTV into an IPTV service offering (e.g., services on the public Internet, and services offered by telco operators and end-users). The foundation of this work is an open service infrastructure that facilitates this form of integration. The authors discuss the set of service enablers that make up the infrastructure and present a working prototype implementation that serves as a proof of concept of their approach. They also outline four possible scenarios for the future of IPTV, which are based on a trend analysis and form the basis for developing business models made possible by their infrastructure.

INTRODUCTION

Interactive digital TV is a service that is traditionally offered via digital broadcast networks such as DVB (Reimers, 2006), but more recently also became widely available through IP networks. This IP-based branch of interactive digital TV is usually referred to as "IPTV" and emerged as a result of the rapid proliferation of broadband networks (Tadayoni, 2006). IPTV services use the inherent two-way nature of these networks to support interactivity, although the actual level of interactivity depends on how deep it goes into the system (Tadayoni, 2006). IPTV is typically available through managed networks, which means that an operator controls the transmission properties of the network. This allows the operator to control the perceptual quality of its IPTV services, which need to be at least as high as traditional broadcast TV services. Examples of network operators are ISPs, cable television companies, and telecom operators. IPTV offers new market opportunities for such players, who often provide IPTV services as part of a "triple play" service offering (i.e., IPTV plus traditional telecommunications services plus internet access). IPTV usually also involves a video on demand component in addition to linear TV.

While IPTV services extend the traditional TV viewing experience with remote control-based interactivity and typically higher quality content, there is a lot of ongoing work on how to further enhance these experiences, often by adding community and advanced interactivity services ("Web 2.0"-style) to IPTV services. Recent studies show that users seem to value such additions (Abreu & Almeida, 2008; Schatz, Baillie, Fröhlich & Egger, 2008; Miletich, 2008; Boertjes, Klok & Schultz, 2008), although their level of appreciation appears to depend on various parameters, such as the TV content being shown and the interaction modality used (e.g., voice or keypads) (Schatz, Baillie, Fröhlich & Egger, 2008), the types of users involved (e.g., teenagers or retired people) (Miletich, 2008), and the types of services being added (Boertjes, Klok & Schultz, 2008). Examples of such add-ons are "home grown" content of friends, and chat, presence, and recommendation services.

In our work, we also extend IPTV services with community and interactivity services, but we use *third-party services* from *outside* the world of IPTV to accomplish this (cf. (Baca, 2008; Lowet & Khmelinskaya, 2008)). Examples are services from the Web (e.g., e-commerce services or community services), from the world of mobile telecommunications (e.g., billing and accounting services), and services that run within the home (e.g., a networked VCR). As we believe that community services are one of the prime ingredients in an extended IPTV service offering, we call our approach *interactive TV Together* (iTVT) and refer to an IPTV service enhanced with third-party services as an *iTVT service*. Within this model, we think of a content provider (e.g., a provider offering long-tail content) as a special kind of third-party service provider.

Figure 1 shows an example of an iTVT service that serves a group of people at three different

Figure 1. Example of an iTVT service

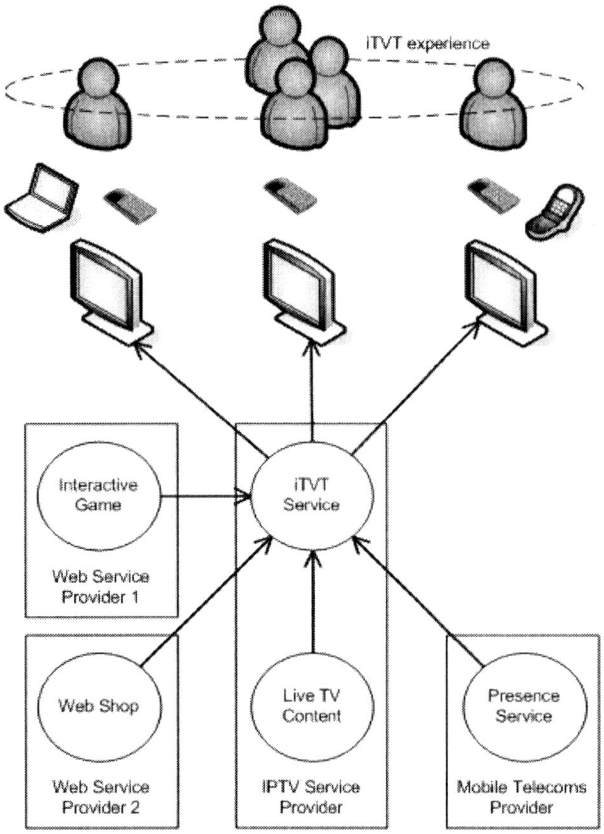

homes. The service consists of an IPTV service and three third-party services from outside the world of IPTV: an interactive gaming service (operated by a service provider on the Web), a shopping service (also operated by a Web service provider), and a presence service (offered by a mobile telecommunications operator). Observe that in this example the IPTV service provider combines the third-party services and its IPTV service into an iTVT service. To simplify the discussion, we will assume this model throughout this chapter, but in general it could also be the responsibility of a separate party to carry out this integration.

The advantage of an iTVT service is that an IPTV service provider can leverage the unique characteristics of the "foreign" third-party services and use them to compensate for the limitations of IPTV services. For example, services from the Web typically support communities, advanced interactivity, and sharing of user generated content, which are features that IPTV services typically lack. Conversely, IPTV services provide high quality multimedia content and provide billing services, whereas multimedia content on the Web is usually best-effort and billing is often problematic. IPTV service providers can thus enhance the interactive TV experiences they facilitate by teaming up with third-party providers from outside the IPTV domain, which means they do not have to realize these services themselves. This reduces the cost of providing and operating iTVT services and creates new business opportunities for third-party service providers. Another advantage is that in the long run "triple play" may no longer be a differentiator for IPTV service providers as more and more of them will be offering this type of service. In this case, the key to differentiation could be the integration of additional (cross-technology) services into an IPTV service (e.g., well-known Internet services), as advocated by our approach. Another advantage is that our approach allows service innovations that take place outside the realm of IPTV to quickly find their way into the world of IPTV, in turn fuelling new interactive TV experiences.

To make iTVT services a reality, we believe we need an *open service infrastructure* that enables third parties to seamlessly add their services to an IPTV service. The major challenge in realizing such an infrastructure is to facilitate *interoperability* as the third-party services we are using are part of different "technology domains" (the Web, IPTV, in-home networks, and mobile telecommunications). By this we mean that, while all based on IP, the services in an iTVT service offering (IPTV and third-party) use different technologies at the service level because they were originally designed for specific types of multimedia services and to operate in specific environments. For example, Web technology typically forms the foundation for best-effort on-demand multimedia services operating on the public internet, whereas IPTV technologies usually focus on the delivery of high quality professional multimedia content in tightly managed environments. These differences are exemplified by the different metadata formats, protocols, and user identities in use in the different technology domains. Overall, the result is a multimedia landscape that is organized in a vertical manner, as illustrated by the four "silos" in Figure 2. The service infrastructure we envision forms a horizontal layer across these technology domains.

In the rest of this chapter we discuss the cross-domain service infrastructure we have developed and explain how the infrastructure's components (service enablers) contribute towards extending IPTV services using third-party services from non-IPTV "silos". We also describe a prototype application that uses the infrastructure and discuss a model for four possible futures for IPTV. We developed the latter as a starting point for understanding the business models that the service infrastructure will make possible, which is part of our future work. We end with a discussion on related work, conclusions, and a brief overview of future work.

We note that while the discussion in this chapter focuses on IPTV, the idea of an iTVT service (i.e., IPTV plus third-party services) can also be applied to non-IP-based interactive TV services, such as DVB services.

Figure 2. Cross-domain service infrastructure for iTVT services

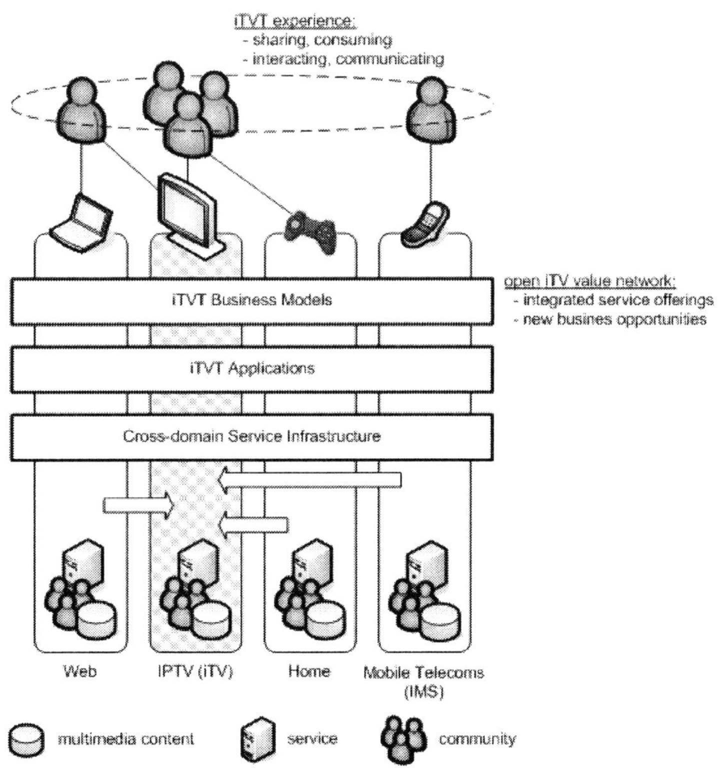

SERVICE INFRASTRUCTURE

The infrastructure we envision consists of a set of *service enablers*, which are building blocks that facilitate iTVT services (cf. the example in Figure 1). Service enablers can be used and operated as services by different parties in different technology domains (e.g., by telco operators or service providers on the web) and also allow application developers to create iTVT services. Service enablers essentially form a horizontal layer across the technology domains we distinguished in Figure 2.

We designed and developed four enablers that we believe are crucial for cross-domain iTVT experiences:

- A *metadata integration service* that unifies the metadata formats in use in different technology domains;
- An *identity management service* that allows third-party services to be integrated into an iTVT service in a user-friendly and anonymous way;
- A *session management service* that controls the establishment and release of the connections involved in an iTVT session; and
- A *personal recommender service* that provides per-user recommendations and facilitates personalized experiences.

iTVT applications use these enablers in an application-specific manner to integrate third-party services with an IPTV service. iTVT applications can run on client devices (e.g., a set top box or a home gateway), on servers in the infrastructure, or both. Client and server-side integration each has its own merits. For example, performing a large part of the integration through client applications provides scalability advantages because most of the intelligence of the system is pushed to the edges. Also, client applications can usually reach services in multiple technology domains at the same time (Internet services, in-home services, and IPTV services), which makes it a natural integration point. An advantage of using server-side iTVT applications is that they provide an increased ease of use: IPTV service providers can simply integrate new third-party services into an existing iTVT service without users needing to install a new client-side iTVT application. As we will see later, our prototype uses both client and server-side iTVT applications. Note that our service enablers are not only used by iTVT applications, but that they also use each other (e.g., the personal recommender service uses the identity management service).

We will discuss our four iTVT enablers in more detail below. One of our main goals in designing them was to use existing standards where possible.

Metadata Integration Service

The iTVT metadata integration service unifies proprietary metadata formats from several technology domains into one format being used internally by all iTVT enablers (TV-Anytime). The metadata integration service bridges the differences between metadata formats in a way transparent to its clients and can handle proprietary formats as well as formats used for internal and external representation of multimedia content (Rein & Winningham, 2008).

In our work, we use the metadata integration service for a cross-domain search engine ("google-style"). This engine can potentially also interact with the personal recommender service to filter out search result based on user preferences. Another potential application is a cross-domain Electronic Program Guide (EPG) that provides metadata on available content sources across technology domains, for instance from both outside the home (e.g., VoD content available from an IPTV service provider) as well as within the home (e.g., content stored on UPnP Media Servers).

Internally, the metadata integration service uses the "data source perspective", known as "federated data orchestration" (Lahanas, 2008), which allows services from different technology domains to maintain their internal formats and *bridges* the differences between them. The advantage of this approach is that it eases the integration of multimedia content from "auxiliary" technology domains into an IPTV service. An alternative approach is to use a single metadata format for all technology domains and force them to migrate to that format, but this requires major modifications in multiple technology domains, which we believe is unrealistic. We chose TV-Anytime (Phase II) (2008) as the internal data format because it is a standard, and de facto the only reliable one. As it is based on XML, TV-Anytime is also an extendable format, which is an important requirement to be able to deal with new metadata languages.

Mapping domain-specific metadata into TV-Anytime is relatively straight forward if the metadata uses well-defined classification schemes. For example, broadcast content usually comes with metadata on the genre of the content, which can then easily be mapped into TV-Anytime. However, for "Web 2.0" content sources this is more complicated because a unique content classification scheme does not exist. Instead, users typically tag their content using their own classification scheme and usually also assign a weight to each of their tags. In order to allow the metadata integration service to also work with this type of metadata, we mapped each tag to the "keyword" language construct in TV-Anytime and mapped the weight of that tag to a numerical value. To achieve the latter, we extended the XML scheme as TV-Anytime can currently only associate three discrete values with the "keyword" construct ("main", "secondary" or "other"). This extension allows tags and their weights to be represented more precisely in TV-Anytime.

In addition to incompatible content classifications across technology domains, there are also a few challenges to be addressed within the Web domain. The first problem is that Web 2.0 providers currently only expose metadata in the form of HTML pages and do not provide any programmatic external interfaces for metadata retrieval. This makes it more difficult to use the metadata repositories of such service providers, which complicates metadata integration. In addition, Web service providers typically use their own proprietary metadata format, which creates an additional interoperability problem, in this case within the Web domain (intra-technology domain). Several standardization bodies are working on this topic, trying to enable content discovery and content description for the (emerging semantic) web, and, in particular, on the Resource Description Format (RDF, 2008). A TV-Anytime based OWL ontology (OWL, 2008) is being created for this purpose, which will allow Web 2.0 service providers to provide metadata in a standardized way.

To bridge the time until these standardization efforts have been realized, we included a parser to the metadata integration service that converts Web 2.0 query results to TV-Anytime compliant metadata, using regular expressions.

Observe that the quality of the metadata associated with a particular content item is an important parameter in metadata integration. In more managed environments (e.g., in the world of broadcasting) metadata is usually of a relatively high quality, but this is often not the case for "Web 2.0" and semi-professional content. Most Web 2.0 platforms do not manage the existence and quality of metadata when new content is being uploaded by a specific user, and if they do so, it is rather in the early stage of development. This, however, is a research topic that is outside the scope of our work.

We implemented the metadata integration service as a web service that can run in any technology domain, for instance within a home environment (e.g., on a DLNA gateway), or at a third-party service provider.

Identity Management Service

The iTVT identity management service is based on OASIS SAML 2.0 (2008) and Liberty Alliance (2008) standards and provides the usual identity management services:

- Storage of personal information such as the credit card numbers of users, their usernames and passwords, home address, and their age;
- Single point of access for users to manage their personal information and for service providers to obtain it;
- Configuration of policies that enable end-users to control who can receive their personal information and under which conditions;
- Invocation of authentication services (e.g., those supported by IMS) to authenticate a user; and
- Generation and handing out of pseudonyms for successfully authenticated users. A pseudonym in this context is a "token" that gives the holder of that token the right to obtain information about a particular user from the identity management service without reauthentication. At the same time, a pseudonym also hides the user's identity, which enables them to remain anonymous.

As is usual for this type of service, the identity management service must be operated by a provider that the user trusts. This provider is usually called an Identity Provider and can be accessed by the service providers affiliated with the Identity Provider. The service providers affiliated with a specific Identity Provider form a so-called circle of trust.

The additional feature of the iTVT identity management service is that we enhanced it with a client that can be run on a gateway device in the user's home (e.g., an IPTV set-top box). This application handles logon interactions with the user (e.g., using an NFC card), requests a set of pseudonyms from the identity management service in the network, and then caches these pseudonyms on the gateway device for later use.

A major advantage of integrating the iTVT identity management service with a home gateway device is that it helps to facilitate a seamless user experience, which is crucial for the acceptance of iTVT services. First of all, the pseudonyms allow for true single sign-on, which is essential for making iTVT services easy to use. Users will not have to enter (or remember) usernames and passwords for the different third-party services in an iTVT service offering, but profit from the single logon which is done when they switch on their IPTV device. This is particularly important in IPTV environments in which input devices will typically be less sophisticated than in the world of PCs, making the entry of personal information often a tedious process. The identity management service enables users to enter and modify such personal information outside an iTVT session, for instance using a PC, and store it at their Identity Provider. Finally, the pseudonym cache that the iTVT identity management service uses allows the gateway device to interact with third-party service providers more quickly, which further improves the iTVT experience.

From the perspective of IPTV and third-party service providers, any identity management service is a crucial enabler for an iTVT service. First of all, it enables IPTV providers and third-party service providers affiliated with a certain Identity Provider to obtain information about users from a single point, which greatly simplifies the setup of business collaborations between them. These business relations can also be changed more easily, which allows the IPTV service provider to respond to market changes more quickly, thus reducing customer churn and increasing user acceptance. Another advantage is that third-party services can charge the user directly, for instance by retrieving the credit card information from the Identity Provider, or they can charge the IPTV service provider they are in a circle of trust with.

By charging the IPTV service provider on a pseudonym basis, it is even possible to entirely protect the user's identity from the third-party service provider during business transactions. This allows users to consume commercial services without exposing their identity. The billing can be entirely performed by their IPTV service provider. Anonymous service usage also allows an IPTV service provider to comfort the privacy concerns of its customers.

Session Management Service

The session management service is responsible for establishing, updating, and releasing iTVT sessions. To accomplish this, the service interconnects user interaction devices, rendering devices (e.g., UPnP renders), and services. As illustrated by the example in Figure 1, an iTVT session may involve services from different technology domains (e.g., streaming services, interactivity services, and telecommunication services), while the interaction and rendering devices in a session may be distributed across multiple homes.

The session management service enables its clients to establish an iTVT session by means of a single request that specifies which services and devices need to be connected together. The service hides the details of the underlying cross-domain interactions, thus reducing the complexity of client applications.

Figure 3 shows the functional components of the session management service superimposed on an example involving two users (Alice and Bob). The main components are Session Control Points, Media Processors, the iTVT Client Loader, and the iTVT Core (we omitted the iTVT Core from Figure 3 to improve its readability).

A *Session Control Point* (SCP) forms the entry point for the session management service and realizes the service's interface. An SCP represents a user identity and manages iTVT sessions on behalf of that user. It allows users to create iTVT sessions, and subsequently add (remote) users, services, and content sources to it. In our system, an SCP maintains a local view on a multi-home iTVT session, which consists of the session's ID, the IDs of the session's users, and a description of the content sources and services involved in the session.

An SCP supports multiple signaling protocols so that it can establish connections with devices within its local home (intra-home), with devices in remote homes (inter-home), and with services and content sources in different technology domains. Our SCPs use SIP and IMS (Magedanz, Witaszek & Knuettel, 2005) to (1) interact with a peer SCP of another home and invite users in that home to join an ongoing iTVT session and (2) to interact with IMS-based services (e.g., an IMS presence service or a video conferencing service). We have developed extensions to the Session Description Protocol (SDP) that allow us to represent iTVT sessions and manage sessions spanning multiple locations and devices. These technologies can also be used to manage inter-home A/V communication and synchronized inter-home web-based services (Lowet D & Shrubsole, 2007). For each content source, an SCP also tracks the streams that flow into the home and the rendering devices to which the flows are routed (possibly passing through a Media Processor, see below).

To add consumer electronics devices within a particular home to an iTVT session (e.g., network-attached storage), we rely on UPnP/DNLA. We assume media sinks are UPnP media renderers or DLNA remote UI clients, which are based upon Web4CE technologies (Dees & Shrubsole, 2007). Web4CE provides a declarative application environment based on existing standards, such as XHTML, CSS TV profile, and DOM level 2, together with various extensions to make it more appropriate to a consumer electronics environment (e.g., handling of remote control key events). By supporting UPnP/DNLA, our SCP can also manage sources for user-generated content (e.g. uploading/live streaming/communication).

A *Media Processor* is an optional component that can manipulate multimedia content that comes

Figure 3. Session management architecture

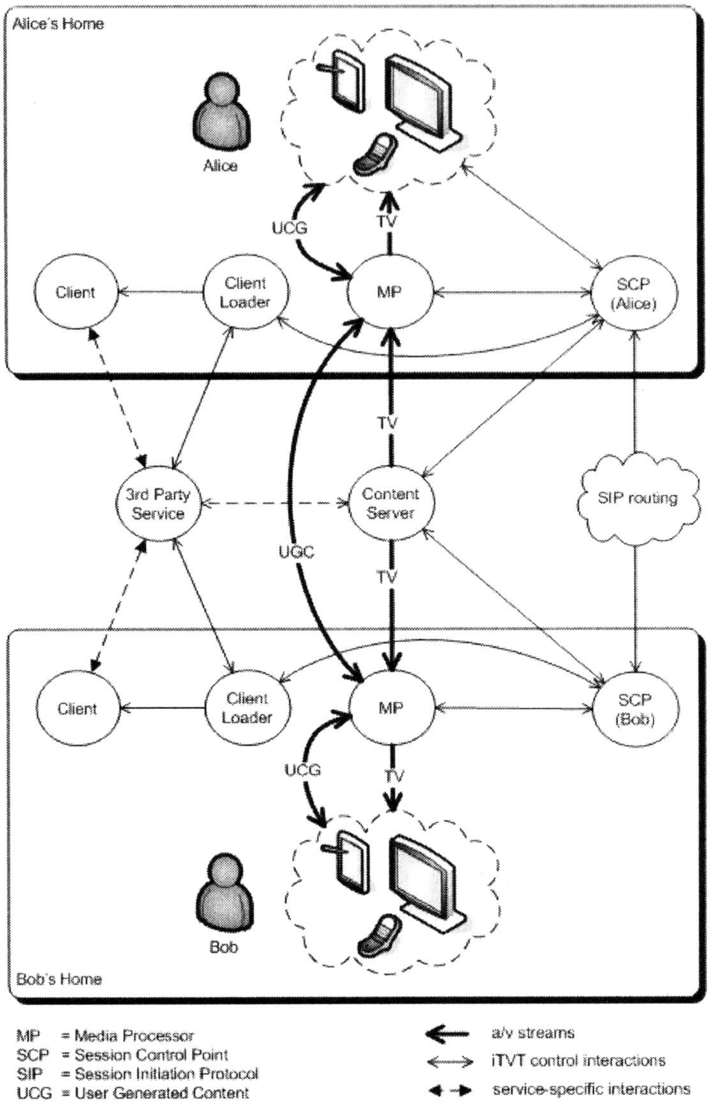

into or flows out of a home. It can for instance downscale high-definition content so that it uses less bandwidth. A Media Processor may also have (application-level) multicast capabilities so that resource-constrained devices in the home (e.g., mobile phones) can easily generate and share live content streams in a multi-user/multi-home iTVT session.

The *Client Loader* is a component that downloads the client logic of a particular third-party service (e.g., in the form of a Java script) from the network and then starts it. The SCP instructs the Client Loader which clients to download based on the requests the SCP receives. The *iTVT Core* acts as a central entry point for client devices (which are based on Web4CE) and contains an EPG service. The core also provides a generic user interface for the session control point, for instance to show SCP-related events on a TV screen (e.g., an invitation to join an iTVT session).

The Media Processor, the Session Control Point, and the EPG Server all register with the home's Service Registry, which is a support facility that enables its clients to discover services in the home. In addition, the devices in the home model their rendering and interaction facilities as services and register them with the Service Registry as well. Mobile devices may also register with the Service Registry if they connect to the in-home infrastructure. UPnP service discovery could be used to realize this service, but actual service discovery mechanisms our outside the scope of our work.

We expect the above components to be deployed in-home, perhaps on a single device such as a home gateway, a network addressable storage device or a home theatre PC. It is also possible that some of the components run in the network rather than in the home (e.g., the Media Processor).

Personal Recommender Service

The personal recommender service provides recommendations about multimedia items for specific users and facilitates the personalization of iTVT services. Personalization is important for several reasons, for instance because it enables the third-party services in an iTVT service offering to better target individual users (e.g., using personalized ads). In addition, it helps users to find interesting content in a timelier manner and helps to reduce information overload (Herlocker, Konstan, Terveen & Riedl, 2004; Resnick & Varian, 1997; van Setten, 2005). The latter is particularly important in an iTVT-like environment, in which the integration of several technology domains results in a steep increase in the number of available services and content. At the same time, this also provides new opportunities because the availability of a wider range of content types (professional, user generated) also makes it possible to extract additional relevant user preferences. This, in turn, can help generate more diverse recommendations, an important issue in content recommendation services (Smith & McClave, 2001). Ultimately, it might even further enhance a user's content consumption experience.

The iTVT personal recommender service uses an *agent-based and user-centric approach*. By this we mean that it represents each user as an agent and that clients can ask that agent for a recommendation about what the user likes, similar to how anyone would ask a question to a user in real life. This is a better understandable model for users, enables the user to be in full control of its preferences as well as who obtains these (Broekens, 2008). Such a user-centric, agent-based approach should allow group-based recommendations, such as in the work by Masthoff (2004), and collaborative filtering mechanisms. Both types of recommendations need cross-user model integration, which might be limited in our system as a result of to the agent-based approach. This is an area of further study.

A second cornerstone of the iTVT recommendation approach is that we explicitly support *third-party recommendation services*, which might reside in different technology domains. The underlying rationale is that these domain-specific recommendation services are usually optimized for specific types of content items (e.g., movies offered by an IPTV service provider or user generated content in someone's home) and that iTVT applications would benefit from being able to use the most appropriate recommendation service. A recommendation service might be coupled to content provider, but it could also run on someone's networked VCR at home.

To add these (domain-specific) recommendation services to our infrastructure, we use the iTVT *recommendation broker*. This is a component that allows providers of recommendation services to register their service with the broker of a particular user. The broker acts as a recommendation-providing agent for that user and therefore enables clients of the broker (e.g., Internet shops) to use those recommendation services the user has registered for in a transparent way. Recommendation requests by clients are forwarded by the broker to the appropriate recommender. In the iTVT architecture, we assume that the recommendation services that have registered with the broker are also responsible for user profiling (i.e., data gathering for the purpose of recommendation). The recommendation broker thus forms the recommendation and profiling access point of a particular user for all components that require recommendations through the iTVT service infrastructure. As a user configures one of the recommendation services registered with his broker as his default recommendation service (e.g., the recommender on his set-top box).

To actually obtain a recommendation, an iTVT application (e.g., an EPG or an Internet shop) sends a recommendation request to the broker of a particular user. This request includes the user's anonymized identity (a pseudonym) and the TV-Anytime metadata of the content item for which the application is requesting a recommendation. The broker translates the anonymous identity to the correct profile ID used by, for instance, the default recommender service. Next, the broker forwards the request to that recommendation service, which returns a recommendation object. This object contains a preference value and a certainty value, both between 0 and 1. The broker also supports an interface that allows iTVT applications to get a recommendation for multiple TV-Antytime metadata objects in one go.

The recommendation broker also consists of a profiling part that enables iTVT applications to create, delete, and update user profiles. For example, when an application wants to update a user's profile based on explicit or implicit user feedback, it sends an update request to the broker. This request includes, as before, the user's pseudonym, the TV-Anytime metadata describing the associated content item, and the updated rating and certainty. Based on the pseudonym, the recommendation broker asks the identity management service for the correct profile ID and routes the update requests to the user's default recommendation service, which updates the user's profile according to its profiling algorithm. This algorithm can therefore be freely chosen by the party that provides the recommendation service. A rating is a number between 0 and 1, where 0 means the user has no interest in the metadata and 1 means the user has full interest. A second interface is available for updating a user's preferences for multiple TV-Anytime metadata objects.

As a result of our user-centric and open approach, we tightly integrated the recommendation broker with the iTVT identity management service, in particular to provide anonymous recommendations. The identity management service couples user identities to pseudonyms, but in this case also associates pseudonyms to profile identifiers and identifiers of recommendation services. As indicated earlier, the broker resolves pseudonyms into profile IDs by contacting the identity management service. Public/private encryption mechanisms ensure that the recommendation broker remains unaware of pseudonym-profile mappings, thus enforcing user privacy.

PROTOTYPE

In order to demonstrate and validate our approach, we have developed an iTVT application that makes use of the iTVT service enablers to provide an iTVT service (i.e., an IPTV service enhanced with third-party services). The application supports a scenario in which two friends (Alice and Bob) play along with an interactive live TV Quiz Show and buy products related to the quiz after the show (cf. Figure 1). The application partly runs on client devices in users' homes (interacting with the identity management service, the session management service, and the metadata integration service) and partly on a server in the infrastructure (interacting with the identity management service and the personal recommendation service).

The IPTV service provider in our prototype is represented by a DVB-IP server, which streams live content to Alice's and Bob's TV sets. The third-party services are a games provider on the Web providing additional questions for the users at home, an internet shop, an IMS presence service for Alice's and Bob's buddies, and a cross-domain metadata search service. Table 1 provides an overview of the technologies we used for the each of the components in the prototype.

Figure 4 shows a few screen shots of the prototype. The setup of the prototype involves two TVs and two PCs, with each TV-PC combination representing a home (Alice's and Bob's). A third PC hosts the streaming service, the third part services, and the enablers (except for the session management service, which runs on both of the client PCs).

The detailed application scenario consists of number of steps, the most important of which are:

1. *Single sign-on.* Bob logs on to the iTVT system using his NFC-enabled mobile phone and signs up for the quiz show. In response to the NFC interaction, the iTVT client application calls the identity management service and uses one of the pseudonyms it provides to log Bob onto the quiz service on the internet.
2. *Inviting an IMS buddy.* Before the quiz show starts, Bob invites his friend Alice to join. Bob uses a TV-based presence client for that purpose (Figure 4a), which is part of the iTVT client application and interacts with the IMS presence server. Bob interacts with the client using his remote control. The session management service sends an IMS-based invitation to Alice's TV, who can then choose to accept it and play along with the quiz (Figure 4b).
3. *Playing the Web quiz.* During the TV quiz, the iTVT client applications in the homes of Bob and Alice receive additional questions from the quiz questions service on the internet. The clients show these questions as an overlay on the game show so that Bob and Alice can answer the questions while watching the TV show (Figure 4c). Bob and Alice use their remote control to provide their answers. The iTVT client application synchronizes the additional questions from the internet with the TV content from the IPTV service provider through so-called "cue points", which are event markers that the IPTV service provider inserts into its content stream (Seeliger and Arbanowski, 2008). The iTVT client fetches a new ques-

Table 1. Technologies used in the prototype iTVT service (3PS = third-party service)

Component	Type	Technologies
Streaming service	IPTV	DVB-IP, cue points
E-shop	3PS, Internet	Web service
Quiz questions service	3PS, Internet	Web service
Presence service	3PS, IMS	IMS presence service, SIP
Metadata search service	3PS, Internet	Web service
Metadata integration	Enabler	TV-Anytime
Identity management	Enabler	SAML 2.0, Liberty Alliance
Session management	Enabler	OSGI bundles, SIP
Personal recommender	Enabler	Web service, Duine
Client application	Application	OSGI, MHP, VLC

tion from the Web service provider whenever it detects a cue point. It uses VLC for rendering the content from the IPTV service provider and MHP to retrieve the cue points from the content stream.

4. *Multi-device interaction.* Alice can also participate in the iTVT service via her Internet tablet and use it to answer the questions. In this case, she uses the browser on her table PC instead of her remote control. The tablet PC shows the additional questions from the internet service, whereas the TV continues to show the TV content. The iTVT client application on the TV retrieves the cue points from the content stream and sends them to Alice's tablet PC via a WLAN connection. The application on the tablet PC subsequently obtains a new question from the internet and displays it.

5. *Cross-domain metadata search.* After several rounds of the game, Alice notices a "joker" icon in the lower right part of her TV (the "Batman"-like shape in Figure 4c). Selecting this icon enables her to quickly carry out a cross-domain search to find an answer to a question she does not know. Under the hood, the iTVT client invokes the cross-domain search engine, which in turn interacts with the metadata integration service to obtain metadata from different technology domains. In our current implementation, we only support Web search engines (Google, YouTube, and Flickr), but we will extend this to other technology domains in the future.

Figure 4. Screen shots: (a) TV and IMS buddy list; (b) invitation to join an iTVT session; (c) game play with buddy scores; and (d) recommendation and buying products. TV content © BNN/Eyeworks.

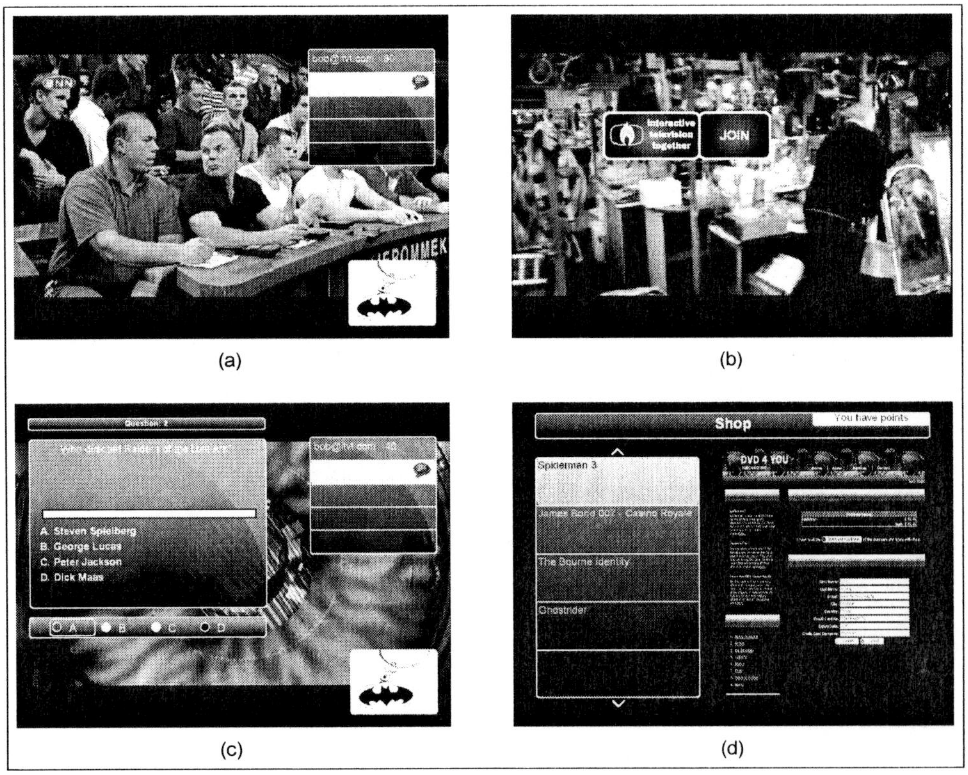

The iTVT client renders the result of the query as an overlay on the TV content.

6. *Buying a product.* After the quiz, the Web provider of the additional questions redirects Alice and Bob to an online DVD shop to allow them to purchase products with the credits they earned. The questions provider uses the pseudonyms it received from Alice's and Bob's iTVT clients in step 1 to transparently log Bob and Alice onto the web shop. The web shop, in turn, interacts with the personal recommender to get a recommendation on products (DVDs) that Alice and Bob might be interested in based on the answers they gave during the quiz (the recommender currently assumes that correctly answered questions express interest in that topic and vice versa). If Alice decides to buy a DVD, the e-shop asks if it should get Alice's personal information. If she agrees, the web shop interacts with the identity management service to obtain Alice's personal information.

IPTV FUTURES

Our open service infrastructure facilitates new business possibilities, both for IPTV service providers as well as for third-party service providers. IPTV service providers can for instance provide a more attractive value proposition, like personal channel packaging or dynamic advertisement insertion (Chorianopoulos & Spinellis, 2007) without major investments in their existing infrastructure, while third-party service providers can more easily play a role in an enhanced IPTV service offer. In addition to established broadcasters, the open service infrastructure also creates opportunities for new players in the media industry, for example for companies from the IT industry (Chorianopoulos & Spinellis, 2007).

To understand which business models will be viable for enhanced IPTV services, we first need to get a grip on the possible paths that the future of IPTV might take. We identified *four such paths based technological, economical, social and judicial trends* and are currently using these futures to develop a few viable classes of business models for iTVT services. These models can be used by existing and potential new players in the IPTV market (e.g., service providers, equipment manufacturers, and system integrators) to develop future-proof business strategies for IPTV. For example, they could be used to decide on how to integrate advertisement services in iTVT services, which may depend on the way advertisers are being charged for their services (e.g., per campaign, per positive response, per reached user, or through one of the many other models available (Anderson, 2006)).

We used an approach inspired by the method of scenario planning (Van der Heijden, 1996) to develop the four future paths for IPTV. The general rationale behind this method is to make multiple possible futures explicit so that it becomes possible to design for some or all of these futures, thus making the design more future-proof. In our case, we used this methodology the scenario planning method to create robust business strategies for IPTV and went through the following steps:

1. We analyzed several trends in the market of IPTV and in society at large from a technological, economical, social, and judicial point of view. Examples of these trends include the Internet as dominant distribution channel, consumers as producers, more fragmented communication, and increased influence of international regulations.
2. We identified trends that are both uncertain (i.e., questionable whether they will become a reality in the future) and have a high impact on the IPTV business.
3. We grouped the trends with high impact and high uncertainty into four possible futures.

Figure 5 shows the result. The horizontal axis indicates the level of "connectedness" of a user, which we define as the degree at which a user wants to feel part of a community and wants to take his personal interests or that of the group as a starting point (Gartner, 2007). The extremes are "collective" and "individual", where the first means that users feel part of a (large) community and as such conform to the community's preferences, whereas the latter indicates that users do not feel part of a (large) community and take their personal interests as a starting point.

The vertical axis indicates the level of "involvement" of a user in an iTVT session. The two extremes are "contributing", which means that users are generate their own content and communicate with other users ('lean forward'), and "consuming", which means that users do not create their own content, but rather 'lean back' and select the content of their interest. In this last case, users thus rely on other parties to provide content that fits their interest.

We named the four scenarios for the future after the type of users that will be typical for each of these futures:

- *Consumator.* In the consummator scenario, users are part of large socio-political groups in society and want to select the user profile they feel most comfortable with. Content suppliers adjust their content offerings to the specific profiles. Professional editorial offices guarantee high quality content that fits the target group. Because of standardization of offerings, chances are high for two like-minded individuals to see the same programs.
- *Cocooner.* Cocooners are users who prefer to select the content that fits their interest most. To accomplish this, their viewing behavior needs to be monitored, such that an electronic system can suggest programs that accurately match their personal taste. Advertisement brokers use this information to serve the individual with relevant advertisements.
- *Clubber.* Clubbers are people with a common interest who use a central platform where they organize themselves. Members view TV programs or movies as a collective. They exchange personal comments, viewpoints, and personal experiences to foster knowledge and growth of the community.
- *Friends-for-friends.* In a friends-for-friends future, the focal point of communication is between devoted friends. Friends love the same programs, but may have different backgrounds, hobbies and may live far away. They use networked systems to keep in touch in an intimate way, see movies together and show each other holiday pictures.

These possible futures (or 'types of usage') are extremes and are meant to make a clear distinction in possible successful scenarios for iTVT services. Hence, it may be perfectly possible, that all four types of usage become a reality. To see how, just imagine a regular evening, where a user comes home from work. He is tired and initially just wants to see the news (Consumator). After the news, this person likes to see that favorite and

Figure 5. Scenarios for the future of interactive TV.

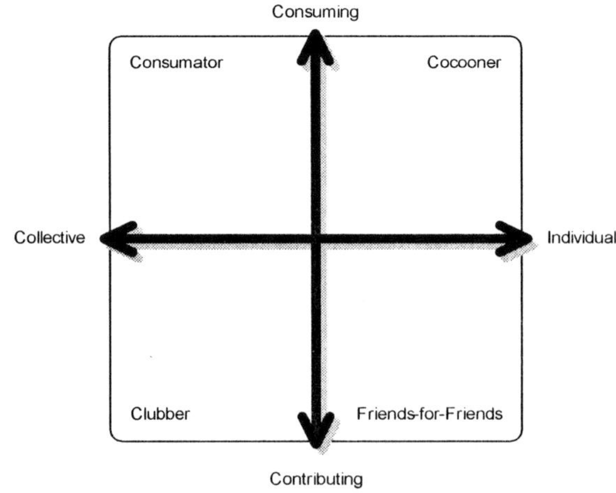

long-awaited movie (Cocooner) and when the movie has finished, he is ready to see how friends are doing (Friends-for-friends). After a short video chat with friends, the user finally sees whether members of the hockey club are still online and see what match they are discussing (Clubber).

RELATED WORK

There exist many systems that integrate "Web 2.0"-style services into an IPTV service (Abreu & Almeida, 2008; Schatz, Baillie, Fröhlich & Egger, 2008; Miletich, 2008; Boertjes, Klok & Schultz, 2008), but to the best of our knowledge there has only been limited previous work on an underlying open service infrastructure that enables applications to integrate *third-party services from outside the world of IPTV* into an IPTV service offering.

One example is the work of Lowet and Khmelinskaya (2008), who propose a browser-based system for making third-party services available on TV sets. Their browsers support CE-HTML and run on TVs. They expose an API that lets third-party service providers develop multi-user interactive TV applications, such as multi-player games. The API enables the service providers to obtain information about users (e.g., the nicknames of their buddies), invite them to a multi-player session, and set up audio and video channels between the participants. Third-party service providers can load their application logic into a browser by means of a Java script. We believe CE-HTML browsers like these could be integrated into our system, for instance by combining it with the identity management service to obtain information about end-users and with the session management service to set up audio/video sessions between multiple homes.

Baca and Holtzman (2008) describe a system that works the other way around: they extended a digital video recorder to call the open API provided by Facebook and get information about the movies that a viewer's Facebook friends like (e.g., in the form of ratings and recently watched movies). The system also posts similar information back into to the Facebook network. This work does not use any underlying service enablers, but could benefit from our enablers (e.g., from the identity management service so that the system can sign on to Facebook in one go).

The Open IPTV Forum (2007; 2008) recently defined an architecture for an open IPTV service infrastructure. Their objectives are similar to ours (facilitating services and content from different technology domains to integrate with IPTV), but we concentrate on IPTV for communities. In addition, our system includes service enablers that are not (yet) supported by the Open IPTV Forum, in particular enablers for metadata integration, identity management, and recommendations. Another difference is that their underlying model does not appear to include third-party service providers, and that additional services need to be operated by either IPTV service providers or by IPTV portals (so-called "Service Platform Providers" in Open IPTV). Finally, their architecture involves a managed version (e.g., for telecom operators) and an unmanaged version (for the Internet), but it is unclear if it is possible to integrate services from both worlds into a single IPTV offering (e.g., an IMS-based voice service and a web shop from the Internet), which is something that our system does allow.

Senot, Toms, Aghasaryan, and Betgé-Brezetz (2007) describe a service enabler for automatically profiling end-users based on how they use IPTV services and 3G mobile video services. The difference with our work is that their system does not explicitly consider third-party services, but focuses on triple play services in which a 3G mobile operator operates both the mobile video service and the IPTV service. As a result, enablers like identity management and metadata integration are less important in their work.

Friedrich, Seeliger, and Arbanowski (2008) focus on extending IMS-based IPTV with IMS services, such as presence and awareness and personalized advertisements. The difference with our work is that we concentrate on social interactive TV and that their service infrastructure does not explicitly focus on combining IPTV with third-party services from outside the world of (IMS-based) IPTV. AmigoTV (2005) is a system from Alcatel that allows telecom operators to add communities, voice, chat, and messaging services to their IPTV service offering. AmigoTV does not explicitly focus on obtaining these services from third parties in other technology domains, although the system can interact with "external" chat services such as ICQ and Yahoo. Abreu and Almeida (2008) and Boertjes, Klok, and Schultz (2008) also propose systems that combine IPTV with additional services, but they too do not allow third-party providers to integrate their services into an IPTV service offering. (Berg, Butterfield, Cosmas, Casagranda, Garrec, Guiraudou, Martinez, Launey, Mazieres, & Milenaso, 2003), (Rauschenbach, Putz, Wolf, Mies, & Stoll, 2004), and (Schatz, Wagner, Egger, & Jordan, 2007) describe systems that combine DVB-based (mobile) broadcast TV with services in cellular networks (e.g., chat services), but their approach is not based on an underlying cross-domain service infrastructure for third-party providers.

Our quiz show application is similar to the quiz application that Luyten, Thys, Huypens, and Coninx (2006) use in their work. The difference, however, is that we concentrate on the underlying infrastructure with the goal of allowing third parties from outside the IPTV domain to integrate their services and content into an IPTV service offering.

CONCLUSION AND FUTURE WORK

We discussed our approach towards extending the viewing experience for interactive digital TV, in particular for IPTV. Our approach is to extend IPTV services with third-party community and interactivity services from outside the world of IPTV, in particular Internet services, mobile telecommunications services, and services in home networks, We call the combination of an IPTV services and a set of third-part services an iTVT (interactive TV Together) service.

In addition to this main concept, we also considered the underlying open service infrastructure that makes iTVT services possible. The main goal of this infrastructure is to facilitate interoperability across different technology domains (IPTV, internet, mobile telecoms, and in-home). It consists of four service enablers (for metadata integration, identity management, session management, and personal recommendations), which enable iTVT applications to integrate services from other "technology domains" into an IPTV service. We found little existing work in this area, even though we think it would offer new business opportunities for several players within and outside the IPTV value network. We discussed the enablers as well as a prototype application that uses them to extend an IPTV service with an interactive multi-player game, multi-device interaction, and a personalized shopping service. Our work is also supported by recent user studies, which indicate that users appreciate such "Web2.0"-style extensions to IPTV services.

Besides our technical work, we also outlined four possible paths for the future of IPTV, which are based on a trend analysis and form the basis for developing business models that our service infrastructure makes possible.

We believe that we have only scratched the surface of cross-domain IPTV and other types of cross-domain content sharing. We are therefore extending our work in the EU project iNEM4U (www.inem4u.eu), for instance to investigate topics like cross-domain communities, context-awareness, synchronization for multi-device interactions, and cross-domain business models for content sharing. We will also investigate the existing enablers in more detail and will carry out small-scale user studies.

ACKNOWLEDGMENT

This work was conducted within Interactive TV Together (iTVT), a joint project of IRT, Logica, NEC Europe Ltd., Philips Research, and Telematica Instituut.

REFERENCES

Abreu, J., & Almeida, P. (2008). From scratch to user evaluation - validating a Social iTV platform. *EuroITV 2008 Workshop on Sharing Content and Experiences with Social Interactive Television*, Salzburg, Austria.

AmigoTV (2005). *AmigoTV: a social TV experience through triple-play convergence.* White-paper.http://www1.alcatellucent.com/publications/abstract.jhtml?repositoryItem=tcm%3A172-195491635

Anderson, C. (2006). *The Long Tail: Why the Future of Business Is Selling Less of More.* New York, Hyperion

Baca, M., & Holtzman, H. (2008). Television meets Facebook: Social Networks through Consumer Electronics. *EuroITV 2008 Workshop on Sharing Content and Experiences with Social Interactive Television*, Salzburg, Austria.

Berg, M., Butterfield, S., Cosmas, J., Casagranda, P., Garrec, D., Guiraudou, M., Martinez, G., Launey, E., Mazieres, B., & Milenaso, D. (2003). *CISMUNDUS: Convergence of Digital Broadcasting and Mobile Telecommunications, International Broadcasting Conference 2003*, Amsterdam, the Netherlands.

Boertjes, E., Klok, J., & Schultz, S. (2008). ConnecTV: Results of the Field Trial. *EuroITV 2008 Workshop on Sharing Content and Experiences with Social Interactive Television*, Salzburg, Austria.

Broekens, J. (2008). Want a Recommendation? Ask my Digital Personality! *EuroITV 2008 Workshop on Interactive Television Commerce and Advertising*, Salzburg, Austria.

Chorianopoulos, K., & Spinellis, D. (2007). Coping with TiVo: Opportunities of the networked digital video recorder. *Telematics and Informatics, 24*(1), 48–58

Dees, W., & Shrubsole, P., (2007). Web4CE: Accessing Web-based Applications on Consumer Devices. *In proceedings of the 16th International World-Wide-Web Conference (WWW2007)* (pp. 1303-1304).

Friedrich, O., Seeliger, R., & Arbanowski, S. (2008). Interactive and Personalized Services for an Open IMS-Based IPTV Infrastructure. *Seventh International Conference on Networking*, (pp. 302-307).

Gartner. (2007). Dataquest Insight. *A Vision of Telecom and Media Firms' Future Relationship with Consumers.*

Herlocker, J. L., Konstan, J. A., Terveen, L. G., & Riedl, J. T. (2004). Evaluating collaborative filtering recommender systems. *ACM Transactions on Information Systems, 22*(1), 5-53.

Lahanas, S. (2008). *Enable SOA Transformation and Cross-Domain Data Fusion.* DM Review Magazine.

Liberty Alliance Project. (2008). Retrieved June 04, 2008, from http://www.projectliberty.org/

Lowet, D., & Khmelinskaya, E. (2008). Social API for CE-HTML devices. *EuroITV 2008 Workshop on Sharing Content and Experiences with Social Interactive Television*, Salzburg, Austria.

Lowet, D., & Shrubsole, P. (2007). Content sharing and experience sharing with Web4CE. *In Adjunct Proceedings of EuroITV 2007: Interactive TV: A Shared Experience: TICSP*

Luyten, K., Thys, K., Huypens, S., & Coninx, K. (2006). Telebuddies: Social Stitching with Interactive Television. *Extended Abstract, Conference on Human Factors in Computing Systems*, Montreal, Quebec, Canada.

Magedanz, T., Witaszek, D., & Knuettel, K. (2005). The IMS Playground @ Fokus – An Open Testebed for Next Generation Network Multimedia Services. *In Proceedings of the 1st International Conference on Testbeds and Research Infrastructures for the DEvelopment of NeTworks and COMmunities (TRIDENTCOM'05)* (pp. 2-11).

Masthoff, J. (2004). Group Modeling: Selecting a Sequence of Television Items to Suit a Group of Viewers. *User Modeling and User-Adapted Interaction, 14*(1), 37-85

Miletich, M. (2008). Five years 'Buntes Fernsehen Engerwitzdorf': The Power of Homemade Television. *EuroITV 2008 Workshop on Sharing Content and Experiences with Social Interactive Television*, Salzburg, Austria.

OASIS Security Assertion Markup Language (SAML). (2008). Retrieved June 04, 2008, from http://www.oasis-open.org/committees/tc_home.php?wg_abbrev=security#samlv20

Open IPTV Forum (2007). *Open IPTV Forum* Whitepaper. Retrieved from http://www.openiptvforum.org/downloads.html

Open IPTV Forum (2008). *Open IPTV Forum Functional Architecture V1.1.* Retrieved from http://www.openiptvforum.org/downloads.html

OWL (2008). *The OWL DL web ontology language*. W3C, http://www.w3.org/TR/owl-features

Rauschenbach, U., Putz, W., Wolf, P., Mies, R., & Stoll, G. (2004). *A Scalable Interactive TV Service Supporting Synchronized Delivery over Broadcast and Broadband Networks, International Broadcasting Conference 2004*, Amsterdam, the Netherlands.

RDF (2008). *Resource Description Framework.* W3C, http://www.w3.org/RDF

Reimers, U. (2006). DVB – The Family of International Standards for Digital Video Broadcasting. *Proceedings of the IEEE, 94*(1), 173-182

Rein, L., & Winningham, T. (2008). *Video metadata model.* The Microformats Project. Retrieved May, 2008, from http://microformats.org/wiki/video-metadata-model.

Resnick, P., & Varian, H. R. (1997). Recommender Systems. *Communications of the ACM, 40*(3), 56-58.

Schatz, R., Baillie, L., Fröhlich, P., & Egger, S. (2008). Getting the Couch Potato to Engage in Conversation: Social TV in a Converging Media Environment. *EuroITV 2008 Workshop on Sharing Content and Experiences with Social Interactive Television*, Salzburg, Austria.

Schatz, R., Wagner, S., Egger, S., & Jordan, N. (2007). Mobile TV becomes Social – Integrating Content with Communications. *International Conference on Information Technology Interface*, Cavtat.

Senot, C., Toms, Y., Aghasaryan, A., & Betgé-Brezetz, S. (2007). Multi-Platform User and Usage Profiling Demonstration: Video-on-Demand Services in IPTV and Mobile Video environments. *Interactive TV: A Shared Experience: TICSP Adjunct Proceedings of EuroITV 2007*, Amsterdam, Netherlands.

Smith, B., & McClave, P. (2001). Similarity vs. Diversity. In D. W. Aha & I. Watson (Eds.), *Proc. of the ICCBR 2001, LNAI 2080*, (pp. 347-361).

Tadayoni, R., & Sigurdsson, H. (2006). IPTV Market Development and Regulatory Aspects. *16th Biennal Conference of the International Telecommunications Society*, Beijing, China.

TV-Anytime (2008). *ETSI TV-Anytime XML Schema. ETSI Technical Specification 102 822.*

Retreived May 2008, from http://www.etsi.org/WebSite/Technologies/TVAnytime.aspx

van der Heijden, K. (1996). *Scenarios*. The art of strategic conversation. John Wiley & Sons.

van Setten, M. (2005). Supporting People in Finding Information. *Hybrid recommender systems and goal-based structuring*. Doctoral dissertation, University of Twente, The Netherlands.

Chapter IV
Social TV from a Computer-Supported Cooperative Work Perspective

Tom Gross
Bauhaus-University Weimar, Germany

Thilo Paul-Stueve
Bauhaus-University Weimar, Germany

Mirko Fetter
Bauhaus-University Weimar, Germany

ABSTRACT

Social TV provides co-located and geographically distributed TV spectators with facilities for jointly watching television and for social interaction. In this chapter the authors discuss Social TV from a computer-supported cooperative work perspective by introducing Social TV, presenting computer-supported cooperative work and its requirements for technological support of social interaction, and by identifying key issues of Social TV concepts and applications—thereby particularly focussing on group awareness, communication, and seamless integration. In particular, this chapter aims to provide users and developers of Social TV systems with concepts and base technology from computer-supported cooperative work and ubiquitous computing as a basis of advanced Social TV.

INTRODUCTION

Social TV applications provide co-located and geographically distributed TV spectators with facilities for jointly watching television and for social interaction before, during, and after watching television. The concept of Social TV can be defined as: 'the increasing integration of television and computer technology to support sociable, computer-mediated group viewing experiences' (Oehlberg et al., 2006, p. 251).

There are many examples for functionality for Social TV such as instant messaging with text as well as speech communication, discussion forums and conferences, online status information, and neighbourhood information like recommendations based on information of spectators with similar viewing history or introduction to spectators with similar interests. These examples clearly show that there is a considerable potential for synergies with other, existing areas of human-computer interaction, computer-mediated communication, and computer-supported cooperative work. Particularly, in the latter area several relevant findings about the users' needs for technological support for social interaction, related requirements for systems and tools, and experience with technology that fulfils these requirements have been made. Looking into the computer-supported cooperative work area has the potential of considerable synergies for both sides: Social TV can benefit from existing experience and knowledge in computer-supported cooperative work, and computer-supported cooperative work can benefit from new findings in Social TV that flow back into computer-supported cooperative work.

In this chapter we briefly introduce some Social TV systems and prototypes that serve as examples where concepts from Social TV and computer-supported cooperative work have overlapping concepts and functionality and therefore room for synergies. We then have a closer look at Social TV from a computer-supported cooperative work perspective—we sketch a typical Social TV scenario, and derive computer-supported cooperative work concepts and functionality that are essential in this scenario. We continue with the a cooperative media space for Social TV that is an example of an advances concept and technology combining computer-supported cooperative work with ubiquitous computing to leverage Social TV. Finally, we present a general framework for advanced Social TV.

SOCIAL TV

Social TV systems allow collocated as well as distributed spectators watching television together. The focus of such systems does not only lie on the presentation of the media to all spectators, but to allow the spectators to keep in touch with each other. This requires keeping the spectators informed about each other, providing them with communication facilities, and possibly with further support for their activity. For instance, Social TV 2 is a prototype system that extends a media center software with communication and ambient awareness facilities to connect distributed spectators (Harboe et al., 2008), and Telebuddies is a system that enhances IDTV set-top boxes with the possibility of annotating media content and user profiles to allow the realisation of enriched interaction with other spectators (Luyten et al., 2006).

The conjoint watching of television can happen at the same time and in real time, but a time-shifted shared experience is also possible. In the following we introduce four typical Social TV systems: a prototype system allowing for a shared television experience in real-time; a low-fidelity and a high-fidelity commercial systems allowing for a time-shifted communication and watching of television; and a system that uses the past media consumption history of the spectators to enhance distribution of media.

AmigoTV

AmigoTV is a television broadcasting application with real-time communication and community support, which has been realised as a prototype implementation. It is based on the assumption that three components are essential for social television applications: personalised content, community support, and rich verbal and non-verbal communication (Coppens et al., 2005, p. 2.). It allows for real-time voice communication over television broadcast content channels. It provides basic presence and buddy management features, via four different online states: available, busy, no speech, and off-line. Images of personal avatars represent buddies on the same television channel that are overlaid over the broadcasted contents; personal image, video, and audio emoticons allow the expression of emotions. The application is controlled via a television remote control and does not require keyboard or mouse input.

The application prototype consists of three components: a messaging middleware for instant messaging session and presence management via the extensible Messaging and Presence Protocol (XMPP (Saint-Andre, 2004)), a Web server for avatar and emoticon storage, and Real-time Transport Protocol (RTP (Schulzrinne et al., 1993)) and Session Initiation Protocol (SIP (Rosenberg et al., 2002)) support for audio instant messaging.

The 'buddy mosaic' view is the community portal of AmigoTV. It displays all buddies' names with the channel they are currently active on, allowing to join them immediately or to start text instant messaging sessions (cf. Figure 1).

Joost

Joost is a commercial online television service for a time-shifted shared television experience. The motivation underneath the Joost system is to combine television based on the internet protocol with social networking functionality, while avoiding content privacy. Joost content is stored on a central server. The Joost clients access this content on the server on demand. The download and local storage of the content is not possible; this makes Joost suitable for business use. Several media groups have a partnership with Joost, so that the service provides over 28.000 shows as of July 2008 (Joost N.V., 2008).

In order to access Joost media content, a dedicated client is needed, which is available for download for two major commercial operating systems. The client allows searching for specific content, browsing by meta-information (e.g., New Shows, Popular Shows), or browsing by content categories (e.g., Cartoons & Animation, Lifestyle, or Music). The content is organised in predefined channels extendable by custom channels. Joost allows inviting friends from within the client via email, integrates a blogging front-end, and provides chat rooms for each channel. The graphical user interface supports full screen and window mode and consists of a content pane displaying the media. All controls and functions are presented as semi-transparent overlays on the

Figure 1. Screenshot of the Buddy Mosaic view of AmigoTV. Source: (Coppens et al., 2004; Coppens et al., 2005).

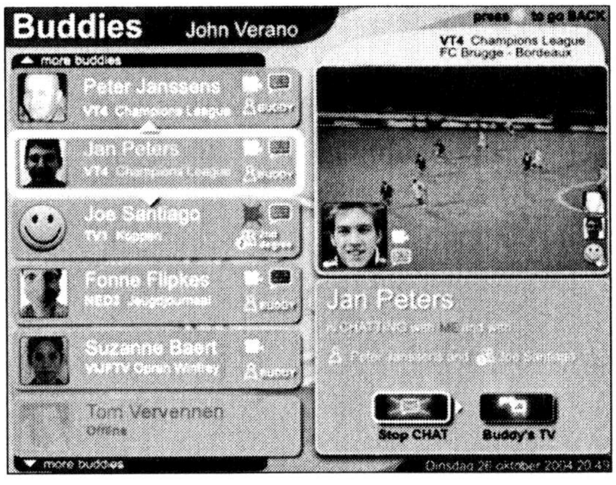

Figure 2. Screenshot of Joost's chat overlay window in front of the media. Source: (Joost N.V., 2008a).

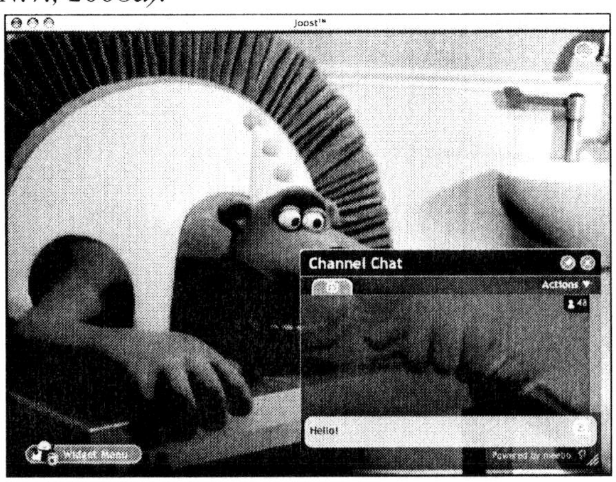

content pane that do not interrupt media playback (cf. Figure 2).

Joost provides interesting base functionality for social interaction around television watching.

Babelgum

Babelgum is a more advanced commercial online television service for a time-shifted television experience. Its client is available for two major commercial operating systems and uses a proprietary peer-to-peer streaming protocol to provide users access to the media contents. The media contents are organised in channels (e.g., News, Trends, or Documentaries). The community concept supported by Babelgum supports special interest groups with media related to a specific topic. Topics range from Food, Wine and Gastronomy in General; over deejay; 80ies cartoons; to surf. Users are provided with information about other active users and can exchange opinions. Film festivals are a special form of communities, where users rate the contents of dedicated categories and can discuss with other users. Smart channels are user-defined auto-updating channels with media content selected by keywords.

Users have the possibility to upload custom media content; to create their own channels and playlists' to comment, tag and rate media contents; and have access to the comments, ratings, and tags of other users. The client operates in full screen mode resembling a television experience on the computer screen. The GUI consists of a full screen pane that displays the media content, and overlay windows for the different domains and functions of the service (cf. Figure 3).

Tribler

Tribler is media viewer and distribution system following a social approach: a social-based peer-to-peer file sharing system for media files. It is built on top of the Bittorrent protocol (Cohen, 2003) and uses the VideoLan VLC media player (VideoLAN Team, 2008) for media playback. It follows a social-based peer-to-peer sharing paradigm that acts on the assumption that 'a similar taste for content can form the foundation for an online community with altruistic behaviour'

Figure 3. Screenshot of Babelgum with My Babelgum overlay window. Source: (Babel Networks Ltd., 2008).

(Pouwelse et al., 2007, p. 129). Its collaborative download protocol 2Fast makes use of social groups to improve the download performance on base of trust.

The graphical user interface of Tribler reflects the social approach to the downloading process by providing access to information regarding the available files and persons in the network, the own profile and library, subscriptions, and friends as partners for the collaborative download (cf. Figure 4).

Core Functionality of these Social TV Systems

The systems presented are exemplary for Social TV. They add a social component to media consumption, but have subtle differences in doing so: one system provides television spectators with synchronised media, real-time communication facilities and information about the television activities of other spectators (AmigoTV), two systems provide facilities for time-shifted watching of television and communication (where the first example provides the basic functionality, and the second one provides extended features), and one system providing facilities for media exchange on base of the user's media consumption history.

Overall these examples show that four core features of Social TV systems can be identified: the provision of media (real-time or delayed), the provision of information about co-spectators and their media consumption, the provision of communication facilities (real-time or delayed), and the provision of services depending on the basis of the media consumption history of the users.

SOCIAL TV FROM A COMPUTER-SUPPORTED COOPERATIVE WORK PERSPECTIVE

In this section we look at Social TV in the context of computer-supported cooperative work. We discuss related concepts of the domain of computer-supported cooperative work and ubiquitous

Figure 4. Screenshot of Tribler with a user list. Source: (Pouwelse et al., 2007).

computing and how they can be applied to Social TV applications. Finally, we derive requirements for cooperative Social TV applications.

Advanced Social TV—A Scenario

In the following scenario we illustrate the identified features of Social TV applications through a typical Social TV application use.

Peter is a student from Germany currently on a student exchange in London. He returns home from an eventful day at the university and wants to relax. He turns on his TV set and is identified by the system. The room settings get adjusted to Peter's preferences when watching TV and his current favourite programme is displayed. Currently on his favourite channel a tennis match is broadcasted. A list of his buddies currently online and watching TV appears on the bottom of the screen. Peter can see his friends Tina and Norbert watching a football match together in Norbert's apartment in Berlin and he asks them if they like to watch the football match together with him (cf. Figure 5). They agree and he joins them.

Peter, Tina, and Norbert are displayed at the bottom of the screen while watching the match. Later, a couple of other friends join them, too. They are added to the bottom of the screen and they all have a discussion about different aspects of the game (cf. Figure 6).

As the discussion gets more specific, the volume of the broadcast is lowered for the participants of the discussion to allow a better conversation. Those who are not involved in the discussion follow the football broadcast with normal volume. They can follow the discussion of their friends in the background what allows them to listen and to join in if they want.

While watching the football match, Peter receives a message from his friend Gordon, who wants to visit Peter. He asks if Peter likes to watch the murder mystery after the football match and Peter agrees. Gordon arrives after the match and they start watching the murder mystery. The settings of the room adapt to the new audience and programme: the lights are dimmed (cf. Figure 7).

The murder mystery turns out to be boring and Peter and Gordon decide to watch something else. They ask Tina and Norbert if they like to join in again. They start the MovieMatch application, which recommends the last episode of a famous chat show Peter has recorded and they all like very much(cf. third image in the bottom row of Figure 8). They agree and the recorded chat show is broadcasted and enjoyed among the friends.

Figure 5. Virtual co-presence of remote spectators

Figure 6. Communication with remote spectators watching the same programme

Figure 7. Adaptation of room parameters to the preferences of multiple co-located spectators

Figure 8. Recommendation based on multiple spectators' interests

Advanced Social TV—Requirements from a CSCW Perspective

In this section we summarise the requirements of a system that is able to provide the services described in the scenario, and identify related concepts in the computer-supported cooperative work and ubiquitous computing domain.

While using the Social TV application in the scenario, the Social TV spectators receive various information about their co-spectators and their situation. The buddy list provides the spectators with presence information about their buddies and with information about their current activity: the programme or movie they are currently watching, and who is watching together. The small real-time pictures in the buddy list provide the users with facial expressions and gestures of their buddies by integrating visual information. The preview of the programmes the buddies are watching provides the spectators with information about what programmes are currently provided and watched. While watching, the real-time images of co-spectators at the bottom of the screen provide information about who is currently watching the same programme and give a more detailed impression of their emotional state and all co-spectators of a programme grouped by their location. This kind of information is referred to as group awareness. The spectator is always able to communicate with his buddies and co-spectators via text or audio chat; the watched programme determines the formation of a communication group. The provision of group awareness and of communication facilities are both key principles of the computer-supported cooperative work domain.

For instance, in the first scene of the scenario the spectator gets identified and logged into the system without further interaction required, and the room settings are adapted to the preferences automatically. In the third scene multiple users are recognised by the system, which then adapts the room settings to the inferred preferences. The audio chat is adapted to support side conversations automatically. This seamless integration of technology helps making the complex functionality of such a system accessible. By taking into account the principles of ubiquitous computing the spectators can enjoy the television experience and do not get distracted by technology such as full-size keyboards or obtrusive microphones, and so .

Group Awareness

Group awareness is described as 'consciousness and information of various aspects of the group and its members' (Gross, 1997, p. 408). It is the mutual awareness of the users and awareness of the workspace, the place of the artefacts relevant to a process. This means that group members should be informed about what happens, who is involved, and how it happens to be able to form a mental image of the group processes. Concepts that support group awareness are informal awareness, social awareness, group-structural awareness, and workspace awareness. We will briefly introduce to these concepts.

Informal awareness is the knowledge about the presence of other persons in the system and their past, current, and future activities. Informal awareness is a prerequisite for people to interact with each other and therefore for joint activities. Co-located people have this kind of information at hand and are able to use it for interaction and planning of further activities (Gutwin et al., 1996). Normally, in face-to-face situations, this kind of information is communicated through non-verbal back-channelling such as eye contact, mimic, or gestures (Gross, 1997; Gross et al., 2005). Group-structural awareness is the knowledge about the internal structure of a group. It comprises information like the roles and responsibilities of single group members, and their status or position as to specific concerns (Gross, 1997; Gross et al., 2005). Workspace awareness is the awareness of the collection of shared artefacts and the interaction of other persons with these artefacts. It is the 'awareness of people and how they interact with the workspace…limited to events happening in the workspace' (Gutwin & Greenberg, 2002, p. 417). Workspace awareness encompasses information such as the identity of users, their actions, the artefacts worked on, or the changes of on artefact (Gross, 1997; Gross et al., 2005).

Communication

Communication has several meanings; the definition that fits best is the third from the Webster's dictionary: 'a process by which information is exchanged between individuals through a common system of symbols, signs, or behaviour' (Merriam-Webster, 2008). Computer-mediated communication is communication between individuals or groups facilitated by computer systems. Computer-mediated communication can be discriminated in asynchronous and synchronous, and uni- and bidirectional, based on text, audio, or video. Various systems with diverging combinations of synchronicity, supported media types, and amount of participants of the communication are available. For instance, e-mail allows individuals and groups to exchange text-based messages asynchronously; internet relay chat allows individuals and groups to bi-directionally exchange text-based messages in near real-time; telephones allow for bi-directional real-time audio communication; and blogs allow for unidirectional asynchronous text-based communication. These are only a few examples of different computer-mediated communication systems; the media support of the different systems converges with the enhancement of technology. Group awareness helps to coordinate the communication among group members minimising unwanted disturbances (Gross et al., 2006a).

Seamless Integration

The paradigm of ubiquitous computing was introduced by Mark Weiser as 'a new way of thinking about computers in the world, one that takes into account the natural human environment and allows the computers themselves to vanish into the background' (Weiser, 1991, p. 66). Central closely related principles of ubiquitous computing are seamless integration of computer technology in the users' everyday life and embodied virtuality:

- The seamless integration of computer technology in the users' everyday life is realised by providing services anywhere and anytime. Users do not have to visit dedicated places of use, but are surrounded by technology. They are provided 'an observed world that is enhanced by the behavioural context of its user' (Roman et al., 2002, p. 74) making the computer 'a tool through which we work, and thus disappearing from our awareness' (Weiser, 1993, p. 76).

- Embodied virtuality is the provision of services through the physical environment letting the computer itself disappear. In contrast to the paradigm of virtual reality, where the user is equipped with technology, such as data gloves or head-mounted displays, to provide access to an artificial world, the users' environment—physical everyday objects—is enhanced to provide services to the user in the real world. The principle of embodied virtuality is often referred to as disappearing computer (Kameas et al., 2003; Schmidt & Van Laerhoven, 2001; Weiser, 1991).

In order to summarise these three concepts from computer-supported cooperative work: group awareness means the general information of other typically remote users in order to have a better shared orientation, communication means explicit conversations among typically remote users, and seamless integration means that technology in general and in particular technology supporting group awareness and communication should be integrated in the activities of the users and not demand too much attention from users.

THE CoMeST EXAMPLE: CONCEPT AND ARCHITECTURE

In this section we exemplify how the above requirements can be put into a concept and implementation of a Social TV system. We describe the concept and building blocks of the CoMeST framework and their interdependencies (Gross et al., 2008). As the above discussion of related work shows, there are several systems and prototypes for Social TV. In this section here we want to show how the specific requirements from CSCW can be realised. For this purpose we introduce our CoMeST framework that consists of platforms developed in our Cooperative Media Lab.

The building blocks of CoMeST are the PRIMI platform (Gross & Oemig, 2005) providing the concepts for group awareness and communication, Sens-ation platform (Gross et al., 2006b) providing the concepts for seamless integration, the MovieMatch Engine that supports the concepts of group recommendations of films, and the SocialTV GUI plug-in that provides the graphical user interface.

We provide an in-depth view of how the subsystems work together to form the CoMeST framework in order to give developers of context-aware Social TV applications a practical overview of the functionality the CoMeST framework provides. Furthermore, we describe how the CoMeST framework can be used to develop cooperative Social TV applications.

Group Awareness and Communication by PRIMI

The Platform for Research on Instant Messaging Infrastructures (PRIMI) is an example of a base platform for developing instant messaging infrastructures. It provides developers with advanced concepts and base technology for presence and availability information and text, audio, and video communication on top of an advanced concept of faces. Faces in PRIMI provide the users of the system the opportunity to organise their presence and availability as well as their online communication distinct for each of their social contexts.

The PRIMIBase system represents the central element of the PRIMI Platform architecture—as it is illustrated in Figure 9—and comprises the PRIMIKernel subsystem and two reference plugin implementations, the PRIMIGUIPlugin and the PRIMISmackPlugin.

The PRIMIKernel forms the core unit of the PRIMIBase system and provides developers with a generic, extendable basis for their own developments. The plugin based architecture allows the integration of instant messaging features into other applications—like in our example a Social TV system—and the rapid development of new instant messaging applications from scratch alike. The PRIMIKernel takes responsibility for basic tasks like mediating awareness information, bridging between multiple instant messaging protocols, handling multiple logins, etc. Furthermore, it supports advanced concepts as for example for integrating sensor information sampled from hard- and software sensors, which can be used to infer more information about the users status and so adapt the application.

The PRIMIGUIPlugin is a reference implementation that offers basic instant messaging functions as well as advanced functions like for example multiple logins into different accounts. The PRIMISmackPlugin is the reference implementation of a ConnectorPlugin founding on the Extensible Messaging and Presence Protocol (XMPP) (Saint-Andre, 2004) providing at instant text communication, buddy management and mediation of presence awareness information. It can communicate with any XMPP server (e.g., Openfire).

So, PRIMI can be extended or completely modified by the implementation of new SourcePlugin, GUIPlugin, or ConnectorPlugin components. Among these extensions so far are for example research prototypes focusing on the selective disclosure of information (Fetter et al., (accepted); Gross & Oemig, 2006) or advanced methods for managing contacts (Fetter & Gross, 2008; Gross & Fetter, 2008).

PRIMI is implemented in the Java programming language (Sun Microsystems, 2007).

Sensor-Based Seamless Integration by Sens-ation

The Sens-ation platform is a service-oriented middleware designed to support developers of context-aware and ubiquitous environments. It provides support for integrating different types of sensors and actuators, and inferencing mechanisms, handles various types of sensor information, and provides means to control actuators. Sens-ation abstracts from technical details of sensors and actuators, such as data formats and protocols, and is able to process data in concurrence. Multiple clients in different locations have access to all sensor data in near real-time and a history of past sensor data allows for complex deductions.

The core platform of Sens-ation consists of six different subsystems: the Adapter subsystem,

Figure 9. Components of the PRIMI platform

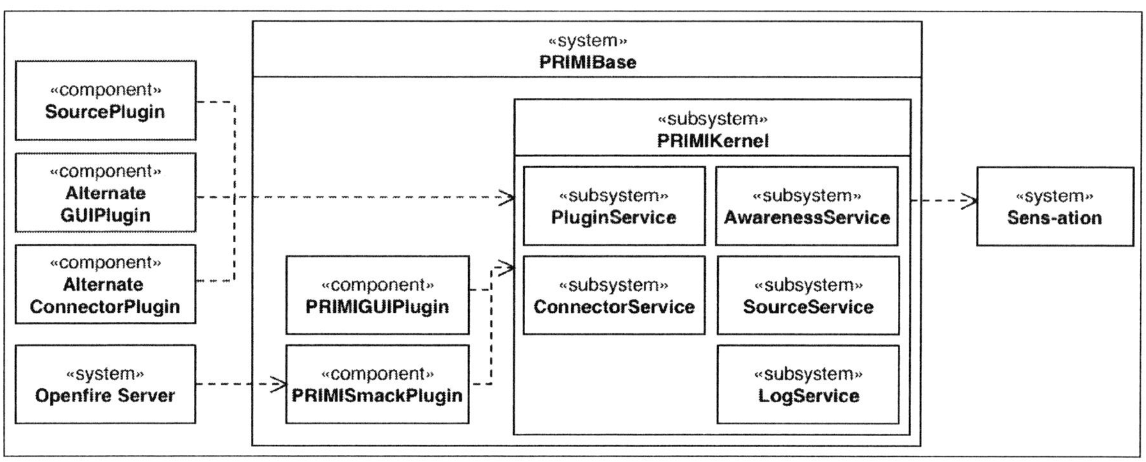

the Management subsystem, the Persistence subsystem, the Inferencing subsystem, the Publisher subsystem, and the Gateway subsystem. (cf. Figure 10).

The Sens-ation platform receives data from the real world through Sensor hardware and software (e.g., light sensors, temperature sensors, online presence sensors). Adapters connect sensors to the platform allowing the submission of sensor data. A variety of protocols allows the submission of sensor data even from distant locations. The central Management caches the sensor data for fast access, stores it for later use in the Persistence subsystem, and provides it to the Inferencing subsystem and to the Publisher subsystem. The Inferencing subsystem allows the deduction of high level information from sensor data. The Publisher subsystem allows the notification of Clients or Actuators with sensor data and data from Inferencing subsystem that matches specific criteria submitted earlier in form of a subscription. Clients and Actuators have access to all data available in the platform through the Gateway subsystem that provides a great variety of different protocols for local access and access over distance such as Web services, XML-RPC (eXtensible Markup Language Remote Procedure Call (Winer, 1999)), or TCP-IP sockets.

Advanced Social TV in CoMeST

The CoMeST framework is an integration of PRIMI and Sens-ation with some additional subsystems to provide advances concepts and functionality for Social TV.

The CoMeST Client consists of PRIMIBase and the SocialTV GUI Plugin. PRIMIBase is the central subsystem of the client. It manages presence and availability information of users received from the Sens-ation subsystem and provides instant messaging facilities using the XMPP server subsystem of the CoMeST Server. The specialised SocialTV GUI Plugin provides the client GUI of the CoMeST framework, and

Figure 10. Components of the Sens-ation platform

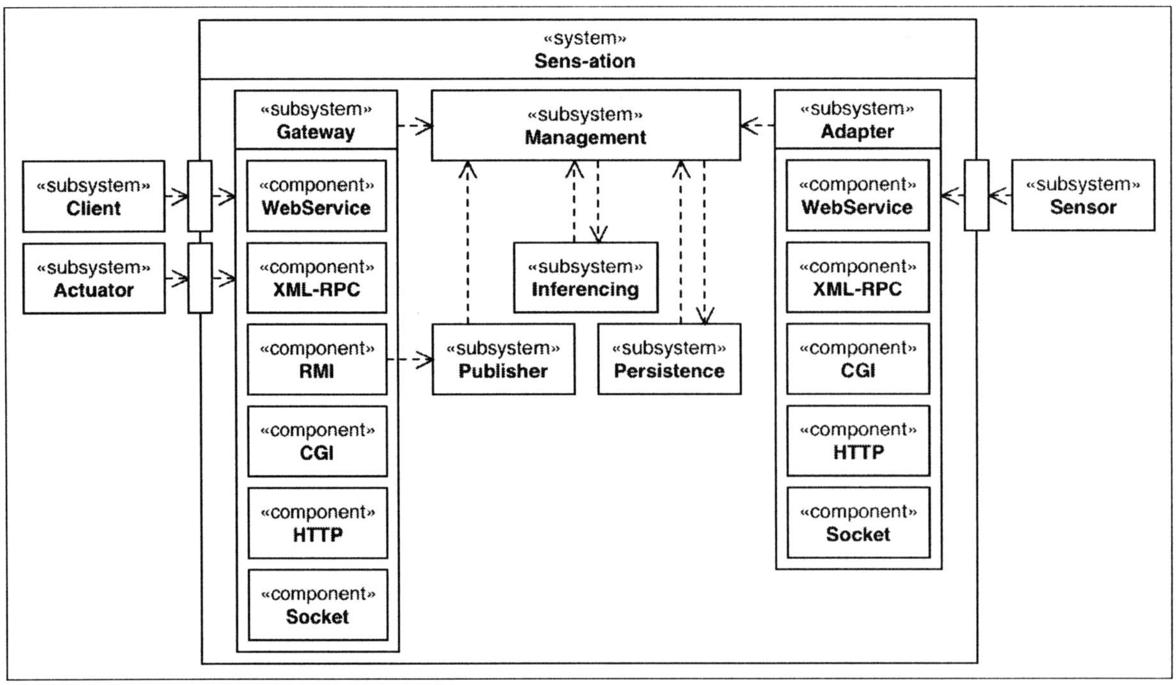

provides a Sensors and an Actuators subsystem for the integration of Sens-ation sensors and actuators. The CoMeSTGUI integrates the instant messaging facilities, provides the user with information about other spectators and groups of spectators, displays media information it receives form the Receiver component of the SocialTV GUI Plugin. The Receiver integrates media sources such as Digital Video Broadcasting (DVB), Digital Versatile Disc (DVD), or Internet Protocol Television (IPTV). The Sensors subsystem is made up of several different Sensor components that provide the Sens-ation subsystem of the CoMeST Server with information: a SpectatorSensor provides information about the physical presence of co-located spectators by matching their mobile phones Bluetooth ID against a lookup table, a ConversationSensor measures the sound intensity in frequency band of the human voice, an AmbientLightSensor measures the luminance of the environment, a ProgrammeSensor receives information about the current programme watched by the spectator from the Receiver of the SocialTV GUI Plugin. Presence and availability information is received by corresponding sensor components from PRIMIBase. The Actuators subsystem provides an actuator that controls the AmbientLightActuator and is driven by the Sens-ation subsystem of the CoMeST Server. Both the Sensors and the Actuators subsystem are open for future extension.

The CoMeST framework is a distributed client-server design. It consists of a CoMeST Server, and CoMeST Clients (cf. Figure 11). The CoMeST Server is built around the Sens-ation platform. It is the central information processing unit of the framework. It receives sensor data from the sensor components of the CoMeST Clients and provides them with inferred information. Three different inference engines provide the locations of all spectators, compute the ambient preferences for the locations, and deduce recommendations with help of the MovieMatch subsystem. The MovieMatch subsystem of the CoMeST Server provides movie and programme recommendations for users and groups of users. It uses sensor

Figure 11. The CoMeST framework components

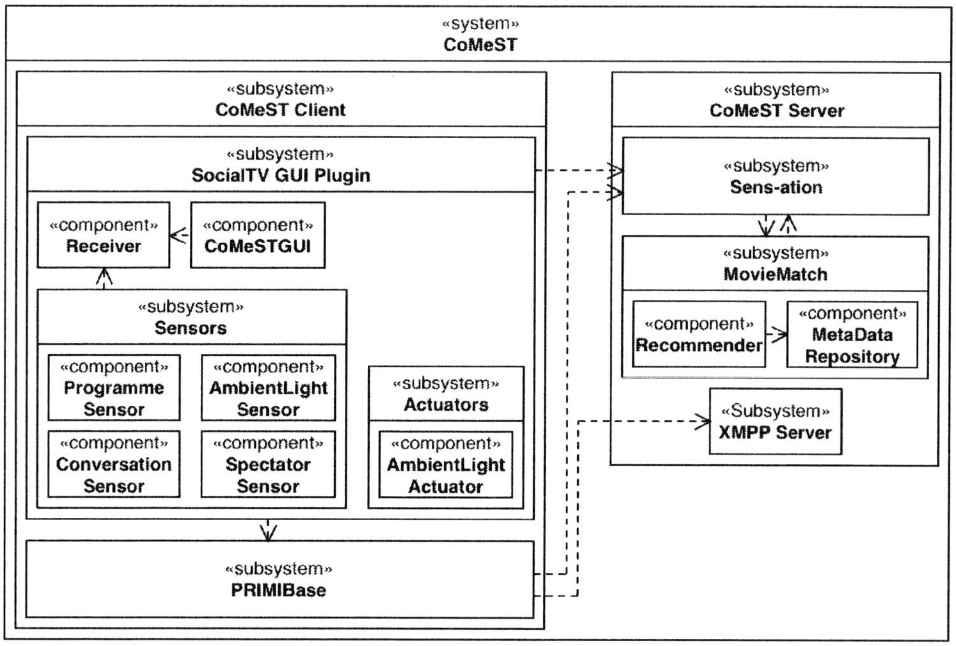

information provided by the Sens-ation to create substantial user profiles of television spectators. Sensors capture detailed information about the viewing habits of a user and submit these as event data to Sens-ation. Additional meta-information for classification of the movies and programmes are provided by the MetaDataRepository that uses data from The Internet Movie Database, which is available for personal and non-commercial use (IMDb, 2008). The recommendations of the Recommender are based on the viewing history of every participating user with respect to detected preferences and aversions, rhythms, and past group constellations. MovieMatch uses concepts and technology for analysing the user behaviour, profiling, and comparison of user profiles from MatchBase (Gross et al., 2006a). An XMPP Server provides the CoMeST Clients with instant text messaging services.

The CoMeST framework augments Social TV applications with respect to the principles of group awareness, communication, and seamless integration. It provides facilities for handling any kind of awareness information in the form of sensor events and provides advanced functionality for communication. Hard and software sensors gather data from the users and their environment in different locations. Inference engines deduce context information that is used to adapt the awareness information provided to the user, to adapt the environment, and to make recommendations for one or more user. Its modular architecture makes it flexible and open for future extensions; new sensors, inference engines and actuators may be added easily.

THE DESIGN SPACE OF TIME, SPACE AND CONNECTION FOR SOCIAL TV

Three dimensions—time, space, and connection—have been found to be helpful to fathom the design for systems that leverage social interaction and seamless integration (Fetter & Gross, 2007) and will be transferred to the field of Social TV.

Communication and interaction in Social TV systems is often designed to meet a very specific scenario: Singular persons watching one TV show together at the same time but at different places. In the following we will discuss how this scenario should be expanded to better meet user expectations and needs. Further we will exemplarily show how the above-mentioned dimensions can help to further explore the design space of Social TV and give guidelines on how to support interaction along these dimensions.

Time

Time is central to our daily social life. People have different schedules, rhythms or even live in different time zones and manage to arrange their social live around these structures. Social TV should meet these requirements. We therefore argue that Social TV systems should take this into account and support synchronous as well as asynchronous social interaction (Dix et al., 1993, p. 447), as users may not always be able to watch TV together at the same time, yet do not want to miss the enriched experience of Social TV. In the following will we give an example of how this can be achieved.

Besides audio, video, and text communication for synchronous interaction, shared video annotation (audio, video, text) can give someone, who watches a show or movie when others have already seen it, the possibility to share the common experience. So in an asynchronous scenario, people should be able to leave visible cues or audible markers in the video, to share their excitement about a scene in the movie or an opinion towards a statement in a show. The system could record the communication of a group in an overlay channel and replay that channel—if desired—together with the recorded TV show for the ones who missed it. The explicit annotation by setting markers in a movie, that is watched alone, can also help to share the experience with friends that see the film afterwards, if those annotations are shared. In both cases asynchronous backchannels can foster interaction that helps the person who watches the movie with annotations later, to share the opinion towards the movie, or towards the annotations. So, social interaction is also possible in a time-relaxed form.

Space

Besides time, space plays a major role in the design space for systems that support social interaction. As Harrison and Dourish (1996) discuss and later Dourish (2006) reflects, space is immanent to computer-supported cooperative work systems, yet can be explored from different perspectives. We want to focus on two perspectives: the real physical space we inhabit and the concept of virtual space we can immerse in. The first perspective is also reflected in the time/space matrix for classifying groupware according to their application in co-located or remote settings. The second perspective is illustrated by Lombard and Ditton (1997) in their conceptualisation of 'presence as transportation' also referring to television in particular. Social TV systems can support communication that seamlessly bridges physical and virtual space.

This can happen by offering support for communication and interaction that accounts for physical co-located settings (i.e., a group of people watching TV together in one location); for remote settings (i.e., individuals watching TV together over distance), and for mixed settings (a group or individual at one location communicates and interacts with groups and individuals at remote locations). A major challenge for all systems—and especially for those supporting mixed settings—is to create awareness about who is interacting with whom.

Further systems should also take into account virtual remote space. Watching together in Social TV is often directed at watching together the same movie or show. In real life we can observe situations, where people decide to watch TV at the same time, yet in different rooms of the apartment to see different contents. These people nevertheless might want to have the feeling of being connected, and want to spontaneously exchange thoughts about what they are watching with the other spectators in the other room. So, watching together might not always mean watching the same program. Social TV systems should also account to these situations and offer means to communicate in such situations.

Connection

Finally connection is the third of the three central dimensions. By connection we refer to the 'closeness of a relationship between each two people in terms of friendship, trust, and privacy' (Fetter & Gross, 2007, p. 46). Exploring the design space along this dimension and taking the findings into account is viable for designing social translucent systems (Erickson & Kellogg, 2000). We suggest that Social TV should allow for interaction among people that share different intense connections, by meeting the concerns of people that are closely connected as well as people that are only loosely coupled and the shades in between.

In order to achieve this, Social TV systems should mediate mutual presence and availability at different levels. For close friends or family members, giving away personal information at a finer grained level, is often adequate. In such close connections it is more likely that users feel comfortable about disclosing information (e.g., about their TV watching rhythms and movie preferences), sharing more private annotations or allowing audio and video as standard communication channels. For weak-ties, instead of a detailed listing of detailed shows and schedule of each spectator, it is more appropriate ands sufficient to show an indicator of the overall similarity in terms of movie taste or preferred TV times. This general similarity can then be used to make contact with other spectators, and possible exchange more details later on. Furthermore, only public annotations are shared and text messaging might be the preferred communication medium at first. Consequently, mechanisms that support such settings without overwhelming the user with to much configuration effort are desirable.

Supplementary the user should be supported with building up and maintaining such connections. While Social TV can help to foster existing relationships, there also should be means to build up new connections with people that for example share the same movie interests.

CONCLUSION

In this chapter we introduced some prominent examples of Social TV applications. We discussed Social TV from a computer-supported cooperative work perspective: we presented a scenario, derived requirements, and provided general information on group awareness, communication, and seamless integration for Social TV. We introduced CoMeST as an example of a sophisticated based technology providing concepts for group awareness, communication, and seamless integration. Finally, we discussed the overall design space of time, space, and connection of Social TV. In fact, group awareness, communication, and seamless integration should always aim to support spectators in time, space, and connection and thereby offer a shared TV watching experience.

REFERENCES

Cohen, B. (2003). *Incentives Build Robustness in BitTorrent. 1st Workshop on Economics of Peer-to-Peer Systems*, Berkeley, CA.

Coppens, T., Trappeniers, L., & Godon, M. (2004). AmigoTV: Towards a Social TV Experience. *2nd European Conference on Interactive Television - EuroITV 2004* (pp. 163-168), Brighton, UK.

Coppens, T., Vanparijs, F., & Handelkyn, K. (2005). *AmigoTV: A Social TV Experience through Triple-Play Convergence*. Paris, France: Compagnie Financiere Alcatel.

Dix, A. J., Finlay, J., Abowd, G. D., & Beale, R. (1993). *Human-Computer Interaction*. Englewood Cliffs, NJP: Prentice-Hall.

Dourish, P. (2006). Re-Space-ing Place: 'Place' and 'Space' Ten Years On. *Proceedings of the ACM 2006 Conference on Computer-Supported Cooperative Work - CSCW 2006*, Banff, Canada. pp. 299-308.

Erickson, T., & Kellogg, W. A. (2000). Social Translucence: An Approach to Designing Systems that Support Social Processes. *ACM Transactions on Office Information Systems, 7*(1), 59-83.

Fetter, M., & Gross, T. (2007). Time, Space, Connection: Scaling Ambient Intelligence. *UP-GRADE - The European Journal for the Informatics Professional, VIII*(4), 44-49.

Fetter, M., & Gross, T. (2008). Contact Management on the Wall: A Card-Game Metaphor for Large Displays. *Proceedings of the Second International Conference on Tangible Embedded Interaction - TEI 2008*, (pp. 247-250). Bonn, Germany.

Fetter, M., Gross, T., & Zeller, B. (2008). *Disclosure Templates: Vorlagen fuer die selektive Freigabe persoenlicher Informationen (Disclosure Templates: Templates for the Selective Disclosure of Personal Information; in German).* Mensch & Computer - 8. Fachuebergreifende Konferenz fuer interaktive und kooperative Medien - M&C 2008, (57-66). Luebeck, Germany.

Gross, T. (1997). Towards Flexible Support for Cooperation: Group Awareness in Shared Workspaces. *Proceedings of the Eighth International Workshop on Database an Expert Systems Applications - DEXA'97*, (pp. 406-411). Toulouse, France.

Gross, T., Braun, S., & Krause, S. (2006a). MatchBase: A Development Suite for Efficient Context-Aware Communication. *Proceedings of the Fourteenth Euromicro Conference on Parallel, Distributed, and Network-Based Processing - PDP 2006*, (pp. 308-315). Montbeliard-Sochaux, France.

Gross, T., Egla, T., & Marquardt, N. (2006b). Sensation: A Service-Oriented Platform for Developing Sensor-Based Infrastructures. *International Journal of Internet Protocol Technology (IJIPT), 1*(3), 159-167.

Gross, T., & Fetter, M. (2008). CoDaMine: Communication Data Mining for Feedback and Control in Ubiquitous Environments. *Proceedings of the Sixteenth Euromicro Conference on Parallel, Distributed, and Network-Based Processing - PDP 2008*, (pp. 539-546). Toulouse, France.

Gross, T., Fetter, M., & Paul-Stueve, T. (2008). Towards Advanced Social TV in a Cooperative Media Space. *International Journal of Human-Computer Interaction, 24*(2), 155-173.

Gross, T., & Oemig, C. (2005). PRIMI: An Open Platform for the Rapid and Easy Development of Instant Messaging Infrastructures. *Proceedings of the 31st EUROMICRO Conference on Software Engineering and Advanced Applications - SEAA 2005*, Oporto, Portugal. (pp. 460-467).

Gross, T., & Oemig, C. (2006). From PRIMI to PRIMIFaces: Technical Concepts for Selective Information Disclosure. *Proceedings of the 32nd EUROMICRO Conference on Software Engineering and Advanced Applications - SEAA 2006, Cavtat*, Dubrovnik, Croatia. (pp. 480-487).

Gross, T., Stary, C., & Totter, A. (2005). User-Centered Awareness in Computer-Supported Cooperative Work-Systems: Structured Embedding of Findings from Social Sciences. *International Journal of Human-Computer Interaction, 18*(3), =323-360.

Gutwin, C., & Greenberg, S. (2002). A Descriptive Framwork of Workspace Awareness for Real-Time Groupware. *Computer Supported Cooperative Work: The Journal of Collaborative Computing, 11*(3-4), 411-446.

Gutwin, C., Greenberg, S., & Roseman, M. (1996). Short Paper: Supporting Awareness of Others in Groupware. Conference Companion of the *Conference on Human Factors in Computing Systems - CHI'96,* Vancouver, Canada. (p. 205).

Harboe, G., Metcalf, C. J., Bentley, F., Tullio, J., Massey, N., & Romano, G. (2008). Ambient Social TV: Drawing People into a Shared Experience. *Proceedings of the Conference on Human Factors in Computing Systems - CHI 2008*, Florence, Italy. (pp. 1-10).

Harrison, S., & Dourish, P. (1996). Re-Place-ing Space: The Roles of Place and Space in Collaborative Systems. *Proceedings of the ACM 1996 Conference on Computer-Supported Cooperative Work - CSCW'96*, Boston, MA. (pp. 67-76).

IMDb. (2008). *The Internet Movie Database (IMDb),* http://www.imdb.com: IMDb.com Inc. (Accessed 11/7/2008).

Joost N. V. (2008). *Joost -What's Joost?,* http://www.joost.com/whatsjoost.html: Joost N.V. (Accessed 1/5/2008).

Kameas, A., Bellis, S., Mavrommati, I., Delaney, K., Colley, M., & Pounds-Cornsih, A. (2003). An Architecture that Treats Everyday Objects as Communicating Tangible Components. *Proceedings of the 1st IEEE International Conference on Pervasive Computing and Communications - PerCom 2003,* Dallas-Fort Worth, TX. (pp. 115-122).

Lombard, M., & Ditton, T. (1997). At the Heart of It All: The Concept of Presence. *Journal of Computer-Mediated Communication, 3*(2).

Luyten, K., Thys, K., Huypens, S., & Coninx, K. (2006). Work-in-Progress: Telebuddies: Social

Stitching with Interactive Television. Extended Abstracts of the Conference on Human Factors in Computing Systems - CHI 2006, Montreal, Canada. (pp. 1049-1054).

Merriam-Webster, I. (2008). *Merriam-Webster Online,* http://www.m-w.com/. (Accessed 21/3/2008).

Oehlberg, L., Ducheneaut, N., Thomton, J. D., Moore, G. E., & Nickell, E. (2006). Social TV: Designing for Distributed, Sociable Television Viewing. *Proceedings of the European Conference on Interactive Television - EuroITV 2006,* Athens, Greece. (pp. 251-262).

Pouwelse, J. A., Garbacki, P., Wang, J., Bakker, A., Yang, J., Iosup, A., Epema, D. H. J., Reinders, M. J. T., van Steen, M. R., & Sips, H. J. (2007). Tribler: A Social-Based Peer-to-Peer System. *Concurrency and Computation: Practice and Experience, 20*(2), (pp. 127-138).

Roman, M., Hess, C., Cerqueira, R., Ranganathan, A., Campbell, R. H., & Nahrstedt, K. (2002). A Middleware Infrastructure for Active Spaces. *IEEE Pervasive Computing, 1*(4), 74-83.

Rosenberg, J., Schulzrinne, H., Camarillo, G., Johnston, A., Peterson, J., Sparks, R., Handley, M., & Schooler, E. (2002). *SIP: Session Initiation Protocol - RFC 3261,* http://www.ietf.org/rfc/rfc3261.txt: The Internet Society. (Accessed 11/7/2008).

Saint-Andre, P. (2004). *Extensible Messaging and Presence Protocol (XMPP): Core - RFC 3920,* http://www.ietf.org/rfc/rfc3920.txt: The Internet Society. (Accessed 11/7/2008).

Schmidt, A., & Van Laerhoven, K. (2001). How to Build Smart Appliances? *IEEE Personal Communications, 8*(4), 66-71.

Schulzrinne, H., Casner, S., Frederick, R., & Jacobson, V. (1993). *RTP: A Transport Protocol for Real-Time Applications - RFC 1889,* http://www.ietf.org/rfc/rfc1889.txt: The Internet Engineering Task Force. (Accessed 2/10/2006).

Sun Microsystems, I. (2007). *J2SE 5.0.* http://java.sun.com/j2se/1.5.0/. (Accessed 31/7/2008).

VideoLAN Team. (2008). *VideoLAN - Free and Open Source Software and Video Streaming Solutions for Every OS!,* http://www.videolan.org/: VideoLAN Team. (Accessed 10/7/2008).

Weiser, M. (1991). The Computer of the 21st Century. *Scientific American, 265*(9), 94-104.

Weiser, M. (19*93). Some Computer Science Issues in Ubiquitous Computing.* Communications of the ACM, 36(7), 75-84.

Winer, D. (1999). *XML-RPC Specification.* http://www.xmlrpc.com/spec. (Accessed 12/12/2007).

Chapter V
Television Content Enrichment and Sharing:
The Ambulant Annotator

Dick C. A. Bulterman
CWI: Centrum voor Wiskunde en Informatica, The Netherlands

Pablo Cesar
CWI: Centrum voor Wiskunde en Informatica, The Netherlands

Jack Jansen
CWI: Centrum voor Wiskunde en Informatica, The Netherlands

Rodrigo Laiola Guimarães
CWI: Centrum voor Wiskunde en Informatica, The Netherlands

ABSTRACT

This chapter reports on the Ambulant Annotator, a middleware extension for Personal Digital Recorders (PDR), in the form of a lightweight authoring tool, which allows the viewer to personalize television content and share it with others. Traditionally, social interactive television research has focused on the provision of synchronous communication mechanisms between distributed peers in the form of direct communication channels (text or audio chats) or distributed control (joint television watching experience). This chapter considers a broader approach that enhances the connectedness between users by providing video sharing capabilities. The Ambulant Annotator empowers viewer-side enrichment of multimedia content in the form of video fragmentation, fragments annotation and enrichment. Once the user has created his personalized enriched version of the video content, the Ambulant Annotator provides mechanisms to share it with his social network by using asynchronous communication technologies. The video manipulation mechanisms presented in this chapter does not modify the original video material, but are encoded as separate overlays in such a way that Digital Rights Management (DRM) restrictions on content reuse are respected.

INTRODUCTION

Social interactive television research has, mostly, focused on providing the user with synchronous communication mechanisms while watching television. This chapter considers instead a broader definition of social television (Chorianopoulos, 2007), in which asynchronous communication methods such as media sharing are available. In order to provide a more connected experience, the shared media can be enriched and personalized by a number of media overlays such as audio commentaries (Cesar et al., 2008). We termed the end-user process of generating such enriched content as the *Authoring from the Couch* paradigm (Cesar et al., 2006) as a means to differentiate it from more formal multimedia authoring methods (Bulterman & Hardman, 2005). *Authoring from the Couch* corresponds to lightweight authoring (Kirk et al., 2007), while watching television content in order to enrich existing material and with the intention of sharing the resulted media content. This chapter discusses in detail this new paradigm and its innovative capabilities. The chapter is structured as follows. First, related work in the areas of multimedia authoring and multimedia sharing are reviewed. Then, a detailed description of the *Authoring from the Couch* paradigm is provided followed by a discussion on the current system architecture. Finally, the last section of the chapter proposes a number of extensions to the *Authoring from the Couch* paradigm to be explored in the future.

RELATED WORK

Contrary to the traditional role of television viewers as passive spectators, there are strong indications that there is room for more active television watching (Ursu et al., 2008; Agamanolis, 2008). However, according to recent research (Jensen, 2005; Chorianopoulos, 2008), it would not be the flavor of interactivity with complex applications that require concentration, long time spans, and distraction from television viewing. Instead, they point out that television applications should provide new interactivity models particularly designed for television. "Lazy interactivity is a low-attention-span paradigm designed for television viewers" (Jensen, 2005). That is, interactive applications intended for quick decision, short attention spans, handheld remotes, and instant gratification. Lazy interactivity thus requires a simpler interface involving minimal consumer effort. The contribution of this chapter is to provide a lazy interactive paradigm for content manipulation and sharing. On the one hand, the *Authoring from the Couch* paradigm empowers the user with manipulation functionality over television content, thus leveraging the user potential impact on the content he is consuming. On the other hand, the paradigm provides the user with sharing capabilities for communicating using media with his social network. This section is divided into two research trends: multimedia authoring systems and video sharing systems.

Multimedia Authoring Systems

Traditionally, the production and distribution of digital television content follow a one-way approach. First, high-quality material is captured using professional audiovisual devices (analogue and/or digital). Then, this raw material is converted into a digital format, if necessary, and edited using tools which operate at a low syntactic level, for example manipulating video as a sequence of frames and streams of un-interpreted audio. The television programs are then authored by aggregating the different media elements into one presentation, in which the layout characteristics of the visual elements are defined, and the interactivity points handled. Finally, the produced content is delivered through broadcasting stations to the end-user's device for consumption.

All in all, authoring systems for interactive digital television assume the user to be seated in front of a workstation on the broadcaster side and in a professional mindset. These systems have tackled the authoring complexity problem in different ways. While some commercial systems artificially reduce authoring complexity by limiting the available features (e.g. JAME Author[1], Cardinal Studio[2] and AltiComposer[3]), several research systems have attempted to provide more comprehensive support for a wide range of authoring needs (Bulterman et al., 1998; Guimarães et al., 2008). Apart from the functionality limitations, systems for interactive digital television are generally technologically-centered, that is, the design of standards, infrastructures and authoring tools normally prioritizes the technical aspects over the real authors' needs. Thus, authors need to spend a reasonable amount of time learning new technological features as they are incorporated in the authoring tools.

Contemporary research and the explosion of new products and new services show a trend in which the viewer is not the end of the content value chain. Instead, the viewer is considered to be an active node in the production-distribution-consumption chain. More recently, triggered by the widespread availability of digital recording devices as well as display and rendering devices, the end-user has become a potential content producer and content distributor. User-Generated content websites, web-based authoring tools, and viewer-side enrichment solutions are examples of this new trend.

The establishment of web technologies had major implications on content creation paradigms and methods: from the traditional professional creation of content to a more fresh and immediate user-generated content. The success of web sites that distribute users' content (e.g., YouTube[4]) has demonstrated that end-users want to change the way they find and consume audiovisual media. At the same time, Web-based authoring tools, such as Yahoo!'s JumpCut[5], generally require a less professional mindset than traditional authoring tools and make users find this an easier way to author. Other solutions for remixing (Shamma, 2007) and repurposing (Pea, 2004) multimedia content exist as well. Finally, end-user enrichment systems, such as YouTube Annotations and the work reported by Pimentel et al. (2008) for the Brazilian digital television environment, are capturing the attention of the broad public.

Content Sharing

Despite many criticisms on the passive nature of television watching, the social uses of television have also been documented. In particular, the use of audiovisual content as a placeholder for starting and sustaining relationships is an everyday experience for the majority of television users (Cesar et al., 2008b). For instance, in the living room people watch television together and outside home people talk about last night's football match; people even call each other to recommend an interesting program. We can categorize systems that allow users to share audiovisual material in the following groups: online video sharing, Web-based television, interactive television, enriched instant messaging, and sharing on the move.

Lately, online video sharing has become a primary activity in the World Wide Web. Some examples include YouTube, Yahoo! Videos[6], and MySpace[7]. These systems provide the user an interface for uploading, searching, viewing, rating, and recommending videos. In this chapter we are especially interested in the interface for recommending video content, which normally is performed by sending an e-mail with a link to the video or by embedding the video into a social web site such as Facebook[8]. While online video sharing systems are, in theory, intended for user-generated content, currently a number of Web-based television systems such as Joost[9] and Lycos Cinema[10] are becoming popular. Web-based television solutions are targeted for watching television programs using the Internet infrastructure. From this chapter's perspective the most interesting functionality, apart from recommending videos, is to watch television together across distances. Similar solutions exist in the form of Peer-to-Peer television such as Tribler (Fokker et al., 2007).

At the same time, traditional interactive television systems are upgrading their infrastructure in order to provide social activities. We can term such solutions as social interactive television. For example, CollaboraTV from AT&T (Harrison & Amento, 2007) permits to record the viewer's comments while watching a television program. Then, such comments are replayed when a friend is watching the video using avatars to identify who has said what. Synchronous solutions, such as Motorola's SocialTV (Harboe et al., 2007; Harboe et al., 2008) and Alcatel's AmigoTV (Coppens et al., 2005), have been proposed as well. While the previously described media-based systems are starting to provide communication and sharing functionality, communication systems are starting to provide media capabilities. For example, enriched instant messaging solutions such as MessengerTV[11] from Microsoft and Zync! (Shamma, 2008) from Yahoo! now provide the option of sharing synchronized videos while chatting.

Finally, with the popularity of powerful mobile phones, users can anytime, anywhere capture and share their personal experiences by exchanging messages. In this case the media capturing and sharing are performed on the move. There have been several studies on the motives of people to share media material, mostly photographs, using mobile phones. Taylor & Harper (2002) observed the social norms around gift-giving and the demand of reciprocity with mobile phones. Another study categorized the uses of photo sharing in two dimensions: social/individual and functional/affective (Kindberg et al., 2005). The results indicate that affective usage was the most predominant. House et al. (2005) provide a similar categorization: creating and maintaining social relationships, personal and group memory, self expression, self presentation, and for practical purposes. In most of the cases, the usage was mainly incidental and not a planned activity (Kirk et al., 2007).

AUTHORING FROM THE COUCH PARADIGM

This section discusses the *Authoring from the Couch* paradigm. First, a working definition is provided together with an example. Then, a more detailed description of the functionality such paradigm should provide is introduced. This section is concluded with an elaboration on the implications this innovative paradigm brings.

Definition

Authoring from the Couch is a lightweight approach in which the authoring task is performed incidentally, while watching television content, with the final intention of sharing a personalized version of the television content with selected members of the users' social network. Just consider the following example. Mark and Katrina are a couple with two children. During a business trip to Vancouver, Katrina watches an interesting 60-minute documentary about South America, where the family is going this summer. She decides to share relevant parts of this presentation with her family back home in Amsterdam. This sharing consists of identifying the video, overlaying a personal navigation structure highlighting portions of interest, and adding a number of voice-over annotations to some of the fragments — such as 'I want to go there' or 'we have to visit this spot'. She then sends the family a message with a pointer to "her" version of the video. Note that such enrichments do not generate a new version of the video; they create a (set of) content wrapper(s) containing a link to the base video, along with a navigation map and a set of - possible personalized - annotations. Katrina uses asynchronous sharing: the 9-hour time difference, plus the differentiated content, makes synchronous sharing (such as via a chat-based system) impossible.

Functionality

As shown in the previous example, the *Authoring from the Couch* paradigm requires non-destructive manipulation capabilities together with sharing functionality. More precisely, the requirements of such paradigm include (Bulterman, 2007; Cesar et al., 2008):

- **Video fragmentation:** A facility that allows a user to define one or more ranges of clips within a base media object. These fragments can be explicitly or implicit exposed to parties with whom the user shares content.
- **Fragment annotation:** A facility to add user-generated notes, comments, or descriptions to a particular media fragment.
- **Enrichment**: A facility to add new temporal links, overlaid media or voice intro to a baseline object. These enrichments may be layered - that is, a particular media object may expose a history of enhancements by various parties.
- Enrichment sharing: a facility to send recommendations to others. This can be done, for example, via an e-mail with a link to the media or as an embedded HTML fragment on a social website.

Implications

Currently, the *Authoring from the Couch* paradigm allows users to enrich a fragment of television content by including personalized overlays (e.g., audio commentary and text), and share it with others through recommendation messages (Cesar et al., 2008). Such innovative paradigm builds upon work on authoring systems, online video sharing systems, and mobile multimedia studies. Nevertheless, there are a number of relevant implications of the paradigm such as Digital Right Management (DRM) and ownership implications, media personalization, and lifecycle of the multimedia content.

From the ownership viewpoint recommendation messages should be encoded in such a way that DRM restrictions on content reuse are respected. Thus, there is a need, in terms of content modeling, for a mechanism that separates the enrichments and the original content. This way, when recommending content, the user only sends the description of the enrichments and not the original content itself. The receiving partner, thus, needs to own or acquire the original material on which the enrichments were performed. Apart from DRM considerations, keeping the enrichment as a separate description file provides two major benefits. Firstly, the already overloaded online video servers do not need to host several encoded copies of the same movie, but only the movie plus several enrichment files. Secondly, viewers can decide which enrichments to watch, if any. Our paradigm, because the video manipulations are non-destructive, allows viewers to select to turn off all the enrichments over a piece of video or select only those that are relevant for him at a given situation.

The traditional view on media personalization has been the one from the content producer based on his notion of viewers. *My media* is a term often associated with media I might like, and recommender systems have been built to provide the user hints about what that media might be. On the other hand, the *Authoring from the Couch* paradigm positions the user as the central component. Thus *My Media* now becomes my personal annotations and enrichments on the media I am consuming. Such shift on the viewpoint has implications on the recommender systems we are aiming at. This chapter argues that apart from traditional recommender engines based on profiles and/or on community likes and dislikes, improved direct recommendation tools need to be addressed.

The final implication, and probably the most interesting one, is the implicit new multimedia content lifecycle that arises from the *Authoring from the Couch* paradigm. Where in the past television content only flows from the broadcaster to the user, now the user is transformed into an active node in the lifecycle of the content. The user, thus, is not a passive entity anymore but can literally *personalize* the media he is consuming and share such personal media with others. We can foresee that the lifecycle of television content does not end at broadcast time, but it starts there.

THE AMBULANT ANNOTATOR: SYSTEM ARCHITECTURE

The Ambulant Annotator (Cesar et al., 2006) is an extension of traditional television receivers that allows the user to fragment, enrich, and share television content. It is implemented in an extended PDR (PDR+). The PDR+ has access to the conventional broadcast content, to high-density disks, and to multimedia content located on the web. In addition, a number of devices at home can be connected to our PDR+ via the home network. These devices include the conventional television set, the remote control, and personal devices. The conventional television acts as a rendering component; it has a broad presentation path (i.e., multimedia content), but a restricted interaction path (i.e., remote control). Other devices, such as a tablet augmenter have a broader interaction path, so they can provide enriched interactivity. For sharing capabilities, the Ambulant Annotator uses the Internet for uploading content in a blog or for sending email messages. In addition, it includes a SMS generator, specifically Wireless Application Protocol (WAP) Push functionality, for recommending enriched television content to mobile phones for people on the move.

The enrichment system is divided into two differentiated components: server and client side. The server side is located in the home-server PDR+. It implements the functionality drafted in the previous section: content fragmentation, content enrichment and annotation, and message generation for content sharing. The client side, on the other hand, resides in the interactive devices (e.g., PDA). It corresponds to the user interface of the enrichment system and it exposes to the user the functionality of the server side. Server and clients communicate using a well-defined interface via the home network. The system, thus, provide a number of levels of interactivity such as passive, when watching the content, and proactive, when enriching the media content.

For television watching, we have integrated the Ambulant Player[12] in a Mac-mini-based television receiver. The open source Ambulant Player, implemented in C++, provides a complete implementation of the Synchronized Multimedia Integration Language (SMIL) language (Bulterman, 2004). It is a multi-platform player that can run on Linux, Macintosh, Windows, and various handhelds. Note that while a software media player is used to manage the presentation of enriched content, the user is presented a conventional television-centric view of the content. In addition, the system included the Ambulant Annotator. The Ambulant Annotator, implemented in Python, is an extension to the Ambulant Player that acts as a DOM interface to and from the player. The Ambulant Annotator can handle the actions sent by the interactive devices and affect the player behavior. Simple examples are to play or pause the presentation, more complex examples include the fragmentation and enrichment of television content. Finally, in our architecture, each of the interactive devices (e.g., PDA) communicates with the Ambulant Annotator using XML-Remote Procedure Call (XML-RPC). This way each interactive device implements the user interface and the functionality using its native language. Functionality provided by the interactive devices includes free-hand pen strokes, audio recording capabilities, and handwriting recognition for enrichment of media content.

When video manipulation takes place, the SMIL file is used to control the interactive display at the client. The SMIL file is transformed into a TV-Anytime[13] description at the point that one or more recommendations are transformed into a recommendation message. At no time is actual content integrated, manipulated or shared among users. All communication is done via an abstracted SMIL file or as a set of TV-Anytime markup. Hence this is a non-destructive solution for the video manipulation challenge. One very important advantage is that viewers might select to see the manipulations performed by a specific user or turn-off all enrichments at once. The SMIL language was selected as the final format of the content because of many reasons: it is a container format, thus the base content is not modified; it provides a rich set of media timing and activation primitives; it has a simple and flexible layout model; it provides both system- and user-test attributes for content control; and it can be used together with other languages such as Scalable Vector Graphics (SVG) form, for example, ink annotations.

RESEARCH DIRECTIONS

Currently, a prototype implementation of the *Authoring from the Couch* paradigm exists in the form of the Ambulant Annotator. We have conducted a number of user studies (Cesar et al., 2008) that indicates that its innovative features appeal the test participants. Nevertheless, there are a number of topics that should be further explored. We can differentiate three themes: lifecycle of enriched multimedia documents, personal versus shared media, and synchronous television watching. The following sections elaborate on them.

Enrichment of Enriched Presentations: Enrichment Lifecycle

One essential topic to further investigate is a version control system for the recommendations. Because a new version of the enriched document is created after the lightweight authoring process, we might want to keep track of the different versions of such file. Once an enriched presentation is shared it starts a new lifecycle in which further enrichments can be produced, for example, as a reply to the first recommendation. Therefore, it is important to provide mechanisms for controlling the different available versions, to know who has produced them, when and how they were created, and to know for whom they were produced. However, more investigation on people's practices is necessary in order to understand how users manage their recommendations. Then, it will be possible to frame how recommendations might evolve along time and how some others are transient in nature. Moreover, for a large amount of annotations indexing and searching mechanisms can be also relevant.

Personal vs. Shared Media

Differently from the standard design of digital television programs, in which the viewer is a passive entity that only selects media content, the *Authoring from the Couch* paradigm empowers users to enrich and share fragments of television content. Taking into account that a person would like to share different views depending on the recipient party; the viewer would not tell the same story around media to his family and to his friends. Thus, there is a need to provide the tools for (1) generating different enrichments for different sharing situations and (2) provide the mechanisms for selective viewing of enrichments.

Synchronous Television Watching

Traditionally, television control commands (e.g., change channel) are negotiated verbally among collocated viewers. On the other hand, when the viewers are distributed but still want to have a joint television experience across distance, there is a need to define policies regarding governance and people's roles to support content control and social communication. A similar problem arises when collocated viewers have their own control devices at hand, so multiple devices will be controlling the television content at the same in the same living room. Moreover, appropriate mechanisms to distribute the same content at the same time to distributed viewers are needed. When someone watching media content invites other people to watch together, the PDR should enable a way to invite them through the return channel, and a multicast tree could be created to send the content flow to all users that joined the shared experience. Inclusion of standards such as IP Multimedia Subsystem (IMS) in consumer electronic products might provide the tools for such problem. Finally, we can consider that several users might perform the authoring task in the *Authoring from the Couch* paradigm in a collaborative manner across distance. This scenario seems to be a particular case of works in the Computer Supported Cooperative Work (CSCW) field, so further investigation of the state of the art in such field needs to be done.

Awareness Information is fundamental for making intelligent decisions on how the communication channel between people should take place. For example, if a person is not available, asynchronous communication mechanisms can be applied. Such awareness information should be non-obtrusive and should respect privacy concerns. Instant messaging engines provide limited awareness information in the form of buddies list with presence status. More innovative approaches in television research include the use of physical artifacts (Harboe et al., 2008) and interfaces showing what other friends of the users are viewing at one particular moment in time (Boertjes et al., 2008). In any case, privacy concerns have to be taken into consideration (Hess, 2008). In relation to the *Authoring from the Couch* paradigm, awareness information can be utilized in two directions. On the one hand, collaborative editing of television content can occur, if other parties happen to be available at the edition process. On the other hand, presence information can determine the best manner to share enriched videos based on the context of the recipient party.

CONCLUSION

From the social interactive television perspective in mind, this paper explores how asynchronous communication mechanisms can be exploited for increasing the connectedness between television viewers across distances. The Ambulant Annotator together with the *Authoring from the Couch* paradigm, presented in this chapter, have a number of implications for future design of social interactive television systems. Firstly, there is a need to shift the focus from a media-centric approach to a user-centered approach. *Personal media* should not be only considered as the media I might like, but as my personal view on the media I consume. Thus, in this chapter we propose a system that empowers viewer-side enrichment of television content. Moreover, due to indexing and archival purposes, and DRM restrictions, such enrichments should not be encoded together with the video material. Secondly, direct recommendation mechanisms of personal media need to be provided for allowing the user to share his personal media with others. Finally, we can foresee a shift on the traditional lifecycle of television content, in which broadcast content is just the beginning. Where users personalized the content and share them with others, thus re-starting its lifecycle all over again. This chapter has elaborated on all these issues based on our experiences designing, implementing, and testing the Ambulant Annotator system.

ACKNOWLEDGMENT

Dick Bulterman and Jack Jansen are also affiliated with the Vrije Universiteit in Amsterdam. This work has been funded by the NOW BRICKS PDC3 project, the ITEA Project Passepartout, the FP6

IST project SPICE and the FP7 IP project TA2. Development of the implementation of this work within the open source Ambulant Player has been funded by the NLnet foundation. We are grateful for this support.

REFERENCES

Agamanolis, S. (2008). At the Intersection of Broadband and Broadcasting: How Interactive TV Technologies can Support Human Connectedness. *International Journal of Human-Computer Interaction, 24*(2), 121–135.

Boertjes, E., Klok, J., & Schultz, S. (2008). ConnecTV: Results of the Field Trial. In *Adjunct Proceedings of EuroITV.*

Bulterman, D. C. A., & Hardman, L. (2005). Structured Multimedia Authoring. *ACM Transactions on Multimedia Computing Communications and Applications, 1*(1), 89-109.

Bulterman, D. C. A., Hardman, L., Jansen, J., Mullender, K. S., & Rutledge, L. (1998). GRiNS: A GRaphical INterface for Creating and Playing SMIL Documents. In *Proceedings of the WWW Conference.*

Bulterman, D. C. A., & Rutledge, L. (2004). *SMIL 2.0: Interactive Multimedia for Web and Mobile Devices.* Springer-Verlag, Heidelberg.

Bulterman, D. C. A. (2007). User-Centered Control within Multimedia Presentations. *Multimedia Systems Journal, 12*(4-5), 423-438.

Cesar, P., Bulterman, D. C. A., & Jansen, J. (2006). An Architecture for End-User TV Content Enrichment. *Journal of Virtual Reality and Broadcasting (JVRB), 3*(9).

Cesar, P., Bulterman, D. C. A., Geerts, D., Jansen, J., Knoche, H., & Seager, W. (2008). Enhancing Social Sharing of Videos: Fragment, Annotate, Enrich, and Share. In *Proceedings of the International Conference on Multimedia.*

Cesar, P., Chorianopoulos, K., & Jensen, J. F. (2008b). Social Television and User Interaction. *ACM Computers in Entertainment, 6*(1).

Chorianopoulos, K. (2007). Content-Enriched Communication: Supporting the Social Uses of TV. *The Journal of the Communications Network, 6*(1), 23–30.

Chorianopoulos, K. (2008). User Interface Design Principles for Interactive Television Applications. *International Journal of Human-Computer Interaction, 24*(6), 556-573.

Coppens, T., Vanparijs, F., & Handekyen, K. (2005). *AmigoTV: A Social TV Experience Through Triple-Play Convergence.* Technology White Paper.

Fokker, J., Ridder, H., Westendorp, P., & Pouwelse, J. (2007). Psychological Backgrounds for Inducing Cooperation in Peer-to-Peer Television. In *Proceedings of EuroITV* (pp. 136-145).

Guimarães, R. L., Costa, R. M. R., & Soares, L. F. G. (2008). Composer: Authoring Tool for iTV Programs. In *Proceedings of EuroITV* (pp. 61-71).

Harboe, G., Massey, N., Metcalf, C., Wheatley, D., & Romano, G. (2007). Perceptions of Value: The Uses of Social Television. In *Proceedings of EuroITV* (pp. 116-125).

Harboe, G., Metcalf, C. J., Bentley, F., Tullio, J., Massey, N., & Romano, G. (2008). Ambient Social TV: Drawing People into a Shared Experience. In *Proceeding of the SIGCHI Conference on Human Factors in Computing Systems* (pp. 1-10).

Harrison, C., & Amento, A. (2007). CollaboraTV: Using Asynchronous Communication to Make TV Social Again. In *Adjunct Proceedings of EuroITV* (pp. 218-222).

Hess, J. (2008). Community Awareness in Social TV Environments. In *Adjunct Proceedings of EuroITV*.

House, N.V., Davis, M., Ames, M., Finn, M., & Viswanathan, V. (2005). The Uses of Personal Networked Digital Imaging: an Empirical Study of Cameraphone Photos and Sharing. In *Extended Abstracts on Human Factors in Computing Systems* (pp. 1853-1856).

Jensen, J. F. (2005). Interactive Television: New Genres, New Format, New Content. In *Proceedings of the Australasian Conference on Interactive Entertainment* (pp. 89-96).

Kindberg, T., Spasojevic, M., Fleck, R., & Sellen, A. (2005). I Saw this and Thought of You: Some Social Uses of Camera Phones, In *Extended Abstracts on Human Factors in Computing Systems* (pp. 1545-1548).

Kirk, D., Sellen, A., Harper, R., & Wood, K. (2007). Understanding Videowork. In *Proceeding of the SIGCHI Conference on Human Factors in Computing Systems* (pp. 61-70).

Pea, R., Mills, M., Rosen, J., Dauber, K., Effelsberg, W., & Hoffert, E. (2004). The DIVER Project: Interactive Digital Video Repurposing. *IEEE Multimedia, 11*(1), 54-61.

Pimentel, M. G. C., Cattelan, R. G., Melo, E. L., & Texeira, C. A. C. (2008). End-User Editing of Interactive Multimedia Documents. In *Proceedings of ACM DocEng* (pp. 298-301).

Shamma, D. A., Shaw, R., Shafton, P. L., & Liu, Y. (2007). Watch What I Watch: Using Community Activity to Understand Content, In *Proceedings of the International Workshop on Multimedia Information Retrieval* (pp. 275-284).

Shamma, D.A., Bastea-Forte, M., Joubert, N., & Liu, Y. (2008). Enhancing Online Personal Connections Through the Synchronized Sharing of Online Video. In *Extended Abstracts on Human Factors in Computing Systems* (pp. 2931-2936).

Taylor, A. S., & Harper, R. (2002). Age-Old Practices in the 'New World': a Study of Gift-Giving between Teenage Mobile Phone Users, In *Proceeding of the SIGCHI Conference on Human Factors in Computing Systems* (pp. 439-446).

Ursu, M. F., Kegel, I. A., Williams, D., Thomas, M., Mayer, H., Zsombori, V., Tuomola, M. L., Larsson, H., & Wyver, J. (2008). ShapeShifting TV: Interactive Screen Media Narratives. *Multimedia Systems Journal, 14*(2), 115-132.

ENDNOTES

1 http://jame.tv/
2 http://www.cardinal.fi
3 http://www.alticast.com
4 http://www.youtube.com/
5 http://jumpcut.com/
6 http://video.yahoo.com/
7 http://www.myspace.com/
8 http://www.facebook.com/
9 http://www.joost.com/
10 http://cinema.lycos.com/
11 http://messengertv.msn.com
12 http://www.cwi.nl/projects/Ambulant/
13 http://www.tv-anytime.org/

Section II

Who, What, Why and How?

Methodology and Audience Studies

Chapter VI
Sociability Heuristics for Evaluating Social Interactive Television Systems

David Geerts
IBBT / K.U.Leuven, Belgium

ABSTRACT

In this chapter, the author introduces 12 heuristics for evaluating the sociability of social interactive television systems. He first introduce the social uses of television, documented by many scholars in media studies, as well as the relatively new concept of sociability in new technologies. He then explain how the heuristics are based on a thorough analysis of several user studies the author performed with several social interactive television systems, as well as of literature that reports lab studies and field studies with such systems. Finally, each heuristic is presented and explained in more detail, along with some more specific guidelines. Geerts hopes these heuristics will enable designers and evaluators of social interactive television systems to make sure they support the social uses of television and create sociable systems.

INTRODUCTION

Evaluation is a critical part of the user-centered design (UCD) process of new technologies. It is best done as soon as possible – to uncover problems when it is still easy and cheap to fix them – and as often as possible, in an iterative manner. Within Human-Computer Interaction (HCI), the multidisciplinary field that is concerned with studying how humans use interactive systems and how to design systems to improve this use, a wide range of methods exist to evaluate if a system is adapted to human needs. These methods are usually classified in three categories: user-based methods, model-based methods and expert-based methods (Jacko, 2003). Of all these methods, heuristic evaluation (part of the expert based methods) is low-cost and easy to perform (Nielsen, 1994), so it is very popular with practitioners. Such a heuristic evaluation is performed by one or more experts that use a number of general usability principles or heuristics to form an opinion on the user friendliness of an application (e.g. Nielsen, 1994, Norman, 1988, Shneidermann & Plaisant, 2005). Most of these heuristics however are meant for computer based applications, and even though some of these guidelines could be transferred to interactive television, they are not always applicable due to the specific nature of television (e.g. being a lean-back medium). Thomas & Macredie (2002) have argued that, if there are no specific guidelines, designers for interactive television would be forced to apply existing HCI guidelines. Because of the essential difference between both media, they feared that these designers would create a mix of ad hoc, company specific standards that are mostly incompatible and inconsistent, based on HCI knowledge stemming from an obsolete economical, social and technological environment. They say that to design new guidelines, the general principles of HCI are best taken as a starting point, but adapted to the properties of interactive television and the characteristics of the traditional television viewer.

There are already some guidelines to design user friendly interfaces for interactive television (e.g. Daly-Jones & Carey, 2000, Lu, 2005, Chorianopoulos, 2008), but they usually focus on the usability of interactive television systems, and do not take the social nature of television into account, or only very limited. Another problem with some of these guidelines is that they are tailored to certain applications, e.g. Electronic Program Guides (EPG), which means they are not widely applicable to other interactive TV systems.

As discussed in the previous paragraph, existing guidelines for interactive television are usually related to usability: "The extent to which a product can be used by specified users to achieve specified goals with effectiveness, efficiency and satisfaction in a specified context of use." (ISO 9241-11). However, for applications being used in a social context such as the social television systems that are presented in other chapters throughout this book, evaluating only usability is not enough. Even if these applications are evaluated to improve their usability, it doesn't mean that the social interactions they are supposed to enable are well supported.

Recently, the term 'sociability' is being used to indicate these aspects that support and enhance social interaction with and through new technologies and applications. Having social interactions with new technologies is not new at all. Even before, but especially since the advent of the internet, people have used new technologies to find ways to communicate, discuss and form networks with each other. Several research areas have arisen that focus on this phenomenon such as Computer Mediated Communication (CMC) and Computer Supported Cooperative Work (CSCW) (Grudin, 1994). However, specifically evaluating how well these social interactions are supported is only recently an area of research.

In this chapter, we will first discuss the social nature of watching television and how this impacts interactive television. We will then introduce the concept of sociability, and look at existing sociability heuristics for online communities. Finally, we introduce our own heuristics for evaluating the sociability of social interactive television systems.

THE SOCIAL USES OF (INTERACTIVE) TELEVISION

According to McQuail (1998), media use is as sociable or as solitary as a person wants it to be. Whereas certain media like books are typically solitary, television is typically very sociable. He cites different researchers that have studied the social uses of television and summarizes some aspects of the 'sociability' of mass media, e.g.

maintaining peer-group relations, using media as a basis for conversations and promoting social contact, replacing missing role models by media personalities and structuring interaction within the family (McQuail, 1998).

One of the studies cited by McQuail that is particularly interesting is the ethnographic research by James Lull (1980) who classifies the social uses of television in a typology. Lull distinguishes structural and relational uses of television (see Figure 1). Structural aspects are concerned with the way television has an impact on the structure of people's lives. Relational aspects consider the influence television has on relationships between people.

Lull divides the structural social uses of television into environmental and regulative aspects. With environmental aspects, he refers to the use of television as background noise. In a lot of families, the television set is turned on and used as a way to bring life into the house. The television can also be used as a companion during household chores, to make them seem less tedious. In a general way, Lull also sees the television as a source of entertainment for the family as an environmental aspect. The regulative aspects of television refer to the use of television to punctuate time and activities. As a lot of television shows are aired at the same time every day, it's easy to use this as an indicator of the time, or as an indication that dinner will be served. When family members talk to each other is also regulated by the scheduling of television programs. The same goes for external activities or communication outside of the family (e.g. phone calls to friends or relatives).

When we look at environmental and regulative aspects, there are some interesting difficulties relating to interactive television. As interactive television demands some interaction from the user, it's not enough to merely put on the television and let it serve as background noise or companionship. Also, as video-on-demand might become more widespread, it's a lot more difficult to structure time and activities around certain programs. The

Figure 1. Social Uses of Television (Lull, 1980)

Social Uses of Television

Structural

Environmental (background noise; companionship; entertainment)

Regulative (punctuation of time and activity; talk patterns)

Relational

Communication Facilitation (experience illustration; common ground; conversational entrance; anxiety reduction; agenda for talk; value clarification)

Affiliation/Avoidance (physical, verbal contact/neglect; family solidarity; family relaxant; conflict reduction; relationship maintenance)

Social Learning (decision-making; behavior modeling; problem solving; value transmission; legitimization; information dissemination; substitute schooling)

Competence/Dominance (role enactment; role reinforcement; substitute role portrayal; intellectual validation; authority exercise; gatekeeping; argument facilitation)

news for example does not necessarily start at seven, but can be put on whenever the viewer(s) want. These conclusions are important to elaborate on in a later phase. They might lead us to design implications to make sure these functions can still be served by interactive television, possibly in a different form.

Relational aspects according to Lull are divided into four categories. Communication facilitation means that television can trigger off conversation, or for a child be an entry point into a discussion because he can refer to something he has seen on television. Television in this sense can also fill holes into a conversation or make it possible to clarify personal norms and values. Social affiliation or avoidance means that television can promote but also prevent physical or verbal contact. Family members can cuddle together in front of the television, but the presence of the television can also be an excuse for not talking to each other. It can also strengthen the solidarity in a family or alleviate conflicts. Social learning refers to the possibilities offered by television to learn how to behave socially or solve problems, but also to spread information and knowledge about norms and values to others. Finally, social competence or dominance means that parents can use their authority to decide when children watch television, role patterns are confirmed or imitated or television characters can be verbally assaulted or corrected to demonstrate their own competence.

Regarding relational aspects, again it is already clear that there are services of interactive television that undermine some of these aspects as mentioned above. The Guardian reports that because of the increased choice of television channels on digital television, the shared experience of discussing a television show that was on the day before is becoming less common (Wells, 2004). If on-demand television gets widely spread, that might become an even bigger problem. On the other hand, interactive television offers a lot of possibilities to support or even enhance the relational social uses of television. Interactive game shows make it possible to play along more actively than is possible with normal television, even with other households across the country, making it easier to strengthen the relations within a family. On-demand programs can help parents in spreading knowledge about norms and values to their children. Child locks can be used to help parents keep control over the moments children watch television, or which television show can be watched.

Although Morley's study "Family Television" (1986), in which he conducts family interviews about their television habits, is limited in scope and focuses more on gender differences than on the social uses of television per se, some of the aspects mentioned by Lull are confirmed. In most of the interviewed families, television is used as accompaniment to domestic activities (cf. environmental aspects), but almost exclusively by women. The remote control serves as the token of power for the father of the family (cf. social competence or dominance). The family members talk to friends and colleagues about television, especially women (cf. television as communication facilitation).

One might argue that these studies are quite old, and the conclusions can not be transferred to modern times given the recent technological developments. However, a more recent analysis of domestic media among Flemish, French, Italian and Swedish children and teenagers (Pasquier et al., 1998) shows evidence that television is still very much a social medium. Pasquier et al. have found that in the four countries studied, most of the families still eat the main meal and watch television together. Although a lot of children (depending on the country, one in four or one in two) have a television in their bedroom, watching television is still a family activity taking place in the living room, especially in France. They conclude that "co-viewing is still the dominant practice everywhere" (Pasquier et al., 1998: 510). When looking at television used as "communication facilitation", as in Lull's typology, the findings show that "television – and VCR – is the main medium which mediates between young people and their parents in all four countries", especially between mothers and daughters. A similar argument for the social role television still plays in the family context has been made as recently as 2007 (Lemish, 2007).

Lee & Lee (1995) have studied the social uses of television as well, but more specifically – and very relevant to our discussion – in order to look at the implications for interactive television. Again, there are numerous similarities with Lull's typology. Lee & Lee discovered four levels of viewing based on the physical proximity of the viewer to the television set and the mental (or physical) involvement with something other than television viewing (e.g. talking to someone, or performing household chores). In level 1, television is the only activity which demands all attention. In level 2 television is one of two activities, e.g. next to reading a book, and the attention of the viewer is divided. In level 3, television is a peripheral activity as accompaniment to another activity, but attention can be turned to the television set when appropriate. And in level 4, television serves as background noise, and the viewer is not necessarily in the same room. Especially the last two levels relate to the environmental aspects of television. Other aspects mentioned by Lee & Lee are the routinization of watching television (cf. regulative aspects), and the general observation that television is still (in 1995) very much a social activity. They also studied the motivations and gratifications for watching television, of which some seem contradictory to interactive television. They especially think that "the desire for undemanding relaxation, involvement in engrossing stories, and the enjoyment of a routinized experience all mitigate the possibilities of new patterns of involvement with television" (Lee & Lee, 1995: 13). They also argue that television as "social grease", meaning the role of television to facilitate interpersonal communication, conflicts with the highly individualized programming possible with interactive television. On the other hand, television as a source for information, self-education and social learning hold some possibilities for interactive television. They conclude that "interactive television is likely to be successful to the extent that it takes into account existing behaviour patterns as well as motivations and gratifications already being satisfied by this enormously successful medium" (Lee & Lee, 1995: 17), indicating the necessity of including social features in interactive television, and ensuring that social interaction with these systems is adapted to the specific characteristics of watching television.

Even though in recent years, many social interactive television systems have been designed (as evidenced by the contributions in this book), little research has gone into adapting these systems to the social uses of television as documented by Lull. One notable and early study has been performed by Arvola and Holmlid (2000) who have studied the qualities-in-use needed for IT-applications that are meant to socialise, by doing a pilot study of a quiz game on interactive television. They detected four qualities-in-use of the game, of which some are divided into subcategories: (1) ease of use, (2) enchantment (on the surface level, the artefact level or the activity level), (3) entertainment (ego challenge or social challenge) and (4) togetherness (immediate togetherness or delayed togetherness). After this pilot study, Arvola (2003) analyzed an add-on to a racing broadcast and an on-demand news application. When the qualities-in-use of the first pilot study and the two new cases were compared, among others regarding their social use, the following three common characteristics were detected and presented as final qualities-in-use[1]: (1) laidback interaction, (2) togetherness (3) entertainment.

Laidback interaction refers to the "lean back" experience of television. Users of interactive television don't want to be too actively involved. Ease of use is closely related to this aspect. Togetherness is the feeling that viewers are actively engaged with other viewers in the room (immediate togetherness), or that they can talk about the program with friends later (delayed togetherness). Entertainment is also an important quality-in-use according to Arvola, with two aspects: ego challenge is the joy of achieving or knowing something, social challenge is the joy of winning from another person.

When we compare these characteristics with Lull's typology, we find some striking resemblances. Immediate togetherness, being 'engaged' with other viewers in the room, can be fit in the category of "social affiliation", as the contact between users (family or friends) is encouraged. It might also be seen as a form of "communication facilitation", because viewers are actively engaged with each other (if not by talking, then by interacting with each other). Delayed togetherness is a form of communication facilitation as well, because the interactive television application acts as a conversation starter. Not only at home, during the interaction, but also afterwards, talking about it with friends, family or colleagues. Finally, the quality social challenge is a form of social competence or dominance, where the viewers can demonstrate their own competence.

The classification by Arvola could already be used to evaluate a social television system, but they are based on a limited study and are not explicitly formulated as guidelines or heuristics. Their practical applicability is therefore limited. Before we introduce our own sociability heuristics for interactive television, based on several user studies, we will first have a look at the concept of sociability and specific sociability heuristics for online communities.

EVALUATING SOCIABILITY

Social interaction is a key concept when talking about sociability. When we look at the meaning of the word sociable (not related to technology), companionship and pleasant social relations are mentioned as central elements (Merriam Webster).

Sociable: 1: inclined by nature to companionship with others of the same species: SOCIAL
2 a: inclined to seek or enjoy companionship b: marked by or conducive to friendliness or pleasant social relations

Sociability in turn is a quality or state that can be appointed to a person if he/she is being sociable, or the act or instance when someone is being sociable.

Sociability: the quality or state of being sociable; also: the act or an instance of being sociable

For a more scientific approach to the concept of sociability, we can look at the famous sociologist Simmel (1949), who has written a seminal paper about the concept of sociability, and defines it as "a feeling for [...] the very fact that one is associated with others and that the solitariness of the individual is resolved into togetherness, a union with others". He says that "sociability in its pure form has no ulterior end, no content, and no result outside itself, it is oriented completely about personalities". This definition explains in more detail that sociability is just about relationships, being together and is completely focused on the act of being sociable, with no other motive than being together and enjoying the companion.

When we transfer the concept of sociability as explained above to the world of technology, sociability can be defined as the quality of a technology to be conducive to pleasant social relations. This is however a very broad definition, and as sociability in new technologies is a relatively new concept, there is no definition yet that comprehensively explains what it entails. Like with usability, sociability manifests itself differently based on the specific platform of the application. Sociability is already being studied within domains as diverse as online communities (Preece, 2000), games (Ducheneaut e.a., 2004, Brown, 2005, Hew et al., 2004), mobile applications (Jans & Calvi, 2007) and interactive television (Rasmussen, 2006). Although there is a lot of research on social interaction in groupware[2] and other work-related applications that support social interactions, we will limit ourselves to non work-related applications as this is a domain in which this kind of research is relatively new.

Jenny Preece has been the first researcher to extensively study how to support and evaluate online social interactions. According to Preece & Maloney-Krichmar (2003) sociability "is concerned with developing software, policies and practices to support social interaction online" (Preece & Maloney-Krichmar, 2003: 605). They argue that the concept of sociability, which they apply to online communities, is very new and, in contrary to usability, still has to be operationalized. They also say a framework for sociability is needed to support designers in creating social applications. De Souza & Preece (2004) have constructed such a framework, the online communities framework (OCF), to analyze and understand online communities, based on semiotic theory and semiotic engineering. The three basic elements relating to sociability in this framework are "people", "purpose" and "policies".

With "purpose", Preece (2000) stresses the importance of a clear purpose of the community that is well stated and linked with other aspects such as a good registration policy. "People" are a key element in online communities, and there are several roles that each play an important part in the sociability of an online community, such as moderators and mediators, professional commentators, provocateurs, general participants and lurkers. Another aspect related to people in communities is the community size. As communities can be neither too small nor too big, reaching critical mass is important, albeit variable from community to community. Finally "policies" are concerned with governing the behaviour of the online community. Topics related to policies are joining and leaving requirements, by-laws, codes of practice for communication, rules for moderation, issues of privacy and trust, and so on.

Based on this framework, Preece (2000) presents sociability guidelines for designing online communities, and derives several heuristics for evaluating not only the sociability, but also the usability of online communities. Some examples of design guidelines are: give the community a clear, meaningful name; decide whether other experts are needed and they can contribute; support personal presence; and encourage empathy, trust and cooperation. When translating these guidelines to more general principles, suitable for evaluating a system, this results in the following heuristics for evaluating the sociability and usability of online communities:

- Why should I join this community?
- How do I join (or leave) this community?
- What are the rules of this community?
- How do I get started reading and sending messages?
- Can I do what I want to do easily?
- Is the community safe?
- Can I express myself as I wish?
- Why should I come back?

These heuristics are clearly tailored to evaluate the usability and sociability of online communities, and can not be used for evaluating interactive television applications. Therefore, we set out to create specific heuristics to evaluate the sociability of social interactive television systems. In the following section, we will first shortly describe how we created these sociability heuristics, and will then introduce and discuss each specific heuristic and some more detailed guidelines.

SOCIABILITY HEURISTICS FOR SOCIAL INTERACTIVE TELEVISION

We performed several user tests with social interactive television systems[3] and thoroughly analysed literature that reports the results of user studies in the lab as well as in the field. Based on these studies we used grounded theory and affinity diagramming to create a list of 12 sociability heuristics that can guide the design as well as the evaluation of social television systems. The focus of these heuristics is on sociability. Usability issues are not directly addressed in these guidelines, as already mentioned there are plenty of other general usability heuristics or specific guidelines for interactive television. Of course, these sociability guidelines should be used together with usability guidelines for interactive television, as a social television system should not only be sociable, but also usable.

The heuristics presented here are not the only or the first guidelines for social television systems. Several other chapters in this book present guidelines based on experience with a single system or on related literature, and Chorianopoulos (2007, 2008) addresses several social issues of interactive television in his interface design principles as well as his discussion on content-enriched communication. Although these guidelines and principles are very valuable, we think our heuristics have the added benefit of being based on several user studies with diverse social interactive television systems, and are explicitly formulated to enable performing a heuristic evaluation with them.

This section is organized as follows: for each heuristic, we will first discuss the background that explains the research the heuristic is derived from. We then present the heuristic, along with a short explanation that should help the evaluator to use the heuristic during a sociability evaluation. Finally, we will present a couple of specific guidelines as instances of that heuristic, so the evaluator has some examples of how this heuristic manifests itself in different systems. However, these specific guidelines are just examples, and should not limit the evaluator in assessing the system to be evaluated.

Communication Modalities

One of the basic functions of most social television systems is supporting communication between remote television viewers. The first issue that arises when implementing communication tools is whether to enable text chat or voice chat, or maybe even both. This issue is currently actively debated (see e.g. Geerts, 2006, Schatz e.a., 2008, Weisz e.a., 2007, Tullio & Harboe, 2008), and is still not concluded. However, the most recent research indicates that offering both modalities is the preferred option for users. This allows them to choose the communication modality that best fits the situation and their own experiences. Older users are less adept at chatting and multitasking, and prefer voice chat, whereas younger users prefer text chat (Geerts, 2006). Certain circumstances are more beneficial for text chat (e.g. when the children are sleeping), others more for voice chat (e.g. when moving around the room) (Tullio & Harboe, 2008). However, if only one option can be chosen, due to technical or budgetary reasons, one has to carefully select the right modality, based on the intended user population and context of use.

Whatever communication modality is chosen, users clearly like a wide range of communication options. Several research studies (e.g. Harboe e.a., 2008) have shown that when restricting communication channels, e.g. by only offering pre-fabricated messages, users are unhappy about it, and try to find other ways of communicating (by taking the phone). This means that users need full communication possibilities. This doesn't mean that other ways of communicating should be abandoned. Lightweight messaging such as emoticons, small generic messages or gestures that convey interest in a certain show all enrich the way users can communicate with each other through the system. Social television systems that focus more on sharing content should also consider implementing communication features.

Finally, when implementing text or audio chat, the communication should be optimally supported. Indicating who is talking, giving feedback when a reply is being typed or offering a simple way to control the audio, all ensure that communication between users goes smoothly.

Heuristic #1: Offer different channels and levels for communicating freely. Enable voice chat as well as text chat if possible, otherwise use voice chat for a broad user audience including people with little chat experience or text chat for a specific audience including people with chat experience. Allow communication on different levels, from low-activity quick responses such as emoticons, gestures or automatic replies, to free-form communication. Make sure the communication process is optimally supported so it can go smoothly.

Examples of specific guidelines:

- Enable free-form communication using voice chat or text chat, and if possible both
- When having multiple people in the same room, provide an easy way of indicating who is talking
- The sound of the TV should be easily distinguishable from the speech of the participants
- Give the user several ways to send or reply to a recommendation, from pre-defined messages to free-form communication
- Make it possible to send lightweight messages such as emoticons, pre-defined messages or gestures
- Indicate when a reply is being typed
- The activation and de-activation of an audio link should be very easy and quick to do
- Users should be able to add a spoken or written commentary to recorded content they are sending to others

Presence and Awareness

In computer-mediated communication, indicating presence is an important feature that supports awareness and enhances sociability. Providing more information than just presence provides even more awareness for the users. In a social television system, awareness features can enable other viewers to see who is currently watching television, what channel they are watching and even what program. There can be several levels of presence indication, from a simple 'someone is watching television' to the names of the people that are watching television to a list of buddies that each have a status and indicate the channel and program names they are watching. This is an excellent way of indicating to other people that they can talk to their friends or family, whether or not about the current program. Even though there might be some privacy issues (see the separate heuristic about privacy), several user studies have shown that users value this information and usually don't have a problem of showing this information of themselves.

One element of indicating presence, is using a status to convey availability, as found in several instant messaging clients. Taking into account other heuristics that emphasize the minimization of distraction or the support of group privacy, different statuses can be used to convey a user does not want to be disturbed (e.g. because he is watching a movie) or that he is watching television together with several family members. Having a good presence and awareness system helps to show these availabilities. Such a system could not only be active while the television set is turned on, but also use more ambient ways of conveying presence (Harboe e.a., 2008).

A complicating factor that distinguishes social television presence systems from other systems is the social and recreational nature of television. It might not be appropriate for colleagues to be on the buddy list, or to see the programs a user is watching. A group watching television might be dynamic, with people entering and leaving the room, ending up with different people that started watching. In any case, the way the buddy list is implemented should be carefully considered against the other heuristics as well.

Heuristic #2: Use awareness tools for communicating availability. Provide a means for indicating presence, and giving information about the current behaviour of other users. There can be several levels of presence and awareness indication, from a simple 'someone is watching television' to a list of buddies that each have a status and indicate the channel and program names they are watching. These tools should indicate if a user is available to chat or otherwise interact, or if there are special circumstances (such as watching in group) other users should take into account.

Examples of specific guidelines:

- Provide a buddy list, including status information and information about what buddies are watching
- Users should be able to put themselves in a status 'non-disturb'
- Provide an option for some buddies to see a different status than other buddies (e.g. friends and family see your buddy as online, but colleagues not)
- Show if and/or which users are currently watching television

Synchronous vs. Asynchronous Use

Similar to other communication tools, social interactive television can support synchronous communication as well as asynchronous communication. This distinction is similar to the traditional social uses of television, where people can either talk or interact while watching television, or talk about television afterwards or even before watching television. In the context of interactive television, this has also been called direct and indirect sociability (Ducheneaut e.a., 2008) or immediate and delayed togetherness (Arvola, 2003).

Accordingly, social television systems can support synchronous use as well as asynchronous use. For example, a user can send a recommendation to watch a specific show to a friend. When the show is currently playing, and that friend is currently watching television, he can immediately tune to the recommended channel to start watching the show and use the communication features of the system to talk together. However, for several reasons, that friend might not be watching television at the moment. He might just not be at home at the moment, or not watching television, but he might also live in a different time zone on another continent. For these and other reasons, the recommendation should be saved somehow, so that the next time the user starts watching television, the user can see that a show as recommended and choose to start watching on demand, or the next time the show is aired. Apart from this specific example, plenty of other features can be thought of that can be implemented synchronously and asynchronously, or that are specific to one of the two modalities, e.g. chatting concurrently versus leaving comments.

Heuristic #3: Allow both synchronous and asynchronous use. Provide different functionalities for interacting and communicating synchronously as well as asynchronously, so users do not always have to be using the system at the same time, but can also benefit from it when they are using it at different times.

Examples of specific guidelines:

- Enable users to communicate in real time but also allow them to leave notes that can be consulted afterwards
- Make it possible to exchange recorded content that can be watched simultaneously with other users, or at another time
- Make sure recommendations not only can be sent directly, but they are also saved so they can be consulted later, or they show up when turning on the system

Remote vs. Collocated Interaction

Using the "place/time matrix" as often used in Computer Supported Collaborative Work (Grudin, 1994), we can not only classify the use of social television systems as synchronous versus asynchronous (see heuristic #3), but also as same place versus remote place. Even though television can be a solitary activity, the dominant model for watching television has been and mostly still is a social activity, meaning that multiple people are watching television in the same location. This would classify traditional television in the same place/same time square of the matrix. With the advent of social interactive television, however, communicating and interacting around television can also happen at the same time, but in different places, or even at a different time at different places. When multiple people are interacting and watching television in the same room, this is called collocated interaction. When different people are interacting and watching television in different rooms, this is called remote interaction.

Although the previous classification seems rather straightforward, it gets more complicated when there are multiple people in the same room and different people at other locations. This means a social interactive television system can fall in two squares simultaneously, and should cater for both collocated and remote audiences. One example from this is the use of side-conversations where two people silently talk to each other to say something private and/or in order to not disturb the other viewers (Ducheneaut e.a., 2008). This can be either collocated, where a voice enabled system might allow users to temporarily mute their own sound to talk freely, or remote, where a text chat system might provide a separate chat window for talking to someone outside of a group conversation.

A social interactive television system should also be careful not to break any of the two, because this might result in a good experience for the remote users, but not for the relationship between the collocated viewers. An example for this is the use of system menu's or certain features that overlay much of the screen. Even though the user that has initiated the action might not find this a problem, his co-viewers could have been following a television program and be annoyed by his actions. This is a careful balance that should be considered for each function.

Heuristic #4:Support remote as well as collocated interaction. As social interactive television systems can be used with multiple users at the same location (collocated) simultaneous with different users at other locations (remote), it should make sure its functions are appropriate for both situations, and not disturb the interactions between collocated or remote viewers.

Examples of specific guidelines:

- Enable private side-conversations between collocated and/or remote viewers
- The system should be able to handle large groups or a large number of people/groups
- The system should not disrupt other collocated viewers e.g. by imposing a big overlay on screen, but should not take up too much space or use a secondary screen for some activities

Information About Viewing Behaviour

One of the other heuristics discusses presence and awareness, including showing information about a user's current activities, such as what channel or what program he is watching. This can be taken a step further, by using the information of current and (especially) past information in specific functionalities. Information about a specific program can for instance include functionality to switch to that channel directly or, more extensively, a shared television guide can contain information about which television programs are most favoured by a user's buddies and even record popular shows from buddies automatically to a user's personal video recorder. Similarly, a system can provide recommendations based on other buddies' watching behaviour. A history of other viewers' behaviour can also give a good indication of favourable viewing times, to make sure there are friends or family members available for talking or interacting.

Although information from viewing behaviour can be exploited in several ways, the user should still feel in control (see heuristic #6) and also maintain his privacy (see heuristic #7). Users have no problem in disclosing most of their viewing behaviour if this will benefit their friends or family, but when used for commercial reasons, they are less inclined to offer this information.

Heuristic #5: Exploit viewing behaviour for informing and engaging other viewers. Use information from a user's viewing behaviour not only for showing currently viewed programs, but also for creating functions that aid communication, social interaction or recommendations for other users.

Examples of specific guidelines:

- Indicate when other buddies were watching frequently in the past, so viewers can see when it's favourable to watch together
- Show information about the channel and program other buddies are watching
- Provide a shared TV-guide with information about buddies' watching behaviour or recommendations
- Exploit social knowledge for recommending content to watch based on buddies' viewing behaviour
- Provide an overview of the most popular shows of a user's friends

User Control

User control and freedom is one of the ten usability heuristics, created by Nielsen and Molich (1990). Although it is defined by Nielsen as "support undo and redo", giving users an emergency exit when they have made a mistake, it is often more widely interpreted as giving the user control over their actions and system settings (which is actually covered by the heuristic "flexibility and efficiency of use"). Although we do not intend to repeat basic usability heuristics with our sociability heuristics for interactive television, we feel this is such an important heuristic with severe implications for the sociability of such systems, that we think it needs inclusion here.

In several user tests we performed ourselves, as well as those reported in literature, users expressed concerns about certain features for several reasons (privacy, personal preference, experience, ...) and indicated they would like to control different aspects of the systems they used during these tests. This ranged from deciding what information they would like to convey to whom, controlling voice or text chat activity (by blocking certain people or muting audio, locally or remote), how to send and receive notifications, etc. In order to support this, users should on the one hand be able to change system settings to suit their own (or their families') preferences and on the other hand be able to have different options for specific functions (mute or not, accept or not, turn off certain features, etc.). The possibility of tailoring actions and changing settings should however not break the lean-back nature of television (Chorianopoulos, 2008) and not disrupt or distract from the television program (see heuristic #8).

Heuristic #6: Give the user appropriate control over actions and system settings. Users should have sufficient control over their actions and system settings, so they can adapt the system in general or specific features to their needs or to the current situation.

Examples of specific guidelines:

- If providing video communication, make sure it can be turned off
- Settings for notifications, status indications, how to send and receive information, etc. should be controllable

- Allow users to mute the sound of the television program, separately from the sound of the voice chat
- The system should be an optional feature, that can be turned on or off when wanted
- A user should have control over audio activity, and determine to mute their own audio and/or mute the remote audio
- Let users decide who can view their comments or annotations, and if they will be anonymous or can be identified

Personal and Group Privacy

The social nature of watching television has several consequences, one being that users not only want to protect their own personal privacy, but also privacy related to the group situation. Regarding personal privacy, not everyone is keen on sharing information about all television programs they are watching. This has been confirmed in several use studies as an important concern, where users indicate they don't want friends or family to know they are watching e.g. porn or, in contrast, a female oriented program. Interestingly, one study pointed out that women would have less problems in disclosing this kind of information (even if it concerns porn) than men (Regan & Todd, 2004). Another privacy related issue has to do with video communication. Although this hasn't been implemented in any of the systems that have been analysed for creating these guidelines, it was mentioned by some participants as a useful feature. However, most users indicated they would feel it intrudes their private space, or they are not dressed appropriately when watching television. If a system would have a video communication feature, users should be able to activate or deactivate it for privacy reasons.

On the other hand, group privacy relates to the collocated situation in some instances. In a field study of a social television system (Harboe e.a., 2008) as well as in a lab study (Regan & Todd, 2004), a remote user on one end said something that a collocated user on the other end wasn't supposed to hear. This could lead to all sorts of awkward situations, which should rather be avoided. A social television system should therefore use functions such as a group status, inform the user in different ways of an open audio channel or other people that are present, or allowing users to send messages to a different, more private device.

Heuristic #7: Guarantee both personal privacy and group privacy. Make sure the system enables users to ensure their own personal privacy, by choosing what (not) to disclose, as well as group privacy, by taking into account the presence of multiple viewers in a collocated viewing situation.

Examples of specific guidelines:

- Video communication is not advisable as main feature
- It should be possible to temporarily switch of the audio-link
- When an audio-link is active, that should be clearly visible
- Certain channels or programs should be blocked from appearing in the 'now watching' list
- Provide a group status, indicating that multiple people at one location are watching TV
- Give users an option to send content to a more private device (e.g. mobile phone or pc)

Distraction

The main purpose of television remains entertaining or informing viewers with television programs. When new features are added that enable communication or interaction via television, these features should not interfere too much with the content that is being shown. This means that interaction with the system needs to be as easy and fluent as possible, so that attention to the system is minimal, and viewers can pay sufficient attention to the program itself. Especially the genre of a program influences how much attention is needed to the television screen (see also heuristic #10). The more plot-oriented a genre is (like movies or drama), the more attention someone needs to pay to the television screen in order to keep following the program (Geerts e.a., 2008). Users will not want to be disturbed while watching this kind of content, or only minimally. Both text chat and voice chat is taking some attention away from the audiovisual content on screen. Although this usually does not stand in the way of enjoying the experience (Weisz e.a., 2007), the system should be designed so that this distraction is as minimal as possible. Making screen elements not too large, putting them in the visual periphery (taking care not to overlay subtitles or closed captioning), enabling quick access to frequent functions and allowing automatic replies or short lightweight messages to be sent, all helps in minimizing distraction.

A helpful tool for users to decide themselves if they want to be more engaged in interaction or not, is the status don't disturb. Viewers could be put on don't disturb in several ways, such as automatically (e.g. based on the kind of content (films), or when minimizing a program) or manually by the viewers themselves. When in don't disturb mode, there should be some small unobtrusive indication when something is happening, otherwise, they won't notice that someone is trying to chat. However, this notification should be unobtrusive and users should be able to ignore it (without any disturbance, so it should go away automatically) or act on it when they feel like it.

A collocated viewing situation brings up another issue: when one user is interacting with the system, and several menus or other messages appear on screen, maybe even overlaying parts of the content, this can disturb other viewers. Separate screens (e.g. on a mobile device) can make sure that collocated viewers are not distracted while one or more persons are active.

Heuristic #8: Minimize distraction from the television program. Design system features so there is not too much distraction from watching the television program. Specific tools can help users control if they want to be distracted or not. Distraction of other collocated viewers should also be taken into account and minimized.

Examples of specific guidelines:

- Users should be able to put themselves in a status 'non-disturb', even better is to do this automatically (e.g. by detecting the TV-genre)
- When a user is in the status 'busy' or 'don't disturb', incoming events should be less intrusive or even blocked
- Don't provide too many statuses, as it is easily forgotten to change this. An automatic system for changing status would be even better.
- Users should be able to exchange light interactions in a quick and easy way
- Screen elements should be small and in the visual periphery, and not overlay subtitles or closed captioning
- Make sure text chat or comments don't take up too much space
- Frequent system functions should be very easy and quick to do
- Pop-up messages should automatically disappear after a few seconds when not acted upon
- Make it possible to send an automated reply
- The TV-program should not be resized too small, or not at all
- The activation and de-activation of an audio link should be very easy and quick to do

Notifications

When using social interactive television systems, users interact with each other in different ways. This can be explicit communication, having a text chat or voice chat, but it can also be less direct and involve a recommendation that is being sent, or even implicit interactions of users that change the television channel or ignore a sent recommendation. As with traditional usability, where feedback is needed about how the system responds to user actions, a sociable system should also provide feedback about actions or responses of other users. When a user sends a recommendation to a friend, he would like to know if that friend has accepted the recommendation or not. A notification from the system when a user starts watching a certain channel or leaves a certain channel helps in the communication process, so users know if they have common ground to talk about.

Of course, when giving the user a notification, this should not interfere with heuristic#8 that states that distraction from the television program should be minimized, otherwise users can get irritated about these (see the chapter by Boertjes). It's also advisable to make this a controllable feature, so that some users can disable them if wanted, or choose which ones they would like to receive and which ones not. In addition, the number of notifications and the way they are sent can be dependable on the status a user is in (don't disturb) or the program genre he is watching.

Heuristic #9: Notify the user of incoming events and situation changes. When users are requested to respond to an action of another user, notify them visibly or audibly of these incoming events. When there are changes in another user's situation, e.g. when switching channels or when going from watching alone to watching in group, users that are actively interacting with this other user should also be notified of this change in situation.

Examples of specific guidelines:

- When a status is changed (especially to group status), this should be notified to users that are having an active conversation, so they know that people have entered the room and are listening in
- When incoming messages are blocked, there should be a small non disruptive indication that there is chat or other activity
- When a recommendation/invitation is accepted, give feedback to the sender if and how the receiver has reacted (accepted or not)
- When someone enters a channel or leave a channel, this should be notified, especially to users that are actively interacting or communicating with this person

Program Genres

As a social television system will usually be used at the same time as watching a television program, its features should be in some way adapted to the content that is being shown. This is less straightforward than it might sound, as this content is very heterogeneous. Television programs (and by extension any video content) is classified into different genres, that each induce specific audience reactions or user experiences. Television program genres range from movies, drama and soap opera to reality TV, news and sports. One differentiator between these genres, that has an impact on social television systems, is the plot-structure. The more plot structure a genre has, the more attention a user has to pay to it to follow the program. This means that genres with high plot-structure are less suitable for social interactive television systems that support synchronous interaction and communication, as chatting demands some attention from a user as well and interferes with following the plot. News, soap, quiz and sport are those genres during which people talk a lot while watching. For designing features for synchronous social iTV systems, these genres are most suitable. For asynchronous systems (such as for video sharing or leaving notes on top of a video stream), film and news are two genres people talk a lot about and which they would also like to share, while documentaries and music programs also score high. This means that the focus of asynchronous TV systems should be on these genres when designing features for these systems (Geerts, 2008).

Different program genres also have different social uses. According to Lull (1980), soap opera is often used by viewers to learn how to deal with relational conflicts, and discuss this with friends and relatives. This explains why people use this in daily conversations. A similar argument can be made for reality shows. These genres are therefore not only suitable for synchronous communication, because they have low plot-structure, but also for asynchronous systems, because people like to discuss about it. Other genres however show other patterns. People often talk during quiz shows for example, because viewers like to show off their knowledge to each other, but do not often talk about quiz shows. Showing competence during a quiz show is usually done synchronously, so one can answer a question before a contestant, and not by bragging afterwards how many answers you knew. So this genre is mostly suitable for synchronous communication.

Genres also play a role in the platform that viewers use to watch their television content. As users nowadays have several options to watch video content (computer, mobile phone, television, …), they can also choose what content they want to watch on which device. Documentaries and movies are preferably viewed on television, in contrast to the weather or breaking news, which people prefer to watch on their mobile phone. This confirms previous research that states that although the PC is becoming more and more an entertainment system, allowing online video watching (e.g. on YouTube) or even chatting while watching television (e.g. on Joost), the living room is still seen as the most comfortable viewing location, because of the big television screen and comfortable couch (Brereton & O'Connor, 2007). Not only plot structure, but also the length of the content plays a role in choosing a platform to watch this content. In addition, the urgent character of the content (e.g. breaking news) and the social situation (group privacy) also has impact on this decision. A mobile phone is usually preferred for short clips with low plot structure that can be watched in private, and that have an urgent character (high immediacy). E-mail received on PC is preferred for medium length clips with medium plot structure, which may or may not have an urgent character but can also be watched in private. Finally, television is a device on which people would want to receive long clips with high plot structure but without an urgent character that can be watched in company.

Heuristic #10: Adapt to appropriate television program genres. Take into account the properties of television genres, such as the plot structure and social uses of certain program genres, and offer features or settings that are appropriate for these genres. The system should be tailored to television genres that are more suited for the synchronicity of the interaction or for specific platforms.

Examples of specific guidelines:

- Tailor synchronous systems for genres with low plot-structure such as news, soap operas, sports, quiz shows, reality TV or talk shows, and less so for genres such as film, drama, documentaries or news magazines that have high plot-structure
- Tailor asynchronous systems for genres that are being talked about a lot such as film, news, documentary, music programs or soap opera
- Tailor social TV on mobile phones for short clips with low plot-structure that can be private
- Tailor social TV on PC for different sized clips with medium to high plot-structure that can be private
- Tailor social TV on television for longer clips with medium to high plot-structure that are usually not private

Sharing Content

Currently, user-generated content (UGC) is created and shared in large amounts. Although a lot of it is publicly available on sites such as Flickr or YouTube, even more UGC is hidden because it is only shared privately, and often not online but by exchanging videos or DVDs, or just showing visitors vacation snapshots. It is therefore no surprise that in many user tests, people expressed a wish to be able to share content, either created themselves or broadcasted content, which they have recorded. Legal issues aside, there are many benefits to social interaction that warrant social television systems to enable the sharing of content.

Sharing content alone is not sufficient however, users should be given additional options to edit the content, e.g. to cut out an interesting part and only share that. Moreover, sharing content should also be integrated with other features of the system, such as communicating by adding spoken or written comments to the content. As users often have access to different devices to watch content next to television, such as a computer or a mobile device, they should be able to send and receive content on these devices as well, and easily transfer it from one device to another device.

Heuristic #11: Let users share content flexibly. Make sure users can send content to and from different devices to share this with other users. They should be able to edit the content as they wish, and add some form of commentary to it. As much as possible, integrate sharing content with other features of the social television system.

Examples of specific guidelines:

- Make it possible to exchange recorded content
- Allow users to add a spoken or written commentary to recorded content they are sending to others
- Users should be able to add their own personal content to the system, and share, recommend or edit this
- Make sure users can send content to others on different devices
- When receiving clips, users should be able to transfer it to another device (cell phone, computer, television) e.g. by forwarding it

Sharing Activities

Social TV is not only about communication, but also about sharing activities. Examples are watching the same television program at the same time, choosing which programs to watch, recommending a favourite television show, pointing out a funny scene, etc. A social television system should make it easy to start and maintain these activities by offering features that encourage them, such as easy to accept (or decline) invitations to watch together, a shared remote control to pause or fast forward content or a shared program guide that allows several people to choose a television program to watch.

Of course, just offering these features is not sufficient. They should take into account the specific nature of television, some of which are already covered by other heuristics, such as the minimization of distraction or the combination of remote and collocated interaction. There are however also specific aspects of sharing activities. For example, when watching content together, people often want to refer to something that is happening on screen and a delay when typing can make this problematic. A system that enables an accurate way of supporting this will enhance the sociability of communicating about television content. Another heuristic that also has impact on sharing activities is providing awareness, e.g. by letting users know when they are watching TV together, and when they are watching unsynchronized. This helps people to manage this shared activity better.

Heuristic #12: Encourage shared activities. Allow users to easily start and maintain shared activities around the television content, such as communicating, watching together, choosing programs or controlling the content. Make sure sufficient information about these activities is provided so efficient communication and interaction is enabled.

Examples of specific guidelines:

- Provide a way to invite other users, e.g. friends and family, to use the system
- Make it possible to invite others to watch the same channel
- When more than two users are watching pre-recorded content, a shared remote should be used - a master/slave model is preferred, but the 'slave' role should be able to take over the 'master' role - the master is the one that initiated the viewing
- When starting to watch a program that has already started, people should be able to call up a short visual or textual synopsis to catch up so they have common ground to talk about
- Make it easy for users to switch to a channel that another buddy is watching, so they can watch together
- Make a shared program guide available, e.g. for choosing programs together
- Provide enough information about a recommendation that is sent, such as a program summary or a picture in picture video, so the receiving user can make an informed decision to act on the recommendation or not.

CONCLUSION

As social interaction and communication surrounding television is one of the most important characteristics of social television systems, designers of these systems need a way to assess if their system successfully supports this. Insight into the social uses of television, and by extension the social uses of interactive television, can be of help, but this does not offer a direct usable method that is reliable. Although user studies, in the lab or in the field, are a reliable way to assess how real users will interact with and through the system, they are more expensive and time consuming than a low-cost method such as heuristic evaluation (see also the chapter by Bernhaupt e.a.). In this chapter, we have introduced twelve sociability heuristics, that were developed based on several lab and field studies. Using these heuristics, designers can evaluate if their system will successfully support social interaction and communication in the diverse contexts that can surround watching and using (interactive) television. The heuristics can also be used as design guidelines, when making choices for certain features or interface elements in the early stages of design.

Of course, these sociability heuristics are not meant as a substitute for doing user studies, but should be used complimentary to them. As with general usability heuristics, these sociability heuristics can help designers to 'catch the low hanging fruit' and already take into account lessons learned from other systems, so they can focus the users studies on more specific issues that arise when using their new system.

We hope that these sociability heuristics will ensure that current and future social television systems will be well-adapted to users' needs and the social uses of watching television.

REFERENCES

Arvola, M., & Holmlid, S. (2000). IT-artefacts for Socializing: Qualities-in-use and Research Framework. *Proceedings of the 23rd Information Systems Research Seminar in Scandinavia*, Sweden

Arvola, M. (2003). *Good to Use!: Use Quality of Multi-User Applications in the Home*. Licentiate's Thesis. Linköping University, Sweden

Brereton, P., & O'Connor, B. (2007). Pleasure and Pedagogy: The Consumption of DVD Add-Ons Among Irish Teenagers. *Convergence*, *13*(2), 143-157

Brown, B. (2005). Play and Sociability in There. In R. Schroeder & A. Axelsson (Eds.), *Work and Play in Shared Virtual Environments*. Springer.

Chorianopoulos, K. (2008). User Interface Design Principles for Interactive Television Applications. *International Journal of Human-Computer Interaction, 24*(6), Taylor Francis

Chorianopoulos, K. (2007). Content enriched communication with Interactive TV. *The Journal of the Communications Network, 6*(1), 23-30. TCN

Coppens, T., Trappeniers, L. & Godon, M., (2004). AmigoTV: towards a social TV experience. In Masthoff, J. Griffiths, R. & Pemberton, L. (Eds.), *Proceedings from the Second European Conference on Interactive Television "Enhancing the experience*. University of Brighton.

Daly-Jones, O., & Carey, R. (2000). *Interactive TV and electronic programming guides: usability guidelines*. Serco Usability Services.

De Souza, C. S., & Preece, J. (2004). A framework for analyzing and understanding online communities. *Interacting with Computers, The Interdisciplinary Journal of Human-Computer Interaction, 16*(3)

Ducheneaut, N., Moore, R. J., & Nickell, E. (2004). Designing for sociability in massively multiplayer games: an examination of the "third places" of SWG. *Proceedings of Other Players - Conference on Multiplayer Phenomena*, 2004 December 6-8, Copenhagen, Denmark

Ducheneaut, N., Moore, R. J., Oehlberg, L., Thornton, J. D., & Nickell, E. (2008). Social TV: Designing for distributed, sociable television viewing. *International Journal of Human-Computer Interaction, 24*(2), 136-154

Geerts, D. (2006). Comparing voice chat and text chat in a communication tool for interactive television. *Proceedings of NordiCHI 2006*, (pp. 461–464).

Geerts, D., Cesar, P., & Bulterman, D. (2008). The Implications of Program Genres for the Design of Social Television Systems. *Proceedings of uxTV2008*. Mountain View, California, October 22-24. ACM.

Greenberg, S., Fitzpatrick, G., Gutwin, C., & Kaplan, S. (1999). Adapting the Locales framework for heuristic evaluation of groupware. *Proceedings of OZCHI'99*, (pp. 28-30).

Grudin, J. (1994). Computer-supported cooperative work: Its history and participation. *IEEE Computer, 27*(5), 19-26.

Harboe, G., Massey, N., Metcalf, C., Wheatley, D., & Romano, G. (2008). The uses of social television. *ACM Computers in Entertainment (CIE), 6*(1).

Hew, K., Gibbs, M., & Wadley, G. (2004). Usability and Sociability of the Xbox Live Voice Channel. *Proceedings of Australian Workshop on Interactive Entertainment 2004*.

Jacko, J. A., & Sears, A. (Eds.) (2003). *The Human-Computer Interaction Handbook: fundamentals, evolving technologies and emerging applications*. Lawrence Erlbaum Associates

Jans, G., & Calvi, L. (2007). How to develop a mobile 2.0 application. *Proceedings of IADIS International Conference Web-based Communities* (pp. 217-224). Salamanca, Spain

Lee, B., & Lee, R. S. (1995). How and why people watch tv: implications for the future of interactive television. *Journal of advertising research, 35*(6)

Lemish, D. (2007). *Children and television: A global perspective*. Oxford UK: Blackwell Publishing.

Lu, K. Y. (2005). *Interaction Design Principles for Interactive Television*. Master's Thesis, Georgia Institute of Technology.

Lull, J. (1980). The social uses of television. *Human Communication Research, 6*(3)

McQuail, D. (1998). *Mass Communication Theory: An introduction*. London: Sage Publications.

Morley, D. (1986). *Family Television. Cultural Power and Domestic Leisure*. London: Comedia

Nielsen, J., & Molich, R. (1990). Heuristic evaluation of user interfaces. *Proceedings of the ACM CHI'90 Conference*, Seattle, Washington

Nielsen, J. (1994). Heuristic evaluation. In J. Nielsen, & R. L. Mack (Eds.). *Usability Inspection methods*. New York, NY: John Wiley & Sons.

Norman, D. (1988). *The Design of Everyday Things*. New York, NY: Basic Books

Pasquier, D., Buzzi, C., d'Haenens, L., & Sjöberg, U. (1998). Family Lifestyles and Media Use Patterns: An Analysis of Domestic Media Among Flemish, French, Italian and Swedish Children and Teenagers. *European Journal of Communication, 13*(4)

Preece, J. (2000). *Online Communities: Designing Usability, Supporting Sociability*. Chichester, UK: John Wiley & Sons.

Preece, J., & Maloney-Krichmar, D. (2003). Online Communities: focusing on sociability and usability. In A. J. Jacko & A. Sears (Eds.), *Handbook of Human-Computer Interaction*. London: Lawrence Erlbaum Associates Inc.

Rasmussen, T. A. (2006). The Sociability of Interactive Television. *Proceedings of the 4th EuroITV Conference*, Athens.

Regan, T., & Todd, I. (2004). Media center buddies: instant messaging around a media center. *Proceedings of the Third Nordic Conference on Human-Computer interaction NordiCHI '04*. ACM Press, New York, NY, (pp. 141-144).

Schatz, R., Baillie, L., Fröhlich, P., & Egger, S. (2008). Getting the Couch Potato to Engage in Conversation: Social TV in a Converging Media Environment. *Adjunct proceedings of EuroITV2008*.

Shneiderman, B., & Plaisant, C. (2005). *Designing the User Interface: Strategies for Effective Human-Computer Interaction. (4th ed.)*. University of Maryland, College Park: Pearson/Addison-Wesley

Simmel, G. (1949). The sociology of sociability. *American Journal of Sociology, 55*. 254-261

Thomas, P., & Macredie, R. D. (2002). Introduction to the new usability. *ACM Transactions on Computer-Human Interaction, 9*(2)

Tullio, J., & Harboe, G. (2008) Investigating the Use of Voice and Text Chat in a Social Television System. *Changing Television Environments, Proceedings of the 6th European Interactive TV Conference (EuroITV 2008)*, Springer

Weisz, J. D., Kiesler, S., Zhang, H., Ren, Y., Kraut, R. E., & Konstan, J. A. (2007). Watching together: integrating text chat with video. *Proceedings of the SIGCHI Conference on Human Factors in Computing Systems CHI '07*. ACM, New York, NY, (pp. 877-886).

Wells, M. (2004). The TV Revolution Starts Here. *The Guardian*, December 4, 2004, UK.

ENDNOTES

1. Enchantment was defined by Arvola & Holmlid (2000) as being related to excitement and engagement, but was not retained as quality-in-use by Arvola (2003). As enchantment is not directly related to the social use of interactive television, we will not further discuss this quality-in-use.
2. For example, Greenberg e.a. (1999) created a set of heuristics for evaluating groupware applications, based on the locales framework.

[3] Some of which are briefly mentioned or more extensively described in this book, e.g. AmigoTV (Coppens, 2004) or Motorola's Social TV (Harboe e.a., 2008)

Chapter VII
Audience Participation in Television and Internet:
Attitudes and Practices of Young People in Portugal

Celia Quico
Universidade Lusófona de Humanidades e Tecnologias, Portugal

ABSTRACT

This chapter seeks to evaluate the attitudes and practices of media participation amongst young Portuguese aged between 12-18 years, with a particular focus on content creation and sharing through media and information and communication technologies (ICT). Audience participation in television and internet will be addressed, having as basis the results and findings of three empirical studies integrated in the PhD research project of the author, namely: an ethnographical study about the usage of media and ICT usage by 10 families conducted at their own domestic contexts, a quantitative survey about the usage of media and ICT by young people aged 12-18 with a total of 962 respondents and, finally, an evaluation study of a participatory media format which was tested and evaluated by 77 teenagers from three different schools. The main objective is to better understand the attitudes and behaviors of young people in Portugal towards the practices of creation and sharing digital content through media and ICT, providing empirical data about the range and frequency of experiences of content creation by this specific population, as well as their interest and adherence to participatory media formats.

INTRODUCTION

The issue of audience participation has been gaining increasing relevance, not only for the academia and public institutions, but also for the media and communication industries. Nevertheless, even if media and ICT are potentially enablers of such practices, what is the scope and range of participation practices amongst people in their everyday live? More specifically, what is the audiences' actual interest, skills and experience in creating content as well as sharing it? Compared with research on access and understanding, we know very little about the creation of messages or content among the audiences (Livingstone, Van Couvering, & Thumin, 2005).

The evaluation of media participation interest and practices amongst young Portuguese people aged 12 to 18 was one of the nuclear questions explored on the PhD thesis *Audiences 12 to 18 years old in the media convergence context in Portugal: the emergence of a participatory culture* (Quico, 2008) in Communication Sciences at Universidade Nova de Lisboa, Portugal. The thesis' main objective was to identify the main uses of media and ICT by young people aged 12-18 in the current media convergence context in Portugal, more specifically, to evaluate the incidence and relevance of creating and sharing digital content among this specific population and their interest and actual practices of media participation.

The theoretical framework of this research work has its foundations on the Audience Studies field, more specifically, it follows the general orientation of Reception Studies and Cultural Studies that consider audiences as active producers of meaning, as well as the notion that audiences are producers of cultural artifacts which derives from media fan cultures research led by scholars such as Henry Jenkins (1992; 2006a). Instead of asking what media do to people, this research project asked what people do with media: how they use media, how they incorporate media, what functions media have in their everyday life, and what degree of importance is given to media. This research is therefore guided by the conceptualization of audiences as active producers of meaning and cultural artifacts, initially proposed by the Uses & Gratifications approach, Cultural Studies and Reception Studies. Also, Jenkins' definition of participatory culture (1992; 2003; 2006b) is used in contrast with the notion of passive media audiences and also as a tool to explore the complex relations between media producers and consumers occurring in the current convergence context.

The research design of the thesis is based on interdisciplinary approach, drawing on quantitative and qualitative methods in order to gain a diversity of perspectives about the subject under study. The multi-perspective and multi-dimensional forms of comprehension of a given situation are defended by David Morley (2007) who advocates in favour of data triangulation as an alternative to adopt a sole methodology, since the combination of different methods such as interviews, daily diaries and ethnographical studies will allow to compare what people say with what they really do (pp. 82-84).

Likewise, Sonia Livingstone et al. (2005) state that in Media Studies and Audience Studies fields the broad trend is towards the triangulation of qualitative and quantitative methods and a pragmatic multidisciplinarity, with the objective of overcome or compensate the disadvantages of certain methods over others (p. 57). For instance, most surveys are supported on the self-evaluation of attitudes and practices, while qualitative methods such as ethnographical studies even if they allow a deep comprehension of certain subjects, as well as provide context for the findings, they sacrifice the advantages of surveys in terms of the diversity and representativeness of the population surveyed. This way, the best practice in the research methods seek to integrate useful and efficient methods from several sources into a multi-method research design, instead of being directed by specific theories or disciplines (Livingstone et al., 2005).

In summary, the empirical research of this thesis was formed by the following studies, supported in different quantitative and qualitative methods, whose theoretical foundations are further developed in their respective sub-chapters:

- Ethnographical study, conducted with the objective of obtaining a detailed and holistic comprehension of media usage by young Portuguese people and their respective families in their domestic context (sample: 10 families with one or more individuals aged 12 to 24 years, totaling 33 participants)—the observation sessions were videotaped as well as the in-depth interviews with all the family members and the fieldwork took place during February-March 2006 and September-October 2007.
- Quantitative survey, whose objective was to identify the main patterns of usage of media and communication technologies by young Portuguese people, in particular, the main uses given to television, internet and mobile phones (sample: 962 individuals, who answered a self-completion questionnaire)—the fieldwork took place during April 2007.
- Evaluation study of a cross-media format, with the main objective of understanding the interest and relevance of a media format which explicitly asks for the participation and creative contribution of the users (sample: 77 participants from 12 to 20 years old, from 3 different schools)—the procedure included the direct visualization and experience of a TV pilot and a web prototype, followed by the self-completion questionnaire and the fieldwork took place during October 2007.
- Case study Sapo Challenge 2007, an initiative of the group Portugal Telecom for the promotion of digital media literacy aimed at young students 12 to 18 years old, which took place during January until June 2007.

The PhD research received the support of the project *MAU-TVMEDIA: User modeling and viewing clusters: television and new media* (2005-2006) conducted by Universidade Lusófona de Humanidades e Tecnologias, financed by the Portuguese agency Fundação para a Ciência e Tecnologia (POCTI/COM/61029/2004).

BACKGROUND

Young people live in an increasingly complex media culture, observes Kristen Drotner (2008), to whom one of the key challenges presented to scholars is to address this empirical complexity with the requisite theoretical and methodological complexity. Drotner recommends (2008) that researchers approach not one medium but several media, to consider online and offline practices as well as the school and domestic context and, finally, to have in mind that young people are interlacing reception and production processes in unprecedented ways (p. 182).

The notion that audiences produce or create cultural artifacts emerged from the theory of creative production by individuals sketched by Michel de Certeau in *The Practice of Everyday Life* (1984), which inversely to the pessimistic point of view of the School of Frankfurt theoreticians, recognized the modes how individuals can resist and defy the attempts to regulate the production and circulation of meaning in popular culture.

Inspired by de Certeau, but recognizing several limitations such as the lack of ethnographical documentation, Jenkins developed his work about fan cultures: this author offers an ethnographic description of particular groups of fans in *Textual Poachers* (1992), their social institutions and cultural practices, as well as their problematic relations with the mass media. The fan cultures do not maintain a radical distinction between readers and writers, since fans are themselves content producers and not mere consumers—creating fanzines, novels, posters, songs, videos, performances, and other kinds of cultural artifacts—which they distribute and share within their network (Jenkins, 1992). Fan cultures transform the media consumption experience into various forms of creative expression and into the construction of a new community: this mode of usage is designated by Jenkins as participatory culture (1992; 2003; 2006b). Yet, even if the research conducted on fan cultures proves that not every audience is passive, it does not prove that all audiences are active (Jenkins, 1992), the same it's to say, fan cultures are not representative of all the audiences.

Still, Livingstone (2004) notes that in the new media environment, it seems that people get more involved with the content and not as much with the form or the channel—favorite bands, television series or football teams, wherever they are, in every medium or platform -, considering fandom increasingly important, as the audiences diversify and fragment (p. 81). In other words, what once were marginal and marginalized media usage trends are now becoming current media usage practices.

THE USES OF MEDIA AND ICT BY YOUNG PORTUGUESE: ETHNOGRAPHIC STUDY AND QUANTITATIVE INQUIRY

The ethnographical study "Os jovens e os usos dos media e TIC"/ "Young Portuguese and their media and ICT uses" had as main objective to obtain a detailed and holistic comprehension of how young people use media and ICT, as well as their respective families. Some of the themes addressed were time use in everyday life, usage patterns of television, internet and mobile phone by the family members, as well as their attitudes towards media and ICT and how they predict to use them in the near future.

In the Cultural Studies field, Ethnography has been centered on the qualitative exploration of values and meaning in the context of a whole way of life (Barker, 2003). To gain access to the naturalized domains, in Morley's expression (1992), and to their characteristic activities is what Ethnography and other qualitative research strategies were mainly used for in the realm of Cultural Studies and Audience Studies. Morley argues (1992) that an action such as watching television needs to be understood within the dynamics and structure of the domestic context of consumption which it is but a part (p. 173). The limitations of statistically based quantitative survey techniques for this purpose are well established, adds Morey, to whom the statistical techniques are disaggregating, since they inevitably isolate units of action from the contexts that make them meaningful. It should be noted that frequently Ethnography has been invoked polemically against the quantitative communications research tradition, since it deals with the qualitative comprehension of cultural activity in context (Barker, 2003). For David Gauntlett & Annette Hill (1999), Ethnography's contribution to the development of Audience Studies was been extremely important by introducing the everyday context in which media are used and by allowing media consumers to express or demonstrate what they make with media.

Regarding the methodology followed to the present ethnographical study, the option was to conduct a 'quick and dirty' ethnography that could enable to obtain fast and effectively a rich and detailed comprehension of Portuguese families' usage of media and ICT in their domestic contexts. The 'quick and dirty' ethnographical approach is distinct from the conventional ethnography studies since they have a short time span and they are not exhaustive regarding the contexts that they investigate: in other words, they are brief but provide a general and informed sense of the settings (Hughes, King, Rodden & Andersen, 1994).

For the current ethnographical study ten families were recruited, in which at least one of their members had to have between 12 and 24 years: after the observation of their behavior dur-

Table 1. Ethnographical study "Os jovens e os usos dos media e TIC" (2007) Participants

ID:	Data e local:	Participants:	Participants—family:
Family 1 (3 members)	22-02-2006 (wed.) Corroios (Lisboa region)	Pedro (18 years)	Grandfather (73 years) Grandmother (65 years)
Family 2 (2 members)	02-03-2006 (thur.) Póvoa St.Adrião (Lisboa region)	Cátia (22 years)	Mother (51 years)
Family 3 (2 members)	09-03-2006 (tues.) Sacavém (Lisboa region)	Ricardo (24 years)	Mother (55 years)
Family 4 (4 members)	11-03-2006 (sat.) Nazaré (Leiria region)	Nádia (17 years) Filipa (21 years)	Father (50 years) Mother (44 years)
Family 5 (5 members)	09-09-2007 (sun.) Leiria (Leiria region)	Filipe (16 years) Diogo (14 years)	Father (42 years) Mother (42 years) Irmão (8 years)
Family 6 (3 members)	14-09-2007 (6ª) Leiria (Leiria region)	Lúcio (15 years)	Father (53 years) Mother (43 years)
Family 7 (4 members)	14-09-2007 (6ª) Leiria (Leiria region)	Catarina (16 years) Inês (14 years)	Father (39 years) Mother (48 years)
Family 8 (4 members)	15-09-2007 (sab) Nazaré (Leiria region)	Frederico (16 years)	Father (49 years) Mother (47 years)
Family 9 (4 members)	16-09-2007 (dom) Nazaré (Leiria region)	Rita (15 years) Raquel (13 years) Bárbara (11 years)	Mother (36 years)
Family 10 (3 members)	04-10-2007 (sab) Carcavelos (Lisboa region)	Alexandre (14 years) Cristina (11 years)	Father (51 years)

ing leisure time at home, semi-structured group interviews were held, both registered in video for further analysis. Regarding the sample recruitment the strategy followed was to ask to friends and close acquaintances to suggest families who could potentially be interested in joining such study, having to observe the rule that it was necessary that at least one of the members had to have between 12 to 24 years of age. The sample is too small to be considered representative—that was not the intention of this study—but even so there was a concern that the families recruited were from different places and different socio-economical strata. Therefore, the ten families sample can be defined as heterogeneous and diversified group of participants (note: the participants' names were altered, in order to preserve their anonymity, as it is usually done in such studies).

This study was joined by a total of 33 participants, from which 16 were young people aged 11-24 years (9 girls and 7 boys). In terms of distribution by age intervals, from 11 to 14 years the study had 6 participants (4 girls and 2 boys), from 15 to 16 years there are 5 participants (2 girls and 3 boys), from 17 to 18 years there are 2 participants (1 girl and 1 boy) and, finally, from 19 to 24 years there are 3 participants (2 girls and 1 boy).

The fieldwork first phase took place during February and March 2006 and four families were recruited for this purpose. The second phase occurred during September and October 2007, after the quantitative survey results were collected and analyzed, and six families took part. It should be mentioned that this ethnographic study functioned as a probe to gather new themes and questions for the quantitative survey, as well as a lens for some of the survey statistical results, providing clarification and context for the vast amount of statistical data.

By analyzing the media and ICT usage in the domestic context and during leisure time it was possible to gain a broad and integrated perspective of young people's everyday life and their respective families. The study's participants have shown evidence of occupying their leisure time with different types of activities: mediated and non-mediated. Their everyday life is complex, multi-dimensional and dynamic.

After dinner, in five out of the ten families the youngsters went to their room or the family's office in order to spend their time using the computer and accessing the internet. In the other five families, the young people preferred to stay with their parents at the living room: it is curious to note that in four out of these five families at least had one or more girls, while in the case of the five families in which the youngsters went to their room or office in four cases they had one or more boys. Once again noting that this sample is too small to extrapolate general trends for the population, within these groups of participants the socialization with parents after dinner tends to be more a girls' regular practice and going to the room or office tends to be more a regular practice performed by boys.

Television is becoming less important on the media diet of young people: most of the young participants already use computer and internet as the preferred way of occupying their time after dinner at home and also all of the young participants project that in the near future (within five years) they will consume less television than today.

On the other hand, internet is becoming the preferred way to obtain entertainment content in the case of young people, such as music content, movies, television programs and games—downloads via online tools such eMule and BitTorrent are a very frequent practice, specially among young boys. MSN (Messenger) and Hi5 (social network similar to MySpace and Facebook) are significantly important in these young people everyday life, mostly as ways to maintain contact with their friends from home.

Parents are ambivalent regarding internet: they consider that this is a very useful tool, specially for school, but they also have shown their concern about the risks and dangers of internet usage, a source of conflict in some of the families visited in which parents define internet access restrictions at home by specifying a time schedule to use internet or a maximum time of daily usage. Parents are also ambivalent towards television, who have shown their concern about the quality and diversity of programming by the main free-to-air channels, as well as the late schedules in which movies and series are broadcasted, even if they are regular viewers of news and novelas (Brazilian and Portuguese soap operas) in these same channels.

Finally, content creation and its online publication has a quite significant expression in young people aged 12 to 18, both in the simpler form of publishing photos on Hi5 or Fotolog (photo blog), as well as creating and maintaining blogs: out of the 16 young participants 11 claimed to publish their photos no Hi5 or on Fotolog, as well as observed in at least 3 cases. It is also significant the number of youngsters who pursue more traditional creative activities such as playing music, dancing, painting, and drawing.

On the other hand, in order to evaluate what are the main overall patterns of media and ICT usage amongst young people, particularly, the uses given to television, internet and mobile phones it was also conducted a large scale quantitative survey. The need for such a survey to be conducted became more salient after a literature review about the subject, which shown that the statistical data regarding young people media and ICT usage was insufficient, more specifically in the case emergent activities and practices such as multi-tasking or creating and sharing digital content.

The application of quantitative surveys as research method for media audiences was initiated by Paul Lazarsfeld during the 40's, which many historians consider as the founder of quantitative audience studies (Schroder, Drotner, Kilne & Murray, 2003) since he was the leader or promoter of a series of research studies as top responsible for the Bureau of Applied Social Research at Columbia University New York. In these studies emerged the tradition in mass communications of empirical research based on social sciences (Durham & Kellner, 2006), with the application of empirical methods to determine consumers' demands, tastes, opinions and mass media effects. The questions concerning the influence of mass media and their effectiveness in persuasion and attitude change might be properly answered by the advances in 'mental measurement' and statistical analysis, which appeared to offer new and powerful tools for achieving generalized and reliable knowledge of previous hidden process and states (McQuail, 2000). Even considering the limitations of quantitative surveys based on statistical data (Morley, 1992), this is a method that allows researchers to gain a broad perspective about media and ICT usage and to identify patterns and general trends in a large scale. Therefore, quantitative surveys are an effective way to avoid some of the limitations of qualitative based studies, which generate rich findings from a qualitative point of view but that rarely are representative of the population being studied (McQuail & Windahl, 1993).

The references for the design of the current survey were Livingstone's *Young People and New Media* (2002) and the Pew Internet & American Life Project surveys (2005, 2007). Also, Ofcom's media literacy survey (2006) was taken into account as basis for this study. Finally, the ethnographical study previously described provided new themes and questions for this survey, also enabling a pre-test of several of the questions that would appear in the final version of this survey.

The sample is formed by individuals from all over the Portuguese regions, mainly from the Lisboa, Santarém, Setúbal and Leiria regions. The total number of individuals aged 12 to 18 years was 962 subjects, to whom was given an 8 page questionnaire to individually fill out. The fieldwork took place during April 2007.

Concerning the distribution in terms of the gender variable, this sample has 515 female subjects—which correspond to 53,5% of the

Table 2. Inquérito Lusófona MAU-TVMEDIA (2007) Subjects' age distribution

	Frequency	Valid Percent
12 years	5	,5
13 years	8	,8
14 years	137	14,2
15 years	178	18,5
16 years	234	24,3
17 years	180	18,7
18 years	220	22,9
Total	962	100,0

Table 3. Inquérito Lusófona MAU-TVMEDIA (2007) Subjects' gender

	Frequency	Valid Percent
Male	447	46,5
Female	515	53,5
Total	962	100,0

sample—and 447 male subjects—that correspond to 46,5% of the total.

Since this survey is quite detailed and extensive, for the purposes of this paper two groups of questions were selected, specifically addressing attitudes and preferences about media and ICT as well as creating and sharing content activities—the complete survey results are available on the PhD thesis by Quico (2008).

Therefore, regarding preferences and attitudes related to television, internet, mobile phone, the statement "increasingly I want to create my own videos, music, photos, texts, etc on my PC" received the complete agreement of 24,4% of the inquired population and the agreement of 35,1% of these young respondents. Note that from the nine statements listed in this question, this was the second phrase with a higher accumulated agreement percentage (59,5%) and the phrase with a smaller accumulated percentage of disagreement (12,6%).

Regarding sharing content, the option "increasingly I want to share my videos, music, content, text, etc on the internet" obtained 40% of agreement (13,5% total agreement and 26,5% of agreement) and 21,6% of disagreement (7,4% of total disagreement and 14,2% of disagreement), while the option "increasingly I want to share my videos, music, content, text, etc via mobile phone" gathered 33,8% of agreement (11,1% total agreement and 22,7% of agreement) and 25% of disagreement (7,6% of total disagreement and

Table 4. Inquérito Lusófona MAU-TVMEDIA (2007) Attitudes and preferences about Television, Internet e Mobile Phone

	-2 Completely disagree %	**-1** Disagree %	**0** Neither agree or disagree %	**1** Agree %	**2** Completely agree %
I prefer TV to internet	12,7	21,4	45,1	14,5	6,4
I don't know how to get by without TV	11,4	25,0	22,9	27,5	13,1
I don't know how to get by without internet	6,9	15,0	21,2	32,1	24,8
I don't know how to get by without the mobile phone	7,0	11,1	14,6	28,3	39,0
Increasingly I want to create my own videos, music, photos, texts, etc on my computer	4,2	8,4	27,8	35,1	24,4
Increasingly I want to share my own videos, music, photos, texts, etc via internet	7,4	14,2	38,5	26,5	13,5
Increasingly I want to share my own videos, music, photos, texts, etc via mobile phone	7,6	17,4	41,3	22,7	11,1
Increasingly I want to watch television programs anytime and anywhere	7,1	16,4	35,8	25,4	15,3
Internet substitutes Television	8,2	15,0	35,8	25,1	15,9

17,4% of disagreement). In short, these results indicate a more significant interest in content creation in comparison with sharing content.

About the group of options that compare television and internet in terms of preferences, the option "I prefer television to internet" gathered an accumulated total of 20,9% of agreement and an accumulated total of 34,1% of disagreement, with 45,1% of neutral answers. In a similar way, to the option "internet substitutes television", 41% of the participants in this survey agreed with this statement (total sum of complete agreement with agreement) and 23,2% of these disagreed (total sum of complete disagreement and disagreement). These results indicate that the balance is in favour of preferring internet to television in this specific age group.

Regarding creating content, the option was to include a question based on Ofcom's *Media Literacy Audit: report on media literacy amongst children* (2006), which asked 12 to 15-years-old what type of creative activities they had already done or were interested in doing.

As shown on Table 5, a high percentage of young people reported having already made a video using a mobile phone (71,2%) as well as having already used a camcorder for the same purpose (57,9%). The next activity that the survey participants reported to have done was creating a photo album online, executed by 54,4% of these youngsters. Creating and maintaining a blog had 47,4% of positive answers, above the percentage of young people who claimed already having created a web site (35,1%).

The more traditional creative activities, such as writing poetry or prose and making a song, did not interest to 37,7% and 37% of these respon-

Table 5. Inquérito Lusófona MAU-TVMEDIA (2007) Creating content

	Done this %	Interested in doing this %	Not interested %	Don't' know %
Write poetry or prose (short stories, drama, poetry, etc)	44,8	10,7	37,7	6,9
Make a song (lyrics or music)	22,0	24,9	37,0	16,2
Create a web site on the internet	35,1	37,6	18,2	9,1
Make a video with a camcorder	57,9	26,5	11,7	3,9
Make a video with a mobile phone	71,2	12,8	12,3	2,3
Make an online photo album	54,4	25,6	15,7	4,4
Create and maintain a blog on the internet	47,4	23,9	22,9	5,8
Create or modify a computer game	12,6	31,8	37,7	17,9
Modify my computer (modding)	27,4	31,3	25,2	16,1

dents, respectively, even if 44,8% of them claimed having already done the first activity, compared with 22% who reported having created a song. The creative activity with less expression in these young respondents was modifying or creating video games: 12,6% claimed that already have done this and 37,7% said that they have no interest in these activities.

In the Ofcom study (2006) about media literacy in which this question was based, to create a web site was reported by 19% of young people aged between 12 to 15 years across the UK, while 35,1% of young Portuguese people who answered this questionnaire have reported so. Creating or maintain a blog gathered 7% of positive answers in the Ofcom study, while 47,7% of the Lusofona's survey subjects had claimed having done such an activity. To make a video using the mobile phone was reported by 15% of youngsters 12-15 years across UK, compared with 71,2% of the young Portuguese respondents. Finally, 13% of young people in the Ofcom study mentioned already having made a video with camcorder, in contrast with 57,9% of young Portuguese. These very significant differences can be explained by the broad age range of the Portuguese study compared with Ofcom's and also by the difference between the period in which each study fieldwork was conducted: Ofcom's during July-August 2005 and Lusofona's during April 2007.

Also, it should be mentioned the results of the study *Teen Content Creators and Consumers* (Lenhart & Madden, 2005) from the Pew Internet & American Life Project, which found that 57% of young Americans aged 12 to 17 create content for the internet. These "content creators", as the authors of this study designate this population, have reported having done one or more of these activities: create a blog, create a personal web page for school, friends or an organization; share original content online, such as photos, stories or videos; remix content they found online. The content creation activities that are more popular among young Americans aged 12 to 17 are creating original content and work on their personal web page: 33% of the subjects of Pew's study claimed to share their creations online; 32% stated that they already created web pages for groups, friends or for school; 22% reported having remixed content gathered online.

PARTICIPATORY MEDIA FORMATS: CROSS-MEDIA FORMAT EVALUATION STUDY AND DIGITAL LITERACY INITIATIVE CASE STUDY

The departure question to conduct the evaluation study of a participatory media format was the following: what is the interest and relevance for young people aged 12 to 18 of a media format that asks for their participation and creative input? In other words, a sample of young users was asked to provide their opinions about the interest and utility of a media format whose goal is to stimulate content creation and sharing amongst the users.

This evaluation study framework is generically situated in the field of user-centered design (Norman, 1988), also incorporating questions about the usual behaviors and practices of the population under study. The main objectives of this study were the following:

- To measure the users' satisfaction towards the television pilot and web prototype.
- To probe the users' interest in participating in the TV pilot and web prototype, as well as the main motivation factors to participate.
- To evaluate the interest level in themes related to the natural and cultural heritage of Portuguese regions.
- To evaluate what is the skill level and the interest in content creation.
- To verify the level of participation in online communities and social networks.

Total Explorer is a participatory media format which aims to actively engage and build a community passionate for Portugal's natural and cultural heritage, targeting young people aged 12 to 18 years-old. The format specifications included a linear TV documentary with three independent stories; a synchronized interactive TV application; on-demand videos available on a web site which includes other media assets for download (podcasts, music, photos) and for participation (voting, commenting, sharing); a mobile application for mobile phones and PDAs; digital content production workshops in several cities and organized visits to local sites.

The development team efforts were concentrated in producing a linear TV pilot and a web prototype. The TV pilot shows relevant and engaging scientific research about a specific region, as well as local characters and local curiosities. Each of the stories on the TV pilot presented the perspective at least a scientific expert, as well as the perspective of local experts. The subject areas covered History, Geography, Biology, Ethnography, Archaeology and Oceanography. The web prototype enabled full access to the video stories and also to extra videos with the scientific experts and local experts' interviews. Also the users had their own personal space, where they can upload their own content. Participatory functionalities such as voting and commenting were included. The web prototype is available online at <http://bocage.ulusofona.pt/totalexplorer/>, whose functional areas are "Vídeos" and "Pessoal": users can watch, vote, comment videos, as well as read and comment articles about the five main stories in video, know the basics of video creation with a tutorial made specifically for this web site and, finally, create their own personal space that enables them to add friends as well as upload their own videos and create favorite lists.

In terms of the methodology, the option was to apply a self-administered questionnaire to a sample of young users, whose number would not be inferior to 50 subjects. The evaluation sessions had an average duration of 60 minutes, during which the participants had the opportunity to see, experiment and evaluate the media format. After partially watching the TV pilot and using the web prototype, the participants were asked to fill out a survey, mostly with closed-ended questions, but that also included several open-ended questions in order to obtain the respondents opinion in their own terms. The self-administrated questionnaire was composed by 7 parts, with a total of 36 questions organized in the following manner: (1) Respondent characterization; (2) Total Explorer TV

Figure 1 and Figure 2. Total Explorer: TV pilot—story—What treasures can be found in the biggest submarine canyon of Europe?

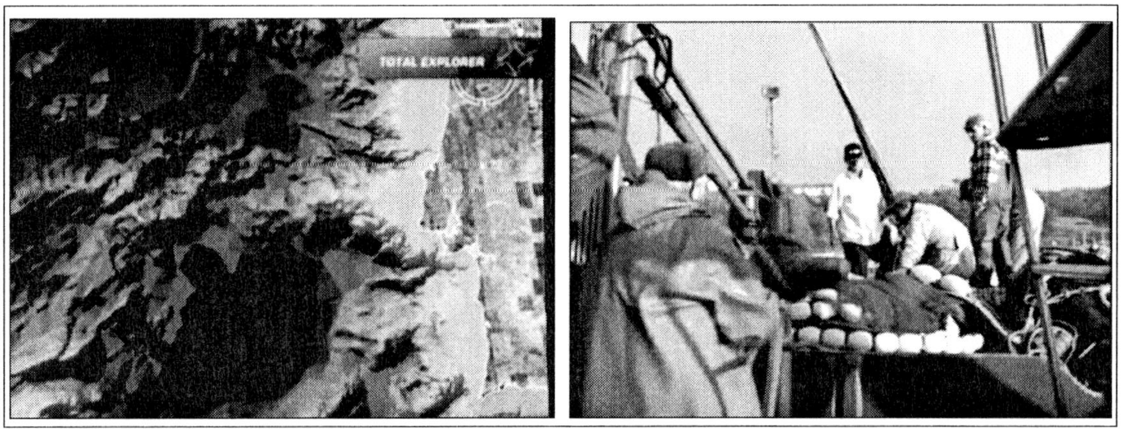

Figure 3. Total Explorer—web prototype, home page

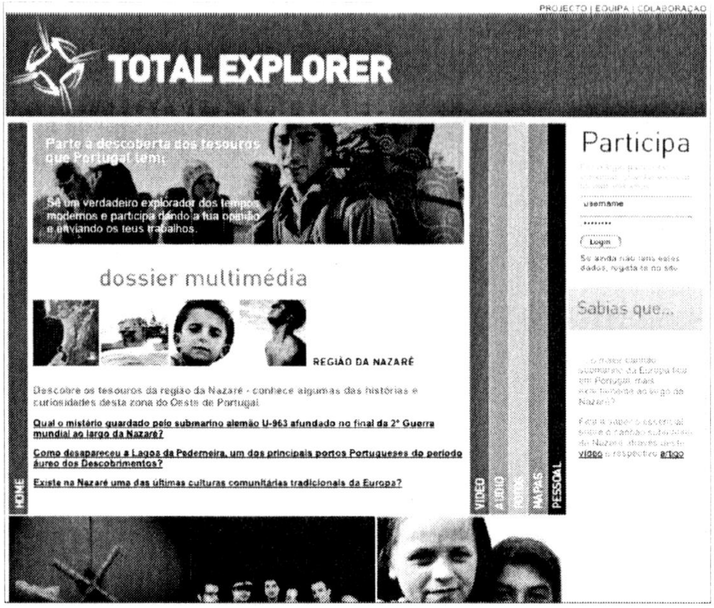

pilot evaluation; (3) Total Explorer web prototype evaluation; (4) Online participation; (5) Online and offline creativity; (6) Television: usage and preferences; (7) Internet: usage and access.

The observation sessions were phased in three distinct moments. After a brief presentation of the study and its goals—as well as the presenting guarantees of the confidentiality of the information -, the sessions would rollout in the following way:

- Each participant chooses one out of the five short documentaries to watch from a DVD, based on the respective titles and brief descriptions (average duration: 10 minutes)
- After this task, each participant uses the web prototype associated to the TV pilot (average duration: 10 minutes)
- Finally, each participant fills out an evaluation questionnaire (average duration: 15-20 minutes)

When the participants entered the room where the sessions were held they would find a home page in their respective computers, in which the sessions' objectives and phases were briefly explained, also containing a link to a descriptive page of the five short documentaries from the TV pilot and another link to the web prototype home page—page available at the following URL <http://totalexplorer.no.sapo.pt/estudo/>.

The fieldwork took place during October 22nd to 25th, 2007, at three schools from the Leiria region, located in the centre of Portugal: Escola Secundária de Domingos Sequeira at Leiria, Externato Dom Fuas Roupinho at Nazaré and Escola Secundária com 3° Ciclo do Ensino Básico at Batalha. The sample is composed by 77 subjects aged 12 to 20, all of them students in these schools: 23 participants from the Escola Secundária de Domingos Sequeira, 14 participants from Externato Dom Fuas Roupinho and 40 participants from Escola Secundária c/ 3° CEB at Batalha. The complete results from this evaluation study are available on the PhD thesis by Quico (2008).

About the questions related with participation in this media format, it was found an expressive declared interest in interacting and participating with *Total Explorer*, in particular, to be involved in school visits to the places displayed and also viewing extra videos associated to the short documentaries, as shown on Table 6.

To this group of youngsters to learn more about their hometown and region is the most significant declared motivation to participate in this format, followed by wining prizes for themselves and learning how to create digital content.

In this line, regarding the general theme of the format–"to know the natural and cultural heritage of Portuguese regions and places"—almost everyone considered it "appealing" or even "highly appealing" (49 and 25 respondents, respectively), with 3 out of 77 participants considering the theme with low appeal. Also related with this topic, the participants declared a clear preference towards TV programs about their own region, compared with the options "other Portuguese regions", "other European countries" and "countries with Portuguese as official language".

Finally, in a nutshell, the respondents were asked if the *Total Explorer* format became a television series and web site would they be interested in watching and participating in this format, 3 out of 4 participants gave a positive answer: 59 out of 77 participants, with 15 participants answering "I don't know/ I don't answer" and 3 participants giving a negative answer. The high percent of positive answers is a solid indicator of the interest and relevance of this kind of format to young people in this age bracket.

Other figures that are solid indicators of the interest and relevance of such a participatory media format for this age group are the average

Table 6. Total Explorer evaluation study (2007) Interest in participating in the media format

n= 77	YES—I'm interested	NO—I'm not interested	Don't know/ Don't answer
To vote on which stories are broadcast on television, before the program is on-air	52	12	10
To choose which story I want to see while the program is being broadcasted (by remote control, web site or SMS)	52	12	10
To watch extra videos about the stories broadcasted during or after the program is over	53	14	5
To vote and comment the videos on a web site	48	16	8
To join competitions that put to the test the knowledge I acquired by watched the TV program	43	26	5
To join competitions that put to the test my talent in creating videos, photos, related with my hometown	39	26	9
To be a part of study or school visits to the places shown on the TV program	63	4	7
To access content related with the TV program though mobile phone (videos, photos, MP3 file, etc)	51	11	12

Table 7. Total Explorer evaluation study (2007) Motivations to participate with original content

n= 77	Leiria	Batalha	Nazaré	TOTAL
Win prizes for me	11	21	7	39
Win prizes for my team	7	16	7	30
Recognition of my work (video, photos, etc) in television	10	15	6	31
Recognition of my work (video, photos) in the web site	7	11	6	24
Spend time with friends and colleagues	6	22	5	33
Know new people who share interest in Portuguese natural and cultural heritage	4	14	6	24
Compete with other teams in creating digital content (videos, photos, etc)	6	13	4	23
Learn to work in a team	3	15	3	21
Learn how to create digital content	14	14	7	35
Learn more about my hometown and region	12	25	6	43
Other reason—which?	1	2	0	3

marks given to the TV and web prototype—positive even if significantly differentiated between the three schools in this evaluation study. In a scale of 0 to 10 points (0 being the lowest mark and 10 the highest mark), the average evaluation mark of the TV pilot was 7,37 points and the web prototype was 7,28 points.

As expected, in the open-ended questions many interesting elements were gathered for this study. Many of the young participants had given evidence of being quite demanding and of having attention to detail by point out some technical failures or less accomplished features of the TV pilot and web prototype: the fact that the web site was not completely functional, the site slow response, the impossibility to send a photo to the web profile after concluding the register process, the background music too high in one of the videos, etc.

In overview, their evaluation was mostly positive: in the open-ended questions, in the case of the TV pilot, most participants valued the acquisition of knowledge, as well as some of the formal aspects of these videos (3D and 2D animations, for instance). In the case of the web prototype, mostly the participants valued the ease of use, the way content was structured and the interface design, even if the videos were the most mentioned feature by the young respondents.

In summary, the results obtained by this study indicate that a significant part of the young participants has strong interest and will to participate in a media format such as the one presented to them. Yet, the sample is too small to extrapolate these findings to the universe of young Portuguese in this age group. This way, the question to be asked is: what if such a participative media format went beyond the laboratory, meaning that, what if a format similar to this was made available to all the Portuguese youngsters during a pre-fixed time span that allowed further research into the interest and practices of creating and sharing content?

This question can be answered by the Sapo Challenge project, a digital media literacy initiative promoted by Portugal Telecom in partnership with the Portuguese Ministry of Education during 2007: this was national competition aimed at young students between 12 to 18 years of age. The second phase of this competition required the participants to create digital content about a Portuguese writer of their preference, in different

Table 8. Total Explorer evaluation study (2007) Interest in watching and participating in a media format such as Total Explorer in the future

n= 77	Leiria	Batalha	Nazaré	TOTAL
Yes	18	31	10	59
No	3	0	0	3
Don't know/ don't answer	2	9	4	15

formats (video, audio, photos, text, etc) and genres (interviews, reports, ads, etc) and publish them in a blog during four weeks. This initiative is further explained and detailed in the PhD thesis by Quico (2008)—it should be noted that the author also joined the team responsible for this project as a multimedia consultant, in charge of defining the competition model and evaluation grid, creating and moderating the official blog of support to the participants, representing the PT group in the evaluation committee which evaluated all the blogs, amongst other activities. The relevance of Sapo Challenge in the context of digital media literacy promotion in Portugal was recognized by Cristina Ponte in the report "Web access of children: Possibilities of internet providers to promote media literacy of children" (2007), presented at the expert conference on European media policy at Leipzig, more precisely to the working group coordinated by Sonia Livingstone.

The quantitative and qualitative results of Sapo Challenge are solid indicators of the interest generated by an initiative that required participants to create and share content. To the second phase of Sapo Challenge 560 teams sent their application, representing a total of 2.800 students and 560 teachers: from the 560 teams, 498 blogs were considered valid and 366 blogs presented at least one content entry. Even if these quantitative results reveal this initiative success, it's mostly the qualitative results that demonstrate the real dimension of the impact it had among its participants. For instance, from March to May 2007 anyone who accessed the Sapo Challenge official blog would visit a true community <http://challenge2007.blogs.sapo.pt/>, formed by young people, their teachers and parents, who shared not only their digital creations but also words of motivation to other participants, critiques to the contest organization, critiques to other participants behavior, in summary, their experience of participation in Sapo Challenge.

CONCLUSION

The evaluation of media participation interest and practices among young Portuguese people aged 12 to 18 was explored in this chapter, as well as their interest, skills and practices of creating and sharing digital content.

Firstly, to create content it was found to be a common practice among the majority of young people aged 12 to 18 in Portugal, who declare to have experience mostly in making videos, taking pictures and making photo albums online, as well as creating and maintaining blogs. The statement "increasingly I want to make my own videos, music, photos, text, etc, on the PC" received the agreement of nearly 60% of the Lusófona's survey about media and ICT usage respondents aged between 12 to 18 years: 24,4% of complete agreement and 35,1% of agreement. From the nine statements about attitudes and preferences included in this survey, this statement was the one that gathered a higher percentage of accumulated agreement, after the statement "I can't go by without the mobile phone", as well as the statement with a lower percentage of disagreement. In the ethnographical study which involved 10 families from different locations in Portugal it was found that creating and publishing content online was a significant activity for the young peoples aged 12 to 18 in this sample, whether publishing photos on the social network profile or photo logs or even creating and maintaining blogs.

Making videos it was found to be a very common activity, with the large majority of young Portuguese people reporting already having made videos with a mobile phone or a camcorder. The large majority of young people 12 to 18 years who responded to the Lusofona's quantitative survey reported filming with a mobile phone or a camcorder; 71,2% of these youngsters claimed already having made a video with a mobile phone and 57,9% used a camcorder for the same purpose. In the participatory media format evaluation study, 46 out of the 77 participants reported having made a video with a mobile phone (59,7%) and 35 already used a camcorder to make a video (45,5%). These percentages are inferior to those registered in the Lusofona's survey, yet the pattern is the same: using the mobile phone is an activity performed by the majority of young peoples in the age group, which is superior to the percentage of youngsters who utilized a camcorder for the same purpose.

Taking pictures and using the social networks sites and photo logs to share photos it was found to be very common practices among young Portuguese people. To make a photo album online is an activity already performed by 54,4% of the participants in Lusofona's survey aged 12 to 18 and 25,6% of these reported having interest in such activity. In the participatory media format evaluation study to take pictures was the most performed creative activity out of 16 listed, with 57 out of 77 participants reporting having done so (74%), while 39 of them reported the usage of photo editing software (50,6%). In the ethnographical study, publishing photos on a social network (Hi5) and photo logs was referenced by 11 out of the 16 participants aged 11 to 24, as well as observed in 3 of the cases.

To create and maintain blogs are activities already performed by approximately half of Portuguese young people within the age group 12 to 18 years. Almost half the young respondents to Lusofona's survey about media and ICT usage mentioned having create a blog, more exactly 47,4%, while 23,9% answered having interest in such activity. These results are consistent with CIES-ISCTE study "Crianças e Jovens: A sua relação com as Tecnologias e os Meios de Comunicação", in which 51,2% of the young respondents aged 9 to 20 years reported having created a blog and 35,7% stated having created a web page (Cardoso, Espanha, Gomes & Lapa, 2006).

Regarding the interest and involvement of young Portuguese people in media formats which explicitly ask them to create and share content, the results obtained by the participatory media format evaluation study indicate that a significant part of the respondents have a strong interest to participate in media formats such as the one presented to them. The large majority of these students aged 12 to 20 years answered having interest in watching or participating in this cross-media format: 59 out of 77 respondents (76,6%), with 3 negative answers and 15 answering "I don not know/ I don't respond". About the motivations to participate, in this study the main factor reported was "to learn more about my hometown and region", selected by 43 out of 77 respondents (55,8%). The material rewards to participate are also important motivational factors for this group, as well as the recognition by peers of the work submitted and learning new skills.

In summary, it was found that content creation is a common practice among the majority of this age group, mainly making videos, taking photos and making online photo albums and, also, creating and maintaining blogs. Also, it was found that a significant part of Portuguese young people have shown interest in participate or effectively participated as media creators in media formats and initiatives.

Yet, even if these results indicate the emergence of a participatory culture in the age group considered for the research work, as defined by Jenkins, there is a clear notion that this exploratory research was only scratched the surface and that many more questions are to be addressed then was previously envisioned. For instance, as these youngsters advance in life how will they occupy their leisure time, more specifically, will their preferential mode of relation with media and ICT be one of consumption or one of creation and sharing? In other words, even if it was found that in this age group emerges a relation mode with media and ICT that is based on participation, that does not imply that's this is a given fact but a trend under development whose real dimension is to determine with more precision in the following years.

Regarding next steps, it should be noted that the subject of participatory culture is currently being further investigated in the "MEDIACULT-PT—Participatory culture and media usage" by a team of Universidade Lusófona, with support from Fundação para a Ciência e Tecnologia. This project, based on the PhD thesis presented in this chapter, deals as main hypothesis with the emergence of a participatory culture as a consequence of new media use, and questions if such participatory culture reinforces mechanisms of social participation in local communities and specific cultural realities and if it is associated to the ways new media promote original forms of use that differentiate from previous ones (Damásio, 2007). Such project will contribute to provide a more nuanced and detailed perspective on the increasingly important topic of audience participation. With globalization, domestication and youth culture, participation is currently a focal point of audiences studies researchers, notes Livingstone (2007), whom also argues for the continued relevance of the critical audience research insights (2008) as our perspective widens from the study of active and interpretive mass television audiences typical of the late 20th century to a broader view of the today's study of interactive, multimedia, dispersed, and mobile audiences. Understanding these trends will hopefully enable both practitioners and researchers in the media and communications fields to improve their relationships with the audiences and to move away, once and for all, from the monologue model of mass media towards the dialogue mode of interactive and participatory media.

REFERENCES

Barker, C. (2003). *Cultural Studies: Theory and Practice*. London: Sage Publications.

Cardoso, G., Espanha, R., Gomes, M. C., & Lapa, T. (2006). *Crianças e Jovens: A sua relação com as Tecnologias e os Meios de Comunicação*, report CIES-ISCTE. Retrieved August 25, 2007, from http://www.cies.iscte.pt/destaques/documents/CriancaseJovens_AsuarelacaocomasTecnologia-seosMeiosdeComunicacao.pdf

Certeau, M. (1984). *The Practice of Everyday Life*. Berkley: University of California Press.

Damásio, M. J. (2007). *Resumo do projecto MEDIACULT-PT/ Projectos de Investigação Científica e de Desenvolvimento Tecnológico em Todos os Domínios Científicos—Concurso 2006*. Fundação para a Ciência e Tecnologia. Retrieved January 11, 2008, from http://www.fct.mctes.pt/projectos/pub/2006/Painel_Result/vglobal_projecto.asp?idProjecto=70893&idElemConcurso=906

Drotner, K. (2008). Leisure Is Hard Work: Digital Practices and Future Competencies. In Buckingham, D. (Ed.), *Youth, Identity, and Digital Media* (pp. 167-184). Cambridge, MA: The MIT Press. Retrieved January 13, 2008, from http://www.mitpressjournals.org/toc/dmal/-/6

Durham, M. G., & Kellner, D. (Ed.) (2006). *Media and Cultural Studies: KeyWorks*. Malden (USA): Blackwell Publishing.

Gauntlett, D., & Hill, A. (1999). *TV Living: Television, Culture and Everyday Life*. London: Routledge.

Jenkins, H. (2006a). *Convergence Culture: Where Old and New Media Collide*. New York and London: New York University Press.

Jenkins, H. (2006b). *Confronting the Challenges of Participatory Culture: Media Education for the 21 st Century.* report to The John D. and Catherine T. MacArthur Foundation. Retrieved January 13, 2008, from http://digitallearning.macfound.org/atf/cf/{7E45C7E0-A3E0-4B89-AC9C-E807E1B0AE4E}/JENKINS_WHITE_PAPER.PDF

Jenkins, H. (2003). Quentin Tarantino's Star Wars? Digital Cinema, Media Convergence and Participatory Culture. In D. Thorburn & H. Jenkins (Eds.), *Rethinking Media Change*. Cambridge, MA: The MIT Press.

Jenkins, H. (1992). *Textual Poachers: Television Fans and Participatory Culture*. New York: Routledge, Chapman and Hall.

Hughes, J., King, V., Rodden, T., & Andersen, H. (1994). Moving out from the control room: Ethnography in system design. In *CSCW'94* (pp. 429-439). Chapel Hill: ACM Press. Retrieved September 13, 2007, from http://delivery.acm.org/10.1145/200000/193065/p429-hughes.pdf?key1=193065&key2=2263690711&coll=&dl=ACM&CFID=15151515&CFTOKEN=6184618

Lenhart, A., Rankin, M., & Smith, A., (2007). *Teens and Social Media: The use of social media gains a greater foothold in teen life as email continues to lose its luster.* Washington, DC: Pew Internet & American Life Project. Retrieved December 19, 2007, from http://www.pewinternet.org/pdfs/PIP_Teens_Social_Media_Final.pdf

Lenhart, A., & Madden, M. (2005).*Teen Content Creators and Consumers*. Washington, DC: Pew Internet & American Life Project. Retrieved September 13, 2007, from http://www.pewinternet.org/PPF/r/166/report_display.asp

Livingstone, S. (2008). Engaging with Media—a Matter of Literacy? *Communication, Culture & Critique*. 1 (2008), 51-62. Retrieved August 21, 2008, from http://www3.interscience.wiley.com/journal/120185561/abstract

Livingstone, S. (2007). Interactivity and participation on the internet: Young people's response to the online invitation to engage in the civic sphere. In P. Dahlgren (Ed.), *Young Citizens and New Media: Learning for Democratic Participation*. London: Routledge.

Livingstone, S., Van Couvering, E., & Thumin, N. (2005). *Adult Media Literacy: A review of the research literature*. London: Ofcom. Retrieved January 9, 2008, from http://www.ofcom.org.uk/advice/media_literacy/medlitpub/medlitpubrss/aml.pdf

Livingstone, S. (2004).The Challenge of Changing Audiences Or, What is the Audience Researcher to do in the Age of the Internet? *European Journal of Communication*, *19*(1), 75-78.

Livingstone, S. (2002). *Young People and New Media*. London: Sage Publications.

Martins, D., & Quico, C. (2007). O desafio da Literacia dos Novos Media em Portugal: o contributo da iniciativa PT Escolas II. In 5° *Congresso SOPCOM* (pp. 2781-2731). Braga: Universidade do Minho—Centro de Estudos de Comunicação e Sociedade.

McQuail, D. (2000). *McQuail's Mass Communication Theory*. London: Sage Publications.

McQuail, D., & Windahl, S. (1993). *Modelos de Comunicação para o Estudo da Comunicação de Massas*. Lisboa: Editorial Notícias.

Morley, D. (2007). *Media, Modernity and Technology: the Geography of the New*. Oxon: Routledge.

Morley, D. (1993). *Television, Audiences and Cultural Studies*. London: Routledge.

Norman, D. A. (1999). *The Invisible Computer*. Cambridge: The MIT Press.

Ofcom (2006). *Media Literacy Audit—Report on Media Literacy Amongst Children*. London: Ofcom. Retrieved January 11, 2008, from http://www.ofcom.org.uk/advice/media_literacy/medlitpub/medlitpubrss/children/

Ponte, C. (2007, May). *Web access of children: Possibilities of internet providers to promote media literacy of children.* Paper presented at the EU 2007 –Expert Conference on European Media Policy—Leipzig, Germany. Retrieved June 8, 2007, from http://www.fcsh.unl.pt/eukidsonline/docs/LeipzigCPONTE.pdf

Ponte, C., & Candeias, C. (2007). Crianças e Internet: que acesso e usos? Que potencialidades e que riscos dessa relação?. Anuário *da Comunicação 2005-2006*. Lisboa: Obercom. Retrieved August 25, 2007, from http://www.obercom.pt/client/?newsId=342&fileName=anuario_2005_2006.pdf

Quico, C. (2008). *Audiências dos 12 aos 18 anos no contexto da convergência dos media em Portugal: emergência de uma cultura participativa?*. Unpublished doctoral dissertation. Universidade Nova de Lisboa—Faculdade de Ciências Sociais e Humanas.

Seiter, E. (2004). *Qualitative Audience* Research. In Allen, R. C. & Hill, A. (Ed.) The Television Studies Reader. London: Rouledge.

Schroder, K., Drotner, K., Kilne, S., & Murray, C. (2003). *Researching Audiences*. London: Arnold.

Chapter VIII
Methods for Involving Users in the Development of Social Interactive TV:
Enhancing Usability and User Experience in Non-Traditional Environments

Regina Bernhaupt
IHCS—IRIT, France

Marianna Obrist
ICT&S Center, University of Salzburg, Austria

Manfred Tscheligi
ICT&S Center, University of Salzburg, Austria

ABSTRACT

Applications for interactive TV (iTV) addressing social aspects will only be successful, if the intended user and user community is taken into account during the development process. Existing methods for evaluating usability and user experience aspects of social interactive TV applications are not really enough to address and focus on the broad range of factors related to social user experience. This chapter presents various forms of user-centered methods enhancing usability and user experience aspects of interactive TV applications in general. Giving an overview on currently used user-centered methods to enhance and evaluate usability and user experience, some selected methods are presented for early and late development stages, reflecting the peculiarities stemming from the non-traditional environment the applications are used in—the home. The chapter will show how the methods can be adopted to focus more precisely on social aspects, especially social user experience. It enables the reader to get an overview on currently used user-centered methods in interactive TV and to learn about benefits and shortcomings of these methods as well as how to choose appropriate methods for their own purposes.

INTRODUCTION

Usability and user experience have become a buzzword in the interactive TV (iTV) industry. As new forms of iTV services are emerging, including applications like live-chatting during watching TV, exchanging personal media content like private home movies via internet, or writing emails, the usability and in general the overall user experience has become a main focus during its development. From the users perspective this kind of applications is difficult, complex and sometimes time-consuming. Navigation within the iTV content and usage of several applications (maybe just for a few times per year) tempt the user to be way more active than in the past. Sometimes users even have the general user experience, that their TV set is no TV set any longer but a time consuming, complex form of a personal computer, instead of an entertaining and relaxing, fun-oriented entertainment device.

Even though, people have always been social in front of the TV by discussing the TV program with other household members, it is still unclear how successful social iTV applications will be for e.g. talking on the phone or playing a game while watching program. Both, usability and user experience aspects, have to be addressed when designing and developing social interactive TV.

Usability is typically defined by three factors: how effective, how efficient and with what level of satisfaction users can perform their tasks (Dix, Finlay, Abowd, and Beale, 2004). A high level of usability is typically achieved using a user-centered development process, developing the product in iterative steps with various forms of evaluation at each development stage. Results of the evaluations are typically informing the further development of the product, system or service. If users are able to perform a task within a certain amount of time, with only few errors and if they are overall satisfied when using the product for their tasks, the system is said to be usable. But usability is just one aspect. Users of a social interactive TV application are using the product to be entertained, to relax, to have fun, to forget about the real world, to interact socially and to have special forms of experiences that go way beyond simple usability. A good example to make the distinction between usability and user experience a bit more clear is the following: Sitting in front of the interactive TV system, a user is playing a game with two of his friends. The goal is to cooperatively solve a game, simply answering questions. The goal can only be achieved if all of the three players reach a certain number of solved questions. The social iTV game would be usable if each user was only given a dialogue box asking if they want to solve the game. If the user answers yes—the game is won. What a bad experience for playing a social game! The user experience of the game lies within solving a number of sometimes difficult questions, trying to figure out where to go, what to do, and exploring the social structures between the players and how they co-operate during playingtheir cooperation during the game: when is it necessary to rely on the other players? When would it be beneficial to solve a quest alone? The players will perceive a new form of experience: social user experience.

User experience of social iTV games, and more general of iTV systems, is different from the concept of usability. Following the fact that a number of qualities of a product go beyond the instrumental aspects of product use and usability, user experience research has evolved (Karapanos and Hassenzahl, 2008; Buxton, 2007; Hassenzahl, 2006). User experience is a broadly defined term, including attainment of behavioral goals, satisfaction of non-instrumental (or hedonic) needs, and acquisition of positive feeling and well-being. Neither a universal definition of user experience nor a cohesive theory of experience yet exists that can inform the Human-Computer Interaction (HCI) community on how to practically design for and evaluate user experience. Researchers are still looking for an adequate and overall definition of user experience (Law, Thora, Hassenzahl,

Hvannberg, 2006; Law, Vermeeren, Hassenzahl, and Blythe, 2007; see also COST Action: MAUSE Group http://www.cost294.org/).

Experience is still something that is defined as being private, with several models describing experience either as a factor-oriented model or using a more holistic approach (Forlizzi and Battarbee, 2004; Arhippainen and Tähti, 2003; Hassenzahl, 2003, 2004, 2006; Kankainen, 2003; Karapanos, Hassenzahl, and Martens, 2008). "An experience is more coalesced, something that could be articulated or named. This type of experience may be characterized by a number of product interactions and emotions, but is schematized with a particular character in one's memory and a sense of completion. An experience has a beginning and an end, and often inspires emotional and behavioral changes in the experience. For example, a dinner party or a news hour seen on television could be classified as an experience" (Battarbee, 2004, p. 162). Battarbee (2004, p.162) defines social user experience as a process of "lifting up experiences to shared attention, where they become part of a social interpretation process that can influence what the experience comes to mean to the individuals and others." During the development of social iTV applications we have to take both aspects into account: On the one hand, usability of the iTV service must be high to allow the user to consume the content, interact with the service, navigate or choose what they want to do. On the other hand, we user experience must be supported, either by working with a factor-model, influencing certain aspects of user experience like emotion, fun, playability—or influencing the design process in general to enable a good overall user experience and social co-experiences especially for the social iTV applications.

Social iTV applications are thus characterized by enabling the user to experience a special form or user experience: the co-experience. Social TV applications in a strict sense would be everything enabling this co-experience, but are typically used when incorporating a special form of communication between users at different locations, or enabling several users in front of the TV to interact in a more social way. More precisely, social TV systems typically integrate some combination of text chat, voice chat or video chat with TV programming and use presence to provide awareness of the status and context of other users of the system (Harboe, Metcalf, Bentley, Tullio, Massey, and Romano, 2008; Boertjes, 2007, p. 1; see also Boertjes, 2007; Harboe, Massey, Metcalf, Wheatley, and Romano, 2008; Oehlberg, Ducheneaut, Thornton, Moore, and Nickell, 2006).

From an industrial perspective social iTV—the new trend of interactive TV—can be summarized as encompassing, among other things, web-based and IPTV programming services that allow viewers to set up personal play lists and communication in real time about the program they are watching; applications that enable online social networking services to extend their reach to television; and services that use online communities as a source of user-generated advertising for distribution on both, broadband and traditional TV. The variety of social TV applications available for the end-consumer is still low, but there is an increasing number of ready to go applications developed within scientific projects and industries (Coppens, Trappeniers, and Godon, 2004). Lommel TV is a project located in a small town called 'Lommel' in Belgium. It's a do-it-yourself community TV channel allowing people to post videos (see http://www1.alcatel-lucent.com/bnd/lommel/). In the SAMBA project—an interactive community TV project in Brazil—the main aim is to create a framework for allowing local communities and citizens (including low income population) to access community-oriented content and services by means of iDTV channels (http://www.ist-samba.eu/). Community TV is also fostered in the CITIZEN MEDIA project by providing an IPTV platform for user-generated content to a local community. TV-centric social services, such as sharing of personal videos and pictures, as well as publishing an own profile on the TV, have been implemented and should enhance the social life and community experiences beyond offline social activities. Moreover, the IPTV platform intends to make people aware about what is happening in their village, bring them together online and in a next step offline again.

Looking at the broad range of social iTV applications, it has to be discussed, how the development process to enhance usability and user experience of interactive TV applications supporting social aspects can be assisted. As social iTV applications are typically used in the living room, the special usage context must be considered as well: how can this usage context be investigated, what kind of methodological variations have to be taken into account, when investigating the context of home—and how can the social user experience be improved?

The following chapter will look at these questions in detail, first describing the special context of the home, then giving a summary of currently applied methods and finally presenting some selected methods in detail, describing their usage during the development of social iTV applications.

NON-TRADITIONAL ENVIRONMENTS

Today's technologies allow users to access information and communication services wherever they are and whenever they want, using a broad range of devices ranging from mobile phones, PDAs or laptops to more fixed devices like set top boxes or computers. Usage of information and communication technologies (ICTs) is heavily influenced by the usage context. During the last decade the context is playing a major role in the development of new products. People on the move using their mobile phone have different needs, than using a mobile phone at the workplace or in their home and Evaluation methods for usability and user experience have to be adopted accordingly (Bernhaupt, Mihalic, and Obrist, 2008). Usage of ICTs like a mobile TV in a public place is different from usage in the home (or even in the office), thus several EU projects are currently focusing on how to develop new forms of ICTs with respect to the usage context, for instance by trying to connect the home and the mobile context within a cross-platform IPTV environment. The use of multiple media in the IPTV environment builds on the idea to support the pre-experiences of people with both devices, the mobile phone and the TV. Thus, existing media experiences with each device are supported instead of replaced by new interaction concepts (Obrist, Kepplinger, Beck, Tscheligi, and Muzak 2008).

However, the main focus for interactive social TV is still on the home environment. The home has been described as a place where people can relax, be comfortable, and be close with other people. The home provides a space for action and interaction in which people freely decide what to do and which technology to use. In terms of research the most interesting question is how people adopt new technologies and how they use these kinds of technologies related to the different aspects of context. The context of use is the environment into which the technology will be placed and used. The home has been described as a living space, defined by emotional, social, educational, and recreational needs of the household members, divided into three components: social, technological and physical space. Obrist, Bernhaupt, and Tscheligi (2006, p. 350) extend these three spaces, using six contextual factors typically related to investigations of the home:

- "**Spatial context:** TV in the living room; multifunctional in all rooms; individual rooms.
- **Temporal context:** TV viewing during the day; week and weekend; season.
- **Social context:** TV used alone; in company; in both contexts; social structure of

the household, e.g. adults, children, older people.
- **Personal context:** Demographic data; prior experience with other technologies; hobbies and interests; TV preferences.
- **Technological context:** Household equipment; technologies used at home.
- **Context of the context home:** Neighbourhood; nation; cultural factors."

What we have learned form our studies (Obrist, Bernhaupt, and Tscheligi, 2006) is that context has to be taken into account during the whole development process. But evaluation methods only reflect the influences of context limitedly. Few methods, like a contextual factor questionnaire at the beginning of the project, focus directly on the context. As usage can depend heavily on the combination of contextual factors, the variation and combination of various evaluation methods is crucial to collect user feedback and in particular users' experiences. To address contextual influences during the product development we argue for an interdisciplinary development team, mixing methodological approaches from several disciplines to enable the development of products that are usable and that convey a high level of (positive) user experience.

For the development of social iTV applications we learned that all methods related to the evaluation of usability and user experience must take contextual influences into account. We thus extended several methodological approaches to include contextual factors and to investigate usage contexts (Bernhaupt, Weiss, Obrist, and Tscheligi, 2007). What we have learned is that during the whole development process the involvement of real users in their real usage context is necessary to achieve a high level of user experience of social iTV applications. Only a combination of several methods can help to really understand the user in the context of home.

USABILITY AND USER EXPERIENCE METHODS FOR SOCIAL ITV

When evaluating usability and user experience, several forms of methods are applicable. In the next sections we want to give an overview of currently applied usability and user experience evaluation methods. We present the methods loosely structured according to the application during early or late development stages.

User-Centered Methods for Early Development Stages

During the development of social iTV applications several methods can be applied in the early development stage. We want to focus on methods that have been already used for the development of social iTV services, as they have been proved to successfully inspire and inform the early development stages. We thus describe in the following: Ethnographic studies, diary studies, cultural and technology probing, social sciences oriented methods like interviews, questionnaires, focus groups and observation and special co-design methods like creative techniques and workshops. For each of the methods we will give the scope of usage, methodological advantages and possible restrictions.

In the field of Human-Computer Interaction (HCI) ethnographic methods are used to better understand how technology is adopted and used in the home context. Crabtree & Rodden (2004), among many others, used ethnographic methods and variations focusing on domestic routines and the role of technology in home life. Ethnographic studies today are an accepted method within the area of HCI to address the possible usage context. When using ethnographic studies for social iTV, the ethnographer must especially focus on the social interaction. But all of these ethnographically inspired studies have in common, that they involve researchers as integral part of the field study.

The presence of a researcher might influence the social communication and social experiences of users. Thus, depending on the focus of the study, ethnographic studies involving the researcher in the field might only be of limited value.

To overcome the limitations of ethnographic studies, diary studies have been used. Participants within the diary study are asked to fill out a (pre-structured) diary, noting special forms of usage or special forms of communication. Diary studies help to overcome the limitations of ethnographic studies as the researcher no longer goes into the field to collect the data, but prepares tools that guide the participant in understanding how to collect the data themselves (Sanders, 2002). On the other side this form of in-situ research method depends on the willingness of the users to fill out the diary in time, to take the diary with them to take notes about usage and it is especially difficult to investigate social situations, as some note taking during conversations or co-usage of products might be inferring the usage process of the product.

Cultural probing provides a variety of ways to apply self-reporting during the ethnographic oriented studies. The method, invented by Gaver, Dunne and Pacenti (1999), uses a so-called probe package including for example post cards, post its, cameras, diaries, maps or other probing material. The participants of the study are asked to fill out the material or use the tools provided to describe the daily usage of products or special usage situations. Technology probing (Hutchinson, Mackay, Westerlund, Bederson, Druin, Plaisant, Beaudouin-Lafon, Conversy, Evans, Hansen, Roussel, Eiderbäck, Lindquist, and Sundblad, 2003) uses a real prototype or product to be evaluated during the study, providing real life insights during the usage of the developed product. Technology probing is especially useful when studying social interactive TV applications, as the adoption of that forms of services is heavily influenced by peoples habits and routines in real live (as shown a field study on ambient Social TV by Harboe, Metcalf, Bentley, Tullio, Massey, and Romano, 2008). Probing approaches can inform the design at early stages of the design process, in the case of technology probing at very late stages of the development. The results of probing are typically qualitative in nature, and especially for technology probing larger samples are often difficult to manage, as the number of functional prototypes if often limited during the development of interactive social TV applications (Coppens, Trappeniers, and Godon, 2004; Obrist, Kepplinger, Beck, Tscheligi, and Muzak, 2008). To design products that people will really use, designers must be aware of user's experiences, emotions, the situation of product use as well as social and cultural influences (Stappers, and Sanders, 2003).

Apart from the above described methods, there is a growing interest in finding new ways to involve and engage users in the very early design process of new applications and services. Co-design is a qualitative research methodology that is often referred to as participatory design or co-operative design approach (Muller, 1993; Spinuzzi, 2005). Participatory design is originated in the Scandinavian approach in the 1970s and 1980s (Ehn, 1992). Its ambition was to empower workers and to enhance democracy and professional development in the workplace (Spinuzzi, 2005; Facer & Williamson, 2004). Participatory design is nowadays being used beyond the domain of workers and their workplace, namely in the extensive domain of end users and HCI, computer-supported cooperative work and related fields (Spinuzzi, 2005). Examples are the projects WebKit and KidReporter (Facer & Williamson, 2004) or CITZEN MEDIA (Obrist, Beck, Kepplinger, Bernhaupt, and Tscheligi, 2008). Especially interesting for social iTV applications is the focus on families, like in the interLiving project and EnComPAs (Westerlund, Lindquist, Mackay, and Sundblad, 2003; Van Rompaey, Hemmeryckx-Deleersnijder, Van Der Meerssche, De Mondt, and Godon 2005b).

Co-design can be defined as the process of first obtaining insight in the natural environment of users who then take on an active role in creating ideas and designing/testing new applications in their natural setting, supported by a multidisciplinary research team. Co-design in particular means actively integrating end users in the development process of a future product by putting them in the role of designer and stakeholder (Van Rompaey, Van Der Meerssche, Godon, Vanden Abeele, Charliers, 2005a) and thus enabling a high level of user involvement (Steen, Kuijt-Evers, and Klok, 2007). In co-design a number of qualitative techniques are used. Next to in-depth interviewing and observation, also cultural probing and technology probes—as described above—are used. Moreover group sessions and co-design workshops are used through an iterative process of evolving and refining ideas together with end users. Co-design workshops were experienced to be very valuable to be conducted in an early design phase in order to involve end users in the idea generation and conceptual phase for new social TV services and applications (Obrist, Beck, Kepplinger, Bernhaupt, and Tscheligi, 2008). This co-design approach also enables users' to collaboratively explore new ideas, get involved in the design of new concepts and have a early kind of co-experience with a newly emerging service or system.

Methods for early design and development phases help in general to understand the usage of the product, especially the influencing factors of social (iTV) usage. These factors can be related to usability and the overall user experience of the social iTV application. Of course the general research question, selection of methods and participants, and interpretation of results, form the basis of realistic and interpretable findings that inform the development.

Influencing factors that are typically neglected in traditional research methods for early design and development are the spatial, temporal, social, personal, technological context and the context of the usage context. Especially the social context and influences on the usage context (context of the usage context) like opinions and information from social communities or groups can only be studied with in-situ research methods like ethnographic studies or probing approaches.

Following the presented co-design approach, user involvement should be understand as a means, where end users are, ideally, involved in the entire development process, e.g. from the beginning, during the iterative design and evaluation process, and also during the evaluation of the designed product or service in the end users' usage context. The ultimate goal of user involvement is to deliver system requirements, which can lead to a system which better matches end users' needs and preferences (Kujala, 2003).

User-Centered Methods for Late Development Phases

During the development of products, several forms of evaluation should take place to steadily inform the (technical) development of the iTV service. To investigate the usability, user expereince and in particular also the co-experience of social iTV applications various forms of methods can be applied. Following we present a selection and give a brief overview of the methods we perceive as most reasonable during the development of social TV applications.

Usability studies in the lab, sometimes referred to as user testing, are performance measurements to determine weather usability and user experience goals have been achieved during the development. Usability and user experience evaluations of prototypes in the lab have several advantages: the usage conditions can be controlled and the prototype must not be fully functional—as tasks can be limited or functions can be replaced by Wizard-of-Oz technique (Dahlbäck, Jöhnsen, and Ahrensberg, 1993). The usability and user experience testing typically includes the following stages: a pre-interview to understand the users' experiences with similar technologies, performance of some tasks, where the difficulty of the tasks is rated, a final interview and some questionnaires. Depending on the research question, usability studies in the lab include several forms of measurements. Eye-tracking can help to discover usage problems (Obrist, Bernhaupt, Beck, and Tscheligi, 2007), bio-physical measurements can be used to investigate aspects like emotions.

The set-up of the studies must either focus on the social usage of the TV system, or should involve several participants interacting with the product (Lull, 1980), or the Co-Discovery technique described by Dumas and Redish (1993) could be used to make it feel more like a social TV context. During a usability test, two test users attempt to perform tasks together while being observed. They are to help each other in the same manner as they would if they were working together to accomplish a common goal using the product. They are encouraged to explain what they are thinking about while working on the tasks. Compared to the thinking-aloud procedure, this technique makes it more natural for the test users to verbalize their thoughts (Dumas and Redish, 1993). Disadvantage of testing social iTV applications in the lab is the missing real context, even if it could somehow be simulated (by testing with more users, make the lab more look like a living room). People might like and use an application in the lab, but when offered other possibilities of entertainment in their home places, they might not adopt usage of the service at all, due to the limited user experience. Evaluation of social iTV applications should thus also include field studies, enabling the study of the product in a real usage situation.

For evaluating a product in the field a variation of technology probes can be used. Technology probes can be used combined with other methods, like ethnographic studies. To support the fast development cycles of products, we typically combine ethnographic studies focusing on general trends with some technology probing of already existing prototypes (Bernhaupt, Obrist, Tscheligi, 2006; Bernhaupt, Obrist, Weiss, Beck & Tscheligi, 2007). The usage of field trials in research helps to discover contextual influences that can not be investigated with other methods. On the other side, field trials require lots of preparation, time and are expensive, as several functional prototypes must be developed and deployed in the participating households.

Evaluation of user experiences during product usage is difficult. The role of evaluation is mainly to inform the design and to improve it at all stages of the design life cycle (Dix, Finlay, Abowd, and Beale, 2004; Stone, Jarrett, Woodroffe, and Minocha, 2005). Different kinds of evaluation may be carried out at different design stages and for different reasons. Evaluation done during the early design stages is undertaken to evaluate the users' requirements, predict the usability of the system, and to assess how well the interface meets the users' needs. Evaluation in a later stage is carried out to check the usability and user experience of a nearly complete system. Several methods and techniques like usability testing in laboratory environments or other in-situ methods, like experience sampling, can be used. The experience sampling method (ESM) is especially suited for evaluation of user experience aspects on an individual level (e.g. users' experiences with a mobile device - Meschtscherjakov, Obrist, Mirlacher, Lankes, and Tscheligi, (2008)). ESM can also be used for evaluating user experience on a social level, in particular to evaluate co-experience aspects (e.g. ESM on IPTV, which is currently implemented and pre-tested within the CITIZEN MEDIA project, to be used in field trials in autumn 2008).

ESM is an in-situ method for collecting information about the context and the content of the daily life of individuals. It aims at catching emotions, motivations and cognitive processes when they actually occur in a real live situation (Larson and Csikszentmihalyi 1983). ESM is typically applied in the area of mobile applications, but is further extended towards other platforms such as described above for the IPTV field and is also made usable for web applications within CITIZEN MEDIA (based on the TUMCAT tool developed by TNO in the Netherlands, http://www.freeband.nl/). Depending on the technical set-up, users are sent questions after using the specific service, or in fixed or variable time steps. ESM helps to discover typical usage problems in real life situations and to investigate the users' experiences while using the product. One of the main disadvantages of ESM is the possible disruption of the user's activity, requiring some attention for answering the questions, which is especially critical for the IPTV ESM application related to the home environment—where we know that the living room is a place for relaxing and often a place for lean-back activities. On the other side the tools functions (like additional cameras included in the device), or the combination of ESM with logging analysis and other qualitative methods such as interviews can provide rich data for the researcher to understand the various aspects of user experience.

In general user experience studies in the field can help understand the usage of social iTV services. Only when understanding the real needs within the households, we are able to develop successfully adopted social iTV services. Within the CITIZEN MEDIA project a combination of several in-situ field methods are used to capture users' experiences with the community IPTV platform. Apart from new creative means, such as a „Smiley Box" developed for collecting users positive and negative experiences in-situ, we are also going to use the experience sampling method directly on the TV for getting real live feedback from end users. These methods are further combined with logging analysis and traditional qualitative methods such as interviews, discussion groups for collecting user feedback, in particular related to their experiences and user acceptance issues. All this user data is used as input for future iterations of the system to make it more fit to the users' requirements and social community needs.

CASE STUDIES FOR THE EARLY INVOLVEMENT OF USERS

Following we are going to describe the application of two methods during the development of social iTV services. The selection, combination and conduction of methods must be carefully, to enable a reasonable investigation of the research topic and to achieve meaningful results. Only with reliable information, an enhancement of usability and user experience is possible.

Creative Cultural and Playful Probing

Creative Cultural Probing and Playful Probing are ethnographic inspired methods and variations of the cultural probing approach, developed to investigate the usage of new forms of technologies and media in the home (Bernhaupt, Weiss, Obrist, and Tscheligi, 2007). The probing method consists of an introduction interview, the ethnographic period of several weeks (including the later described packages) and an in-depth final interview with household members after the study. Creative cultural and playful probing use a package of material to support the self-observation of participating households. The package includes so called creative cards for each research topic. Additionally the creative cards include some play elements (like building a prototype of a future system using modelling clay), and a specially designed game, reflecting the research topic to be investigated.

The method was used to investigate social aspects of interactive TV, especially how people watch TV together in groups, how they would like to personalize the iTV content for the household and if people would like to have a new form of interactive TV experience—we call it "individualized TV", by having the possibility to identify themselves in front of the TV. First results from this trial can be found in Bernhaupt, Wilfinger, Weiss, and Tscheligi (2008).

For the preparation of the study first the research topic was described :To understand how social iTV services, like recommendations based on the user profiles, personalization of services and - contrary—social usage of services are influencing each other. The intended user population (households with several members) was specified, followed by recruiting 40 households, spread over a city and some areas on the country side.

For the development of the creative and playful probing material we addressed at least three creative cards to each research hypothesis. Additionally we developed the playful game, extending a card game called UNO with a new form of activity: a special card, once played is prompting the players to answer some questions related to the research topic. Figure 1 shows the cards and the game. Activities on the cards included questions like: services I would like to receive "just for me" or "for all members of the household". These questions should help to understand what kind of services would have some "social" potential, and what kind of services need the identification of the individual.

The study lasted two weeks (including a trial of a recommendation system) and was concluded by a final interview, investigating the probing material and asking participating household about their general usage behavior on TV, recommendations, privacy and security.

Creative and playful probing are methods enabling the researcher to understand the social usage context, to identify services, that are used social (with other people) and services that are more individual. The study above revealed an interesting social aspect: people would like to become recommendations for the household (to see a nice film together with the kids) but at the same time prefer individual recommendations ("when the kids are to bed, I really like to relax in front of the TV, and would need different recommendations") (Bernhaupt, Wilfinger, Weiss, and Tscheligi, 2008). Recommender systems thus show social aspects that should be included in the development of new social TV services.

Figure 1. Creative Cards used during the study: on the bottom right: A question asking for recommendations that people would like to receive solely or for all members of the household.

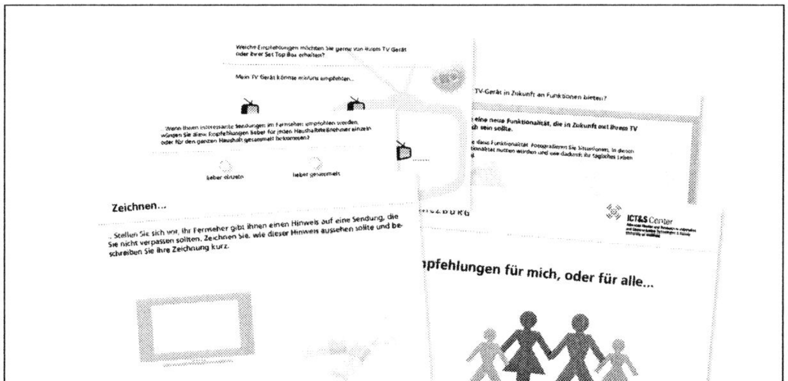

Co-Design Workshops

As described in section 3.1, the co-design approach is increasingly used for early user involvement in a design process of new services and applications. Referring to the methodological framework developed within the EnComPAs project by Alcatel-Lucent (http://encompas.org/), co-design could be divided in four major steps: SCAN (you scan a field/problem), THINK (you think towards/choose a solution), MAKE (you make/implement something) and TRY (you try it out). These steps are based on frameworks out of both software development and design cycles. Following a contextual investigation in the first scan phase, co-design workshops are most applicable in the „Think" phase in order to collaboratively generate new concepts and ideas with end users.

In our research within the CITIZEN MEDIA project, we tried to explore several techniques to make co-design workshops more successful in terms of quality of the results. As part of this research, we developed a card based technique called „Instant Card Technique (short IC-Technique)", which we evaluated and iteratively improved in three successive co-design workshops within two projects: „Future Communication Design" (May 2007), „Game Design " (June 2007) and „Mobile Service Design" (August 2007) Workshops (see Beck, Obrist, Bernhaupt, and Tscheligi, 2008). The IC-Technique allows a simple gathering of user-generated ideas, supports people to express their ideas, considers the way of life of the users, offers a formalized procedure by using the instant card types and results in context scenarios.

Collaborative card-based techniques already have a long tradition in participatory design. The CARD technique (Collaborative Analysis of Requirements and Design) by Tudor, Muller, Dayton, and Root (1993) and its refinement by Muller (2001) set initial steps in the usage of cards for participatory analysis and design. Similarly, Lafrenière (1996) invented CUTA (Collaborative Users' Task Analysis) to provide a simple and low-cost technique for analyzing task flows. Another closely related method is the Inspiration Card Workshop of Halskov and Dalsgård (2006). These cards serve as triggers of ideas in the creation of design concepts. The IC-Technique is to some extent similar to the above presented card techniques, as it also uses cards, which are then further arranged. But the goals of the related card techniques differ from the goal of the IC-Technique. Our technique doesn't solely concentrate on tasks and activities; it has a much broader research focus, including users and the context of use. Furthermore, the IC-Technique aims making users express their ideas. Compared to Muller's layered participatory analysis, the IC-Technique also strives for structuring the research process by classifying the card material. But our technique extends the usage of the cards by using the cards as building blocks for the scenario development.

The five typical steps for using the IC-Technique in a Co-Design Workshop are described below.

1st Step: Think about your Workshop—Goals and Participants

The technique has to be designed carefully by using procedures which are familiar to users, e.g. to take notes on cards and tell stories about activities. An important consideration is how to manage the different roles of the participants during the workshop. In our understanding the researchers typically act as mediators (workshop leaders), the users act as designers. But it can be fertile for the brainstorming and design part of the workshop when researchers act as facilitators by digging deeper and asking further questions.

2nd Step: Select the Workshop Location

We experienced that it is essential for brainstorming and for the overall success of the workshop

Figure 2. Workshop held in a local pub, supporting participants in feeling more familiar with the situation.

that participants feel comfortable and secure in the selected workshop location. Co-Design workshops should therefore preferably be conducted in the real user context (see Figure 2).

3rd Step (Set-Up): Adjust the Instant Cards to Your Workshop Goals

In order to structure the scenario development process, different types of cards are used in the workshop. Six types of basic instant cards are used (see Figure 3): user cards, location cards, time cards, technology cards, activity cards and goals card.

4th Step: Adjust the Workshop Phases to your Workshop Goals

The main workshops phases are: (1) Introduction, (2) Finding a common basis, (3) Scenario development, (4) Final Presentation and (5) Feedback. A detailed description of each phase can be found in Beck, Obrist, Bernhaupt, and Tscheligi (2008).

5th Step: Consider the Outcome of the Workshop, Its Analysis and Further Use for Design

The outcomes of the IC-Technique are user-generated ideas and a set of context scenarios. In general the following material can be analyzed: the developed poster, audiovisual and/or oral presentation of the scenario by the workshop participants on tape, and observations and notes from the researchers accompanying the workshop.

The IC-Technique not only helps to increase the user involvement, but also helps to investigate

Figure 3. Various card forms supporting the different stages

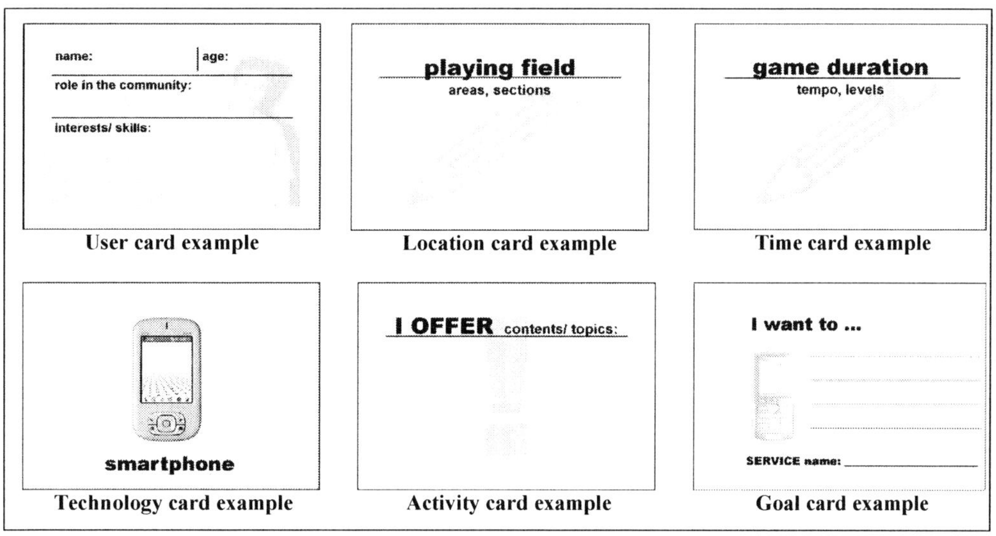

Figure 4. Users select instant cards and take notes themselves in their community location.

Figure 5. Workshop participant explains the game visualized on the poster.

several aspects of design ideas by guiding users during the development of usage scenarios. In particular, the flexible usage of the six card types helps investigate various research aspects, guides users through the idea generation process, selection of ideas and development of usage scenarios for new service ideas and concepts. The presented five step procedure is very flexible, in order to enable the researcher to modify the instant cards towards their specific research aim, for instance to decide about the order of the introduced instant cards into the workshop as well as on the amount of cards.

Even if the IC-Technique has been proved to be very useful to support the idea generation process in co-design workshops, there are still some critical issues which have to be improved, such as the selection of participants for the workshop (to represent the intended user group), as well as the grouping of people in sub-groups during the workshop. We experienced that some dominant participants in a sub-group or the combination of young and old could lead to some problems in the brainstorming and development of new conceptual ideas (some ideas might be overheard or remain unspoken). This social context has to be carefully taken into account. We have recently also used the IC-Technique with children for gathering ideas on the future development of an emotional game including co-creation activities and social interaction, intended to be used on the PC as well as on the TV in a public environment. First results show that the technique is very useful for children above 10 years old. For younger children a more visual approach for the instant card design has to be considered.

We additionally experienced that participants who were previously involved in a field study around the topic of the workshop could express themselves better as newly involved participants. Therefore, we were also thinking about an extension of the recent IC-Technique set up, by combining it with a short pre-experience study using the cultural probes approach in combination with selected instant cards. Moreover, we intend to further improve the IC-Technique for other research areas.

METHODS APPLICABLE DURING THE IMPLEMENTATION PHASE

This section presents a selection of most commonly used methods involving the user in terms of usability and user experience during the implementation phase. Laboratory tests offer a controlled, interruption free testing environment, but do not reflect real life conditions (Rosson and

Carroll, 2002). Normally a usability laboratory is well equipped with audio/visual recording possibilities and facilities for analysis (Dix, Finlay, Abowd, and Beale, 2004). One advantage is that the participant works in an interruption free environment; on the other hand the testing situation misses the context and cannot reflect the real life. In-situ field trials offer a natural evaluation situation, where one can observe behaviors, which could be missed in a laboratory, but the test setting can hardly be repeated. There are many different methods available to perform in-situ user research (see a detailed overview in Fields, Amaldi, Wong, and Gil, 2007; Dix, Finlay, Abowd, and Beale, 2004). With regard to the increasing amount of user research in mobile and private or sensitive contexts, in-situ methods are becoming more and more important (see also Bernhaupt, Mihalic, and Obrist, 2007). Insights on useful evaluation methods for iTV applications, inspired by the field of media studies, is already provided by Chorianopoulous and Spinellis (2006). In the following sections, we will describe two selected methods technology probing and user experience studies in the field, to show how this form of evaluation can inform the development of social iTV services.

Field Studies (Involving Several Users) Including a Variation of Technology Probes

Using field studies to evaluate aspects of usability and user experience helps to understand the real usage of the product in people's daily life. The real usage of products is typically related to a kind of adoption process. Using a (prototype) system in the study, can help understand how people learn to use the system, how usage behavior might change over time, and to what extend the usage of the system can be expected. We want to summarize what we have learned during several field studies related to the introduction of new interactive TV formats, services and technologies related to interactive TV, especially remote controls (Bernhaupt, Obrist, Weiss, Beck, and Tscheligi, 2007).

We used technology probes as the method of evaluation new prototypes of interactive TV services (field trial 2004/05, 2005/06), remote controls (2006) and a recommender system (2007). A field trial is only worth time and effort if the used technology is sufficiently stable to be used by people, with no prior experience. Especially when evaluating user experience, technical instability might heavily influence the user experience, leading to non-realistic results. Our experiences with technology probes showed, that when applying the method the following aspects must be taken into account:

- **Learnability of the system:** Technology probes are especially useful to understand the usage of the prototypes in the field. When people used and learned the system at home, they showed in general less problems, than during (comparative) usability studies in the lab. People really try to use the system, they know their system at home, and the added component (like a new service, a new TV channel with additional information, or just the exchange of the remote control) is less difficult to learn in the real home context.
- **Usage over time:** Field study using technology probing helps to understand the long-term usage of the system, and why and when people stop to use new technologies. The usage of a new remote control can be interesting for some participants in the household, while other ones prefer to use the old one—due to lack of interest or simply due to the additional effort that learning a new remote control would need.
- **Information material:** The trial should be accompanied by information material, to reduce the likelihood of calls at the people installing the system or the researchers. People tend to use the system as a real system,

and they refer to their typical strategies of getting help, if they can not use the system (call the hotline, call the researcher, ask other ones participating in the study, etc.). This kind of support might be given only limited (we had hot-lines available for two of the tests during the normal working times, but the hotline would have been needed during the TV-watching hours—in the evening and during the weekend). Providing some basic instruction manual, or—as preferred by most of the participants—short videos, can help to reduced this problem.

User Experience Studies in the Field (Investigating Social User Experience)

Usability Studies are an adequate method to ensure usability aspects of the system. But social aspects of user experience are difficult to access. One possibility to investigate social aspects are usability studies involving several users in the lab, or field trials and studies, including usability observations or co-discovery techniques. As a realistic social environment can not be reached in a lab based study, a combination with in-situ studies is therefore needed, in particular to capture the social experience. Based on our lessons learned during various field trials (e.g. Obrist, Bernhaupt, and Tscheligi, 2006; Bernhaupt, Obrist, Weiss, Beck, and Tscheligi, 2008) we describe methods combined in a field trial to investigate social iTV.

Ethnographic methods, cultural probing and other inquiry techniques and methods are already available to more actively involved users in a user-centered design. Although, we know from previous research (Bernhaupt, Weiss, Obrist, and Tscheligi, 2007), in particular in the home context, that it is not easy to keep users motivated to participate over a long time. A second aspect is that it is not easy to enter the home context, a private place, as a researcher. Therefore, we applied more

Figure 6. Smiley Box

creative and playful techniques (see section 4.1.), as well as more sociable methods that are useful in particular when taking the social context of TV viewing into account. Following we describe a method used as part of a field study focusing on social user experiences of TV-centric services.

For our long-term field trial for the social community IPTV platform developed within the CITIZEN MEDIA project and deployed in about 60 test households we come up with a new methodological approach, we called „Smiley Box" (see Figure 6). The Smiley Box given to the test participants is a small cardboard box which has to be feed by filled out „smiley" question cards.

The box was intended to:

- Be a nice object participants put in their living room (in analogy to the Set top box they received for using the IPTV application)
- Be an artifact to motivate all household members to contribute (give feedback)
- Address different ages present in households (old and young)
- To make the whole evaluation process more fun for the participants (alternative way to ask questions, not just through a questionnaire).

The first Smiley Boxes have been sent out to 30 test participants in spring 2008. Each household received the empty box and a certain amount of prepared cards (positive feedback cards, negative feedback cards, idea cards, and cards for further information) to fill out and put into the box. In order to keep test participants motivated expressing their opinion they received/will receive additional cards (addressing special factors of user experience in general and sociability in particular) through the post in defined intervals (once a month).

Based on a first analysis of user feedback, the results are twofold. On the one hand, the „Smiley Box“ earned a lot of sympathy from the test participants, facilitating further steps for the researcher to ease the inlet to the users' territory and to establish relationships. On the other hand, the feedback rate was very low (only 8 out of 30 household send back their cards). The results from the analysis revealed that most of the participants were disappointed about the low activities on the IPTV platform, which also reduced their initial motivation to use it, and was also mentioned as reason for not being able to provide more feedback on their experiences through the smiley cards. The feedback received through the smiley box, was reported back to the developer. One participant also attached a long letter to the box explaining the difficulties experienced with the platform, but also the challenges of such a IPTV platform for a local community by supporting social activities. In order to further explore the social issues surrounding the IPTV platform, several other methods are used, such as content analysis combined with personal interviews and home visits. Moreover, the combination of this qualitative feedback with the logging data delivers additional data on the usage behaviour and patterns on the platform.

As next step we further explore the applicability of the smiley approach, in particular the smiley cards, beyond the home context. We already used the smiley cards to gather in-situ user feedback during co-design workshops. This revealed additional positive feedback and strengthened the participant-researcher-communication/relationship. As the method itself is assumed to be accepted by users, but its value in sense of gathering lots of user feedback is still low (connected to low field activities around deployed social community IPTV service), we plan to develop a digital version of the Smiley Box with digital cards, which will be available as feature of the IPTV platform itself.

SUMMARY AND OUTLOOK

Developing social iTV applications addressing new aspects of interactive TV like social communication, preference lists or social recommendations, is a challenging task. The introduction of interactive TV services showed, that iTV service usage is different from what was expected during the introduction in the 1990s (Gawlinski, 2003). People tend to use new forms of services only for a small period of time (weeks to months) and then tend to ask for new services. iTV services in general are thus more entertainment oriented than information oriented. Interactive TV neglected also the social aspects of watching TV, like sitting in front of the TV together, but also recommending TV shows or series to other ones, or simply talking about an ongoing TV show (like calling your friends on the phone, to talk about the great football match currently shown on TV).

Evaluation methods of usability and user experience for social iTV must thus take into account the usage context of the applications, and must enable understanding the possible influencing factors for social TV applications. During the early stages of development so called in-situ methods can help to investigate requirements, limitations and pitfalls of social iTV services. Ethnographic methods, cultural probing, technology probes, creative or playful probing, social-science oriented methods like interviews, can be used to evaluate the usage context and to get new design insights for developing social TV services. Participatory and in particular co-design methods (e.g. co-design workshops) are also used in the early design and development phase to collaboratively explore new ideas for services and applications, create and discuss concepts between end users and towards the design team. During later stages of development, the evaluation of a prototype or system can help to understand usage problems (like using usability studies in the lab environment) and to evaluate the user experience (like using field studies including technology probes, in-situ methods like experience sampling).

Further research should focus on establishing low budget, easy to use and standardized methods for the evaluation of user experience, to enable a comparison of achieved (positive) user experiences. The name for the inhere presented IC-Technique (Instant Card Technique) was consciously selected to outline the „instant" usage of this technique. The specific design of the instant cards intends to accelerate the part of the scenario development process, where the user is involved. Users don't have to invest a lot of time to create new ideas and concepts in form of scenarios. The instant cards serve as a memo pad for the ideas of the workshop participants. Due to their small size, they only allow some short notes of the most important thoughts. The instant cards encourage participants to immediately write down their ideas when they emerge in the scenario development process. The instant cards don't provide masses of detailed information to the workshop' participants; rather they just offer participants small pieces of information as triggers for ideas. They also serve as ready-to-use cards for making selections in order to describe specific user behavior. Moreover, due to their minimalistic design, modifications of the trigger information on the instant cards according to the specific research questions are carried out easily. The IC-Technique will be improved, using this technique in planned co-design workshops within other projects.

In general it could be stated, that it is very important to ensure an early user involvement in the development of social interactive TV. User can provide valuable input in the early conceptual design phase of new social TV services by bringing in their daily experiences, expectations and thoughts. Thereby not only usability aspects, but in particular the users' experiences can be addressed for non-traditional environments, such as the home context. In later development stages it is necessary to extend the methodological set up toward a combination of qualitative and quantitative methods and techniques. Triangulation of methods and constant user involvement during field trials will lead to a fast user feedback to the developers, which can be used for constant and iterative improvements of social iTV services and applications.

REFERENCES

Arhippainen, L., &Tähti, M. (2003). Empirical Evaluation of User Experience in Two Adaptive Mobile Application Prototypes. *In Proceedings of the International Conference on Mobile and Ubiquitous Multimedia,* Norrköping.

Battarbee, K. (2004). *Co-experience: Understanding user experience in social interaction.* Doctoral dissertation. Series ILMARI A51, University of Art and Design Helsinki.

Beck, E., Obrist, M., Bernhaupt, R., & Tscheligi, M., (2008). Instant Card Technique: How and Why to apply in User-Centered Design. *In Proc. PDC2008*, Bloomington, Indiana, USA.

Bernhaupt, R., Obrist, M., & Tscheligi, M. (2006). Usability and Usage of iTV Services: Lessons learned in an Austrian Field Trial. *In Proc. of EuroiTV 2006*, Athens, Greece.

Bernhaupt, R., Mihalic, K., & Obrist, M. (2007). Methods for Usability Evaluation of Mobile Applications. In J. Lumsden (Ed.), *Handbook of Research on User Interface Design and Evaluation for Mobile Technology,* IGI Global. (pp. 745-758).

Bernhaupt, R., Weiss, A., Obrist, M., & Tscheligi, M. (2007). Playful Probing: Making Probing more

Fun. In *Proc. INTERACT 2007,* 10 -14 Sep 2007, Springer LNCS, (pp. 606-619).

Bernhaupt, R., Obrist, M., Weiss, A., Beck, E., & Tscheligi, M. (2008). Trends in the Living Room and Beyond: Results from Ethnographic Studies using Creative and Playful Probing. In *Computers in Entertainment,* 5(1) (March 2008).

Bernhaupt, R., Wilfinger, D., Weiss, A., & Tscheligi, M. (2008). An Ethnographic Study on Recommendations in the Living Room: Implications for Design of iTV Recommender Systems. In *Proc. of EuroiTV 2008*, (pp. 92-101).

Boertjes, E. (2007). ConnecTV: Share the Experience. *In EuroITV 2007 Adj. Proc.*

Buxton, B. (2007). *Sketching user experiences: getting the design right and the right design.* Morgan Kaufmann.

Chorianopoulos, K, & Spinellis, D. (2006). User interface evaluation of interactive TV: a media studies perspective. *In Univ Access Inf. Soc.,* (2006) 5, 209-218.

Coppens, T., Trappeniers, L., &Godon, M. (2004). AmigoTV: towards a social TV experience. *In Proc. EuroITV 2004.*

Crabtree, A., & Rodden, T. (2004). Domestic Routines and Design for the Home. *In CSCW: The Journal of Collaborative Computing, 13*(2), 191-220.

Dahlbäck, N., Jönsson, A., & Ahrenberg, L. (1993). Wizard of Oz studies: why and how. *In Proce. of the 1st international Conference on intelligent User interfaces.* New York, NY: ACM.

Dix, A., Finlay, J., Abowd, G. D., & Beale, R. (2004). *Human-Computer Interaction* (3rd Edition), Prentice Hall Inc.

Dumas, J., & Redish, J. (1993). *A Practical Guide to Usability Testing.* Norwood, NJ: Ablex Publishing Corporation.

Ehn, P. (1992). Scandinavian Design: On Participation and Skill. *In* P. S. Adler, & T. A. Winograd (Eds.), *Usability: Turning technologies into tools.* New York: Oxford University Press, (pp. 96-132).

Facer, K., & Williamson, B. (2004). *Designing technologies to support creativity and collaboration. A handbook from NESTA Futurelab* Bristol: NESTA Futurelab.

Fields, B., Amaldi, P., Wong, W., & Gill, S. (2007). Editorial: In-use, In-situ: Extending Field Research Methods, In: *International Journal of Human Computer Interaction, 22*(1), 1-6.

Forlizzi, J., & Battarbee, K. (2004). Understanding experience in interactive systems. *In Proc. of the 2004 conference on Designing Interactive Systems*, ACM, New York, USA (p. 261).

Gaver, W. H., Dunne, A., & Pacenti, E. (1999). Cultural Probes. *In Interactions, 6*(1), 21-29.

Halskov, K., & Dalsgård, P. (2006). Inspiration card workshops. In Proceedings of the 6th ACM Conference on Designing interactive Systems. *In Proc. DIS '06.* ACM Press, New York, NY, (pp. 2-11).

Harboe, G., Massey, N., Metcalf, C. J., Wheatley, D., & Romano, G. (2008). The Uses of Social Television. In *Computers in Entertainment, 6*(1) (Jan 2008).

Harboe, G., Metcalf, C., Bentley, F., Tullio, J., Massey, N., & Romano, G. (2008). Ambient Social TV: Drawing People into a Shared Experience. *In Proc.s CHI 2008,* (pp. 1-10).

Hassenzahl, M., & Tractinsky, N., (2006). User Experience - a research agenda. In *Behavior & Information Technology, 25*(2), 91-97.

Hassenzahl, M. (2003). The Thing and I: understanding the Relationship Between User and Product, In M. Blythe, C. Overbeeke, A. Monk & P. Wright (Eds.), *Funology: From Usability to*

Enjoyment (pp. 31-42). Kluwer Academic Publishers, Dordrecht.

Hassenzahl, M. (2004). The interplay of beauty, goodness and usability in interactive products. *In Human Computer Interaction*, 19, 319-349.

Hutchinson, H., Mackay, W., Westerlund, B., Bederson, B. B., Druin, A., Plaisant, C., Beaudouin-Lafon, M., Conversy, S., Evans, H., Hansen, H., Roussel, N., Eiderbäck, B., Lindquist, S., & Sundblad, Y. (2003). Technology Probes: Inspiring Design for and with Families. *In Proc. of the CHI 2003,* ACM Press (2003), (pp. 17-24).

Kankainen, A. (2003). UCPCD: user-centered product concept design. In *Proc. of the 2003 conference on Designing for User Experiences.* ACM Press, San Francisco, California, USA.

Karapanos, E., Hassenzahl, M., & Martens, J-B., (2008). User Experience Over Time. Work in Prgress, In *Proc. CHI 2008*, ACM, New York, USA, (pp. 3561-3566).

Kujala, S. (2003). User involvement: a review of the benefits and challenges. *In Behaviour and Information Technology, 22*(1), 1-17.

Lafrenière, D. (1996). CUTA: a simple, practical, low-cost approach to task analysis. *In Interactions,* 3(5) 35-39.

Larson, R., & Csikszentmihalyi, M., (1983). The experience sampling method. In H. T. Reiss (Ed.), *Naturalistic approaches to studying social interaction, New directions for methodology of social and behavioral sciences.* Jossey-Bass, San Francisco, (pp. 41-56).

Law, E., Thora, E., Hassenzahl, M., & Hvannberg, E. (Eds.) (2006). *UX'2006: 2nd International Workshop on User eXperience*, 14 October 2006, Olso, Norway, held in conjunction with Nordichi'2006, [online]. Available: http://cost294.org/upload/408.pdf (accessed: 30.05. 2008).

Law, E., Vermeeren, A., Hassenzahl, M., & Blythe, M., (Eds.) (2007). *Workshop "Towards a UX Manifesto" at the International Conference HCI 2007,* Lancaster, UK, September 3, 2007 [online]. Available: http://141.115.28.2/cost294/upload/506.pdf, (accessed: 30.05. 2008).

Lull, J. (1980). The social uses of television. In: *Human Communication Research*, 6:3, Spring, pp. 197-209

Meschtscherjakov, A., Obrist, M., Mirlacher, T., Lankes, M., Tscheligi, M., (2008). *Evaluating the Emotional Attachment to Mobile Devices & Services,* HCI & Usability Unit: Technical Report 2008.

Muller, M. J., & Kuhn, S. (1993). Participatory design. *Commun. ACM,* 36(6), 24-28.

Muller, M. J. (2001). Layered participatory analysis: new developments in the CARD technique. In *Proc. of the SIGCHI Conference on Human Factors in Computing Systems.*

Obrist, M., Bernhaupt, R., & Tscheligi, M. (2006). Interactive Television for the Home: An ethnographic study on users' requirements and experiences. *In Proc. of EuroiTV'06*, Athens, Greece, (pp. 349-358).

Obrist, M., Bernhaupt, R., Beck, E., & Tscheligi, M. (2007). Focusing on Elderly: an iTV Usability Evaluation Study with Eye-Tracking. In P. Cesar, K. Chorianopoulos & J. F. Jensen (Eds.), *Interactive TV: A shared Experience. 5th European Conference, EuroITV 2007* Amsterdam, The Netherlands, May 24-25, 2007, Proceedings, (pp. 66-75).

Obrist, M., Beck, E., Kepplinger, S., Bernhaupt, R., & Tscheligi, M. (2008). Local Communities: Back to Life (Live) through IPTV. In *Proc. of EuroiTV2008*, (pp. 148-157).

Obrist, M., Kepplinger, S., Beck, E., Tscheligi, M., & Muzak, P. (2008). Usability & User Experience: Preliminary Results from Evaluating an IPTV

Community Plattform. In *Proc. of EuroiTV2008*, (pp. 303-307).

Oehlberg, L., Ducheneaut, N., Thornton, J. D., Moore, R. J., & Nickell, E. (2006). Social TV: Designing for Distributed, Sociable Television Viewing. In *Proc. EuroITV 2006* (pp. 25–26).

Rosson, M. B., & Carroll, J. M., (2002). *Usability Engineering. Scenario-Based Development of Human-Computer Interaction.* San Francisco: Morgan Kaufmann Publishers.

Sanders, E. (2002). *Ethnography in NPD Research. How Applied Ethnography can improve your NPD Research Process.* Retrieved October 9, 2005, from http://www.pdma.org/ visions/apr02/ applied.html.

Spinuzzi, C. (2005). The methodology of participatory design. *In Technical Communication, 52*(2).

Stappers, P. J., & Sanders, E. B.-N. (2003). Generative tools for contextmapping: tuning the tools. In *Proc. of the Third International Conference on Design & Emotion*, 2003 in Loughborough.

Steen, M., Kuijt-Evers, L., & Klok, J., (2007). *Early user involvement in research and design projects—A review of methods and practices.*

Stone, D., Jarrett, C., Woodroffe, M., & Minocha, S. (2005). *User Interface Design and Evaluation.* Amsterdam et al.: Morgan Kaufmann Publishers.

Tudor, L. G., Muller, M. J., Dayton, T., & Root, R. W. (1993). A participatory design technique for high-level task analysis, critique, and redesign: The CARD method. In *Proc. HFES'93.*

Van Rompaey, V., Van Der Meerssche, B., Godon, M., Vanden Abeele, M., & Charliers, K., (2005a). Connecting the family home: Co-designing new technologies for Community Communication. In *Proc ECCR/ECA* Conference.

Van Rompaey, V., Hemmeryckx-Deleersnijder, B., Van Der Meerssche, B., De Mondt, H., & Godon, M. (2005b). Beyond Marketing. Applying qualitative user experience research techniques on social media applications. In *Journal of The Communications Network, 4*(3), 26-30.

Westerlund, B., Lindquist, S., Mackay, W., & Sundblad, Y., (2003). Co-designing methods for designing with and for families, In *Proc. for 5th European Academy of Design Conference,* Barcelona.

KEY TERMS

Usability Evaluation Methods: A set of methods used to evaluate a system, mock-up, or prototype in terms of usability.

User Experience: All aspects of how people use an interactive product, the way it feels like in their hands, how well they understand how it works, how they feel about it while they're using it, how well it serves their purposes, and how well it fits into the entire context in which they are using it

Non-Traditional Environments: Describes environments that are distinct from a work setting, like the home, on the move, in public places.

Co-Design: Means actively integrating end users in the development process of a future product by putting them in the role of designer and stakeholder.

Chapter IX
Inducing User Cooperation in Peer-to-Peer Television:
Deriving Mechanisms from Psychological Theories

Jenneke Fokker
Delft University of Technology, The Netherlands

Huib de Ridder
Delft University of Technology, The Netherlands

Piet Westendorp
Delft University of Technology, The Netherlands
Eindhoven University of Technology, The Netherlands

Johan Pouwelse
Delft University of Technology, The Netherlands

ABSTRACT

Television and the Internet have proven to be a popular combination for both broadcasters and viewers. Because of this popularity they are increasingly facing the consequences of central bottlenecks, which could be overcome by taking a different approach: Peer-to-Peer (P2P) technology. However, P2P systems can only be successful with as much cooperation among as many users as possible. This chapter explains how this cooperation is hard to enforce, and how inducing it might be more successful. Relevant psychological theories are listed that can be used to induce this user cooperation, along with possible applications of cooperation inducing mechanisms for Peer-to-Peer Television (P2P-TV) systems. The authors aim to provide practical criteria along which these mechanisms can be evaluated on their contribution to social activity in P2P-TV systems.

INTRODUCTION

The combination of Television and Internet is expanding traditional broadcasts with flexible, nearly unlimited access to content and narrowcasting of niche content. The movement beyond the channel model frees viewers from fixed schedules and gets them faster and more dynamic access to more content, including large quantities of user generated content. Probably the best example at this moment is YouTube.com, which initiated large-scale production and distribution of user generated content. YouTube.com and similar video distribution systems, however, make use of traditional client-server architectures, requiring the maintenance of large and expensive servers. The main drawback is that they are unsuitable for large amounts of viewers trying to access the same video content at the same time, resulting in reduced quality-of-service (QoS) and availability of content, plus delays and increased download times. An interesting alternative for broadcasters to distribute up to HDTV quality content to millions of viewers in a short period of time is Peer-to-Peer (P2P) technology. An explanation of P2P technology can be found in Answer.com (2008). Instead of a large central server sending the complete file to all interested clients, a P2P system allows broadcasters to reach the same clients by just sending parts of one copy to a limited number of clients and having them distribute the file by exchanging these parts with other interested clients, or 'peers' in terminology of P2P technology, until each peer has the complete file. This illustrates one aspect of P2P technology, namely the injection of content. See section User Cooperation in P2P-TV Systems for more details.

A recent white paper by Cisco (2008) forecasts that during the coming three years P2P traffic will grow at about 30 percent per year with the increasing use of P2P for standard-definition video file exchange, and advent of high-definition video files and television content exchange. Figure 1 shows Cisco's forecasts about consumer Internet traffic up to 2011 (Cisco, 2008) in which P2P clearly dominates and Internet video to PC and TV is rising.

In recent years, the number of initiatives focusing on combining P2P with TV has grown rapidly. Examples are Joost.com, Babelgum.com, Livestation.com, Sopcast.com, PPLive.com,

Figure 1. Global consumer Internet traffic 2007-2011 forecast (source: Cisco, 2008)

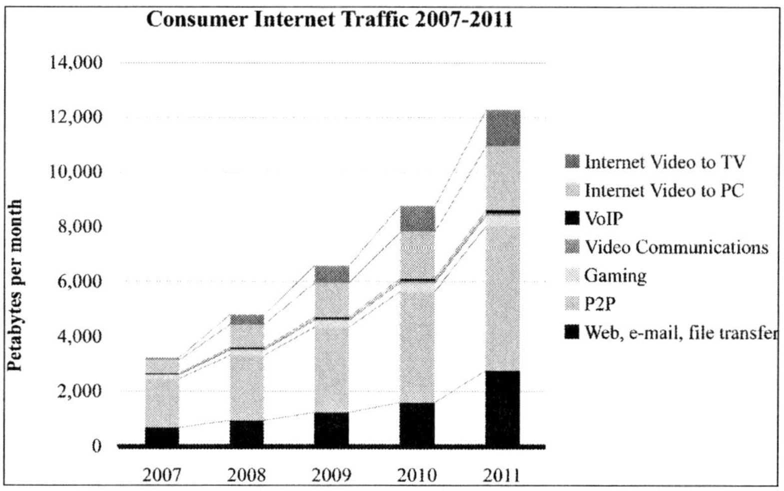

and Tribler.org (Wang et al., 2008). The latter is a system especially developed to investigate means to distribute and share large multimedia data. We will further elaborate on Tribler in the section Tribler and User Cooperation but first discuss the advantages of P2P technology and its challenges.

Both users and broadcasters benefit from P2P technology for a number of reasons. First of all, P2P technology yields low cost of ownership of content, because it aggregates distributed resources through smart interoperability. One-click uploading and near simultaneous distribution to thousands of viewers of user-generated content become possible with hardly any costs involved. Secondly, P2P technology is a promising approach in that television broadcasters do not need expensive and high maintenance central servers. Existing network connections plus infrastructure can be used and any maintenance effort is distributed over the users. It can be used for distributing large files among large amounts of viewers simultaneously.

The above-mentioned benefits of P2P technology only hold if users are prepared to contribute computer resources, maintenance, and time as much as possible. However, for online systems it has been found that only 1% of the users is responsible for almost all contributions, about 9% contributes every once in a while, and 90% only lurks and contributes nothing (Adar & Huberman, 2000; De Valck, 2005; Nielsen, 2006). So, the vast majority wants a high profit at no costs. The same phenomenon is recognized by Beenen et al. (2004) in their work about motivating contributions to online communities, such as MovieLens.org. In P2P networks there is an increased chance of social loafing, because people think it is less likely that they will be evaluated on an individual basis (Goren et al., 2003; Harkins & Szymanski, 1989). P2P technology must deal with this kind of behavior because it lacks a central authority to monitor the network, safeguard the technical and artistic quality of content, and metadata describing the files. As a consequence, the system can be misused deliberately or unintentionally. P2P networks consist of diverse and anonymous users, and most of them usually do not feel responsible to contribute positively to the community, unless a technically enforced tit-for-tat (t4t) mechanism is used as in BitTorrent (Pouwelse et al., 2005).

A technical approach would be regulation or enforcement of more cooperation. However, this is hard to realize, because it would require complex distributed algorithms, in which decisions have to be made by each peer, usually having little global knowledge (Milojicic et al., 2002). An alternative is to look for means to *induce* massive, positive, and voluntary cooperation of users. For this, we need to know what could possibly motivate users to behave altruistically. To build knowledge on human motivations that can help solving this cooperation challenge (social) psychological literature was reviewed using as the main keywords *altruism*, *cooperation*, and *prosocial behavior*. This sometimes touched on theories from evolutionary biology, but other theoretical fields (such as economy, anthropology, game theory, and decision making) have been left out even though we realize that interesting information can be found there too.

This chapter is organized as follows: The section Tribler and User Cooperation elaborates on Tribler and what makes this system social. Thus explaining the forms of user cooperation essential for the success of any fully distributed P2P-TV system. The next section (Relevant Psychological Backgrounds) then discusses theories yielding relevant psychological backgrounds. This is followed by the section Overview of Psychological Backgrounds and Suggested Mechanisms which presents a translation of these psychological backgrounds into possible mechanisms that could induce massive and voluntary cooperation in a P2P-TV system. This section also discusses the mechanisms that have already been applied to Tribler. The final section presents conclusions and suggestions for future work.

TRIBLER AND USER COOPERATION

Tribler is a platform for streaming (live and on demand) and downloading of video content. Figure 2 shows the main screen of Tribler, version 4.1.9. On the left side available files for downloading are shown with thumbnails. Users can double click on a file to download or view it. Tribler's technology depends on the users to inject, moderate, and distribute content. The underlying social structure of friends and taste neighbors facilitates a unique interactivity by features such as recommendations from taste-related users, finding and making friends, speeding up downloads by requesting friends to give idle bandwidth. This social interactivity is crucial as we believe it is the basis for inducing most forms of cooperation that are essential for the success of P2P-TV systems.

Recently, Chorianopoulos & Lekakos (2008) have provided an interesting overview of research findings on sociability in interactive TV. They come to the following definition: Interactive television is a user experience involving at least one user and one or more audiovisual and networked devices. Moreover, there are two social aspects of TV to take into consideration when designing social TV: (1) Presence of the viewers. This may be co-located or at a distance. (2) Temporal aspect of communication. Communication may be synchronous or asynchronous. In the current Tribler software, there is predominantly distance viewing and asynchronous communication. However, in Tribler communication is not as conclusive as described by Chorianopoulos & Lekakos yet; it does not facilitate an open audio channel, instant messaging, or emoticons. Nevertheless, we deem the social interaction, which is essential to P2P-TV systems as described above, to be a form of

Figure 2. Discovered files view in Tribler version 4.1.9 (source: http://www.tribler.org/, May 2008)

communication as well. In future, more advanced modalities of communication and a shift to more synchronous viewing (more streaming instead of downloading) are likely.

User Cooperation in P2P-TV systems

Within Tribler our ideas about how human co-operation can be induced are operationalized and tested in future versions. Within this project the following aspects of cooperation have been identified to be essential for the success of P2P-TV systems:

- **Peer uptime**. When peers stay online for a long period of time, the availability and distribution of content is improved. As soon as users turn off their computers, the video content that is stored on their hard disks would no longer be available to other users.
- **Donate bandwidth**. Users donate their idle bandwidth to others, helping them to increase their bandwidth for speeding up downloads. In our Tribler software, users can request the help of friends to speed up downloads.
- **Injections.** Users add new and unique movies to the system. Content may range from professional videos injected by broadcasters to user generated content.
- **Moderations**. In the absence of a central authority, users check, insert, and update content metadata in order to keep the content and metadata clean, correct, and free from intentional pollution and user mistakes. To this end, the following types of moderation are essential in P2P-TV systems:
    - *Moderation of metadata*: Adding the correct metadata when new content is injected and checking the metadata others have added for mistakes or missing information. Searching and browsing will be more efficient if all information is correct and complete.
    - *Tagging*: Freely choosing annotations for the content (or parts of it). Searching and browsing will become much richer. Fokker et al. (2006; 2007c) discussed the value of tagging video content in Wikipedia.
    - *User-to-user moderation*: Keeping an eye on each other's actions and content. Both showing appreciation and punishing someone for his actions are part of this. EBay.com relies heavily on this type of moderation.
- **Social reach.** Users create and maintain a network of trusted friends. This is based on the assumption that friends are willing to help each other without getting something in return immediately.

These aspects are not easily enforced technically. However, *inducement* of cooperation might be more successful. We believe P2P-TV users can be induced to cooperate voluntarily in an online environment where there are no real face-to-face encounters, nor a central authority to enforce this cooperation, because people *do* help others: next to selfish behavior, altruistic motives abound. The term altruism (a.k.a. prosocial behavior in social psychology) refers to all actions that provide benefit to others but no obvious benefits for the person who helps (Baron et al., 1998). Altruism not only occurs in real life (e.g. donating blood), but also in virtual environments. Some good examples are Wikipedia.org, Flickr.com or World of Warcraft (www.worldofwarcraft.com). Wikipedia is a web-based, free-content encyclopedia, written and moderated collaboratively by volunteers. Flickr is a photo sharing website. Users contribute not only by putting their pictures on the website, but also by annotating and commenting on them. World of Warcraft is a so-called massively multiplayer online game, connecting very large groups of players through the Internet in a persistent game world. In this game collaborations not only take place between friends and other real-life contacts, but also very often spontaneously between strangers (Nardi & Harris, 2006).

Fokker et al. (2006) have made an inventory of mechanisms to induce massive and voluntary cooperation already used in state-of-the-art websites, such as Amazon, Wikipedia, and Flickr. In the next section some of these mechanisms and websites are used as examples of how relevant psychological backgrounds could be applied to P2P-TV systems.

RELEVANT PSYCHOLOGICAL BACKGROUNDS

People often behave seemingly less selfish in a situation that affects others than they would when it only affects themselves. What people decide in these situations sometimes seems hard to explain from a selfish point of view. This section explains the underlying rationale of what seems irrational at first sight. First, theories that can possibly explain altruism are discussed, followed by how they may be applied to P2P-TV systems. We present 13 theories and discuss their usefulness in the context of P2P-TV by suggesting applications to the user interface. The theories are mostly taken from (social) psychology and evolutionary biology. We do not claim to be complete, but believe that this list gives a good overview of human motivations that could underlie altruism. The theories are listed in order of their decreasing frequency of existing operationalizations in state-of-the-art websites (Fokker et al., 2007b).

Costly Signaling Theory

People provide large benefits to others even when they know this help will not be reciprocated, because this costly, altruistic act may benefit the altruist indirectly by gaining status (Grafen, 1990; McAndrew, 2002). McAndrew states: "If one can afford to expend a great deal of money, energy, or time in a manner that seems to be irrelevant to one's selfish interests, then the resources that one has in reserve must be very great indeed." The costly signaling theory is also called competitive altruism (Hardy & Van Vugt, 2006); altruism, if publicly displayed, increases the reputation and status of the altruists, which makes them more attractive interaction partners. Mauss (2000) found strong examples of competitive altruism amongst tribes along the North Pacific coast of North America who, in a ceremony called Potlatch, give away valuable goods to display wealth and status. The other tribe is expected to reciprocate with interest and not being able to reciprocate means a loss of rank or status. According to Rehn (2001) this social event is a ritual which – in the long run – brings about social cohesion and economic benefits. The great expenses made start to make sense if the effects over a longer period of time are taken into consideration.

Applications: It is important to realize that this theory will only work in P2P-TV if people's *behavior and online reputation are visible to others*, as this boosts their self-esteem, status and reputation. Costly signaling can be found in almost all investigated websites and applications, among which SETI@Home. This website generates users' online reputations (consisting of information such as credits and ranking) by registering the donation of resources (e.g. CPU cycles). In message boards and forums, some users make an image of their reputation and use it as an avatar or signature (see Figure 3).

At first sight, costs may seem low in P2P-TV systems, but users make much personal information public, thus making themselves vulnerable to malicious others. It also makes them more trustworthy and interesting to possible interaction partners. It is important to *allow users to stand out from the crowd* by displaying personal information, among which reputation is very important.

Figure 3. A public user profile, often used to show off within message boards in SETI@Home (May 2008)

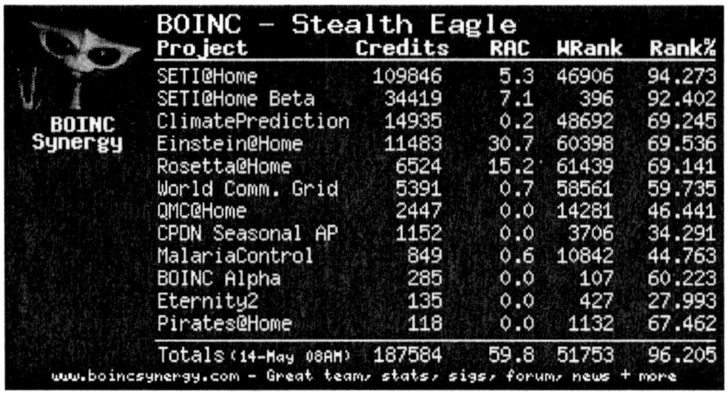

BOINC - Stealth Eagle

BOINC Synergy

Project	Credits	RAC	WRank	Rank%
SETI@Home	109846	5.3	46906	94.273
SETI@Home Beta	34419	7.1	396	92.402
ClimatePrediction	14935	0.2	48692	69.245
Einstein@Home	11483	30.7	60398	69.536
Rosetta@Home	6524	15.2	61439	69.141
World Comm. Grid	5391	0.7	58561	59.735
QMC@Home	2447	0.0	14281	46.441
CPDN Seasonal AP	1152	0.0	3706	34.291
MalariaControl	849	0.6	10842	44.763
BOINC Alpha	285	0.0	107	60.223
Eternity2	135	0.0	427	27.993
Pirates@Home	118	0.0	1132	67.462
Totals (14-May 08AM)	187584	59.8	51753	96.205

www.boincsynergy.com - Great team, stats, sigs, forum, news + more

Social Impact Theory

Latané (1981) explains with his social impact theory how a group influences the decisions that individual members make. Baron et al. (1998) conclude that the likelihood a person will respond to social influence depends on:

1. Strength: the power and status of the group,
2. Immediacy: the closeness of group members in space and time, and
3. Size: as group size increases, 'diffusion of responsibility' occurs more often: each member feels less and less responsible to contribute and puts less effort in it.

Applications: The strength and impact of group size is relevant for P2P systems where people hardly ever feel responsible to contribute. This holds in particular for large online communities in which people usually do not know each other. To make sure that users take responsibility, it should be possible to start and join *small enough groups that are powerful and exclusive*. This way they will value belonging to that group much more and contribute more voluntarily. The *need to belong* is fundamental to a large part of the investigated websites and applications. For instance, MySpace, YouTube, and Flickr all evolve around communities. Figure 4 shows the 'Guitar' group on YouTube. It is the largest group with close to 20,000 members, but it is not nearly the most productive

Figure 4. 'Guitar', the largest but not nearly the most productive group on YouTube.com (May 2008)

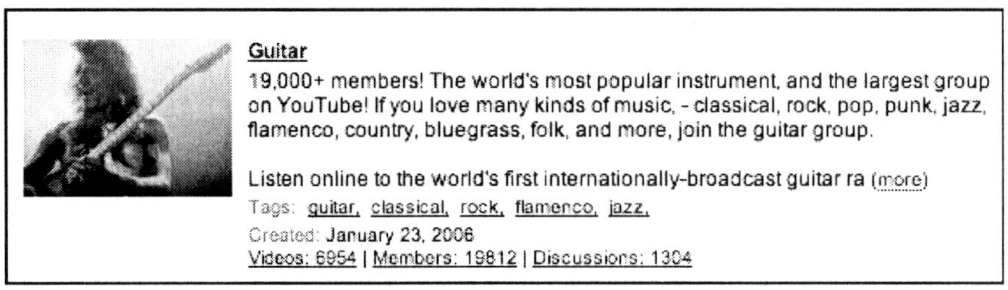

one. It only contains close to 7,000 videos. Groups with higher productions feature close to 28,000 videos and may be significantly smaller.

As for immediacy, Bradner & Mark (2002) have investigated the influence of geographic distance on cooperation, persuasion and deception in a video conferencing community. They found that the ability to persuade another and the willingness to initially cooperate decrease with distance while deception of another person increases with distance. This suggests that *communicating information about the geographic distance between two closely located users* of P2P-TV may positively influence the cooperation between them.

Kin Selection Theory

This theory has its origin in Hamilton's theory of inclusive fitness (Hamilton, 1964). This states that – given a cost/benefit ratio - altruism is more likely with higher genetical relatedness. If two individuals are sufficiently closely related, altruism can benefit reproduction of a set of genes, despite losses to the individual altruist. The benefit of altruism decreases rapidly with declining relatedness, as is exemplified by Hamilton's rule (Brembs, 2001; Hamilton, 1964):

Equation 1

$$C/B < R$$

- C = the cost or effort it takes to help for the altruist
- B = the benefit for the recipient.
- R = relatedness (the chance that two individuals carry the same gene)

Altruistic acts are placed in the context of the survival of genes. The chance that offspring carries the same gene as the parent is ½. The full-sib relationship is just as close. Grandparent grandchild relationship stands at ¼, and so on. So, for anyone to perform an altruistic act to benefit a sib, the benefit must average at least twice the loss to the individual. The benefit must be at least four times the loss for a half-sib and so on (Hamilton, 1964).

However, in online communities genetical relatedness is scarce. The principle could be maintained if the degrees of separation approach is adopted, originating from Milgram's small world problem (1967). He found that any two people in the world are separated within at most six degrees, in which the first degree are the direct contacts, the second degree are the contacts-of-contacts, and every following degree is one step further away. Converting relatedness to DoS would lead to the following equation:

Equation 2

$$C/B < 2^{-DoS}$$

- C = the cost or effort it takes to help a user
- B = the benefit for the recipient.
- DoS = degrees of separation

Applications: In analogy with Milgram (1967) relations can be defined by their *social distance*, as is done by many social networks, for instance in LinkedIn (Figure 5). The *degrees of separation* (1st degree = friend, 2nd = friend of a friend, and so on) could be shown under the assumption that if two individuals are sufficiently close to each other in a network of trusted relationships, altruistic behavior is likely to occur.

Social Exchange Theory

When people give something, they expect something in return. The social exchange theory states that the exchangeable goods need not be of the same type. Esteem is regarded as an exchangeable good as well (Homans, 1958). The goods exchanged usually differ between 'exchange' or

Figure 5. Network statistics for a user of LinkedIn.com, showing for each degree of separation the connections that could be made through trusted professionals (May 2008).

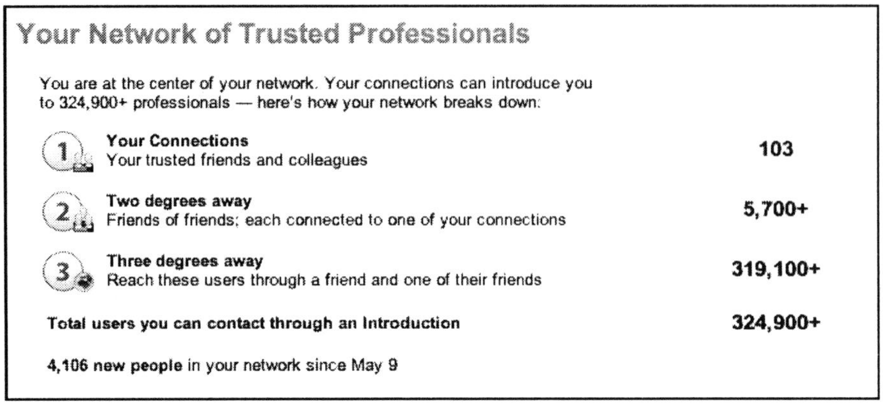

'mutual' relationships and 'communal' or 'reciprocal' relationships (Clark & Mills, 1993). Mutual relationships can be found between people who do not know each other and have never interacted before. Their first interactions are typically based on mutuality and mistrust, accepting no delay in a return on something given. Reciprocal relationships on the other hand, can be found among long-term relationships, friends, and kin. They are based on trust and delay in reciprocity is acceptable. Exactly this acceptability of delayed reciprocity could be exploited in P2P systems to induce cooperation that is flexible over time and resource type. The current protocol used in P2P file exchange only enforces instantaneous bandwidth exchange in the context of one file download. To better exploit available resources within a P2P network, it is desirable to have reciprocal exchange of more resources than bandwidth only and include a long-term memory of what was exchanged with whom, allowing a delayed exchange.

Fokker et al. (2008) hypothesized that trust in reciprocity will decrease with increasing DoS. For lower DoS people will be lenient to the moment when and what is returned. In contrast, expectancy for immediate and equal exchanges will be found among higher DoS. Fokker et al performed an experiment to determine the degree of separation that forms the boundary between mutual and reciprocal relationships. To this end, 33 participants were asked to indicate in a questionnaire from which relationships they 1) are certain of reciprocity, 2) are uncertain of reciprocity, and 3) do not expect reciprocity at all. This was in a setting of an online book exchange community. The results can be found in Figure 6, showing that for friends and friends of friends (DoS=1 and 2) participants are predominantly certain, i.e. they trust, that reciprocity will occur. From a friend of a friend's friend and further away (DoS=3 to N) participants are not certain or don't expect reciprocity at all. The results suggest that trust in reciprocity decreases with the degrees of separation. Furthermore, for an online social network, reciprocal relationships are to be found among first and second DoS while mutual relationships are to be found among the third DoS and further away. Note that the decrease of trust with the increase of Dos is in line with Hamilton's rule.

Applications: In P2P-TV one could implement immediate reciprocity of resources (ideal for mutual relationships) as well as delayed reciprocity (ideal for communal relationships). In other words, *allow for different delays of reciprocity for different relationship types*. To the best of our knowledge, this has never been built in any ap-

Figure 6. Degree of expected reciprocity of DoS=1 to DoS=N. Data are based on 33 participants.

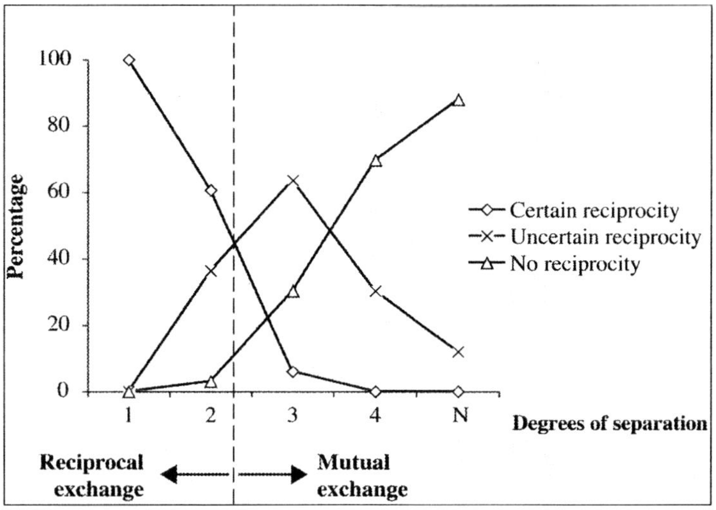

Figure 7. Member details on eBay.com, showing the overall positive, neutral, and negative feedbacks, and detailed seller ratings (May, 2008)

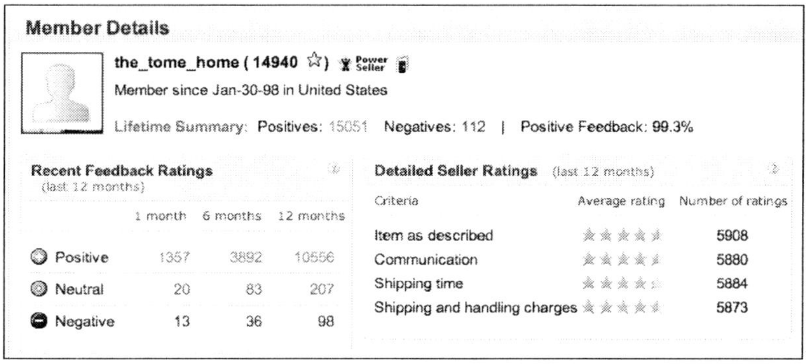

plication or website. An interesting consequence is that users must be able to *set the relationship type,* similar to for instance Orkut, which uses the types 'best friends', 'good friends', 'friends', 'acquaintances', and 'haven't met'.

Another way of applying the social exchange theory to P2P-TV is to provide the means to *express approval or disapproval* in return for received help. This is very similar to eBay's user-to-user moderation involving a transaction between a buyer and a seller, see Figure 7. This user-to-user moderation can also be the basis for forming *virtual friendships*, e.g. users who have repeatedly exchanged resources successfully and positively can decide to become friends.

Social Learning Theory

Most human behavior is learned observationally through modeling behaviors, attitudes, and emotional reactions of others (Bandura, 1971). People observe others, form an idea of how new behaviors are performed and reacted upon, and use this information as guidance in later situations.

Figure 8. Similarity in taste for music between two users of Last.fm (May 2008)

People will adopt modeled behavior quicker if the model is similar to the observer or enjoys a high status and admiration from its peers.

Applications: Translating the similarity principle into P2P-TV implies that *the similarity between users has to be shown,* for instance in their taste for TV-shows. This is comparable to what Last.fm does with taste similarity for music (Figure 8) or Amazon.com's 'Customers who bought this item also bought' for products such as books.

If we want users to set a good example to others, then their altruistic behavior should be *as visible as possible* to others. The more visible someone's behavior is, the more likely someone else will follow example. Yet another application is to create a *role model* with a higher authority or reputation, who can set an example to newbies and explain the social norms. This has proven to work well in systems that use *reputation (or user membership) levels,* as long as the levels are distinctive enough and not too many. Cheng and Vassileva (2005) address the challenge of determining the right amount of membership levels and how they can be reached by users of an online community.

Social Balance Theory

People seek balanced states in their relationships, called interpersonal consistency (Heider, 1958), in the context of dyads as well as in a group of peers. For instance, liking the person/group you are associated with, or disliking a person/group you are not associated with, are both states of balance according to this theory. Imbalance will make people act, either by changing their sentiment about or their association with that person or group.

Applications: Users will feel the *need to belong* to groups in P2P-TV. This is quite basic to social software, like Skype, Orkut, or MySpace. This need can be exploited if the *groups are formed around important aspects of cooperation* in P2P-TV, for instance the 'injection' group. People will be allowed to join this injection group after they have injected fresh content of good quality (both technically and artistically). This is similar to the applications of the social identity theory (see section Social Identity Theory).

Group Selection Theory

The group selection theory states that the vehicles of selection are not just individuals but can also be groups (Wilson & Sober, 1994). The theory suggests to account for behaviors observed in both human and animal societies that appear to benefit the group, even at the cost of individuals.

Applications: It may seem a long way to translate these principles into mechanisms for P2P systems, but it basically implies that we should enable groups to be formed within an application, and that *users should be informed of the group's power and reputation.* Power and reputation are good indicators of a group's health and chances of survival over a longer period of time.

Social Identity Theory

Group membership contributes to the establishment of a positive social identity for the group member (Tajfel & Turner, 1986). A prestigious or restrictive group will reinforce the self-concept of an individual who is admitted to it. People usually try to maintain positive views of the groups they belong to, are positively biased in their judgments of other individuals in those groups, and negatively biased to those outside.

The value derived from joining a group depends on the effort it took to become a member. A group that accepts anyone without restrictions won't boost the member's self esteem. There are hacker groups who ask new users to hack their way into the community, thus proving their skills. The experienced value of belonging to this hacker group is higher than when it would be open to anyone.

Applications: As stated before in the discussion on social impact theory, people have a *need to belong.* If someone belongs to a group within P2P-TV, it is to be expected he is prepared to contribute to that group. To enhance this kind of behavior it is recommended that groups should have *rules for admitting* people, like hacker communities in which newbies first have to prove their skills before getting admitted. Another example is Orkut. People can only join if they were invited. A feeling of *exclusiveness* is created in this way.

It is also important that it is *visible* to which groups a user belongs, because it boosts his self-esteem if he knows others can see he belongs to an exclusive group. The open architecture of P2P networks however, makes it difficult to create and maintain exclusive groups. Therefore, it is important to investigate how much one should invest in such a mechanism.

Theory of Reciprocal Altruism

The theory of reciprocal altruism states that cooperation is conditional: A cooperator will keep helping an unrelated individual, unless the latter refuses to reciprocate (Trivers, 1971). In other words, even if two individuals are not genetically related, they will still help as long as they trust help will be reciprocated. This implies that the interactions are repeated over time.

Applications: If we want two users to help each other repeatedly, it must be possible to *keep track of their given / received help ratio* and *make that ratio visible.* Another consequence is that *users must be able to rebuke each other* for not reciprocating help. Of course, the opposite is just as important: *show approval for reciprocating.* As we discussed before, this is an effective mechanism used in eBay.

Equity Theory

This theory emphasizes fairness in one-on-one relations. People expect a fair share of available resources appropriate to their status (Greenberg & Scott, 1996). The equity theory focuses on whether the distribution of resources is perceived as fair to both individuals. If not, they may experience cognitive dissonance and feel the need to relieve this stress. They will adjust the costs and benefits of their relationships to balance the instable situation. Walster et al. (1978) pointed out that it is exactly the ratio of costs and benefits that determine the perceived fairness, and not just the fair distribution of benefits or costs only.

Applications: For P2P-TV we propose to use these principles by making transactions transparent: *Show how much others contribute in respect to their available resources.* This is in line with BitTorrent sharing ratio, which is based on the tit-for-tat principle: people must give as much bandwidth as they have received. Any benefits of cooperation must be distributed between two individuals according to how much they had given in the first place. Of course this need not only concern bandwidth, but any other exchangeable resource. For applying mixed exchange (e.g. injection of a movie in return for bandwidth), knowledge on what people will accept as a fair exchange is required. Another possibility could be allowing *negotiation* between individuals.

Theory of Group Cohesiveness

This theory explains how small, face-to-face informal social groups exert pressure on their members to uphold social norms. Cohesiveness is defined as the resultant of all the forces acting on members of a group to remain in the group (Festinger et al., 1950). The determinants of group cohesiveness are:

1. Attractiveness of the group to its members,
2. The mutual attraction of those members to each other as individuals, and
3. The mediation of individual goals by the group.

Applications: Even though this theory was formulated for situations of face-to-face encounters, we can apply it to encounters in P2P-TV as well. The more users identify with a group and its members, the more they will feel obliged to uphold the social norms of that community. This means we should *make information about a group and its members* (e.g. the similarity in taste) *and their shared social norms as visible as possible*. Moreover, individuals should have the *means to interact*.

Cognitive Evaluation Theory

Faced with a task, people evaluate it in terms of how well it meets their needs for feeling competent and in control. If they think they will be able to complete the task, they are 'intrinsically motivated' and don't need further external motivation. The cognitive evaluation theory was formulated to explain effects of rewards, feedback, and other external events on intrinsic motivation (Ryan & Deci, 2000). In contrast to operant conditioning (Skinner, 1953) - where desired behavior can be reached by rewards and punishments - the cognitive evaluation theory states that extrinsic motivations can have a counter-productive effect on intrinsic motivation, because they bring about an externally perceived locus of causality. Too many rewards or punishments will have a counter effect if participants perceive them to be controlling, rather than supportive (Frey & Jegen, 2001; Ripeanu et al., 2006).

Applications: The implication for P2P-TV is that *a balance between facilitating and enforcing cooperation has to be found*. However, it seems difficult to advice on applications of this principle. Technical enforcement of cooperation is difficult to achieve anyway, because the network is completely decentralized as was explained before. The minimum requirement should be that P2P-TV systems become platforms in which it is *easy for users to contribute to*, thus satisfying their intrinsic motivation.

Self-Affirmation Theory

This theory discusses how the overall goal of people is to protect their image of self-integrity, of their moral and adaptive adequacy (Steele, 1988). When this image is threatened, people will respond in such a way that is least harmful to their self-worth. There are two ways to respond to such threats: (1) through defensive responses that directly reduce the threat, and (2) through the affirmation of alternative resources of self-integrity. For example, if a user doesn't have the content to be an injector, he can be good at something else, such as performing moderations.

Applications: A P2P system should *provide means for users to specialize*. For P2P-TV that could be on one of the five before-mentioned aspects of cooperation, for instance allow them to become 'top moderators'.

OVERVIEW OF PSYCHOLOGICAL BACKGROUNDS AND SUGGESTED MECHANISMS

The cooperation inducing mechanisms described in the previous section were inspired on state-of-the-art websites and complemented with applicable mechanisms from the discussed theories. Figure 9 presents an overview of the relation between the theories and underlying mechanisms, based on Fokker et al. (2007a). The mapping of the theories to possible mechanisms resulted into four clusters of theories and five clusters of mechanisms. This procedure yields a guideline for translating the theories into new possibilities for cooperation inducing mechanisms. The clustering is best explained with two examples: (1) The mechanisms 'social norms' and 'reputation' are clustered closely because they are applicable based on many corresponding theories such as social learning theory, equity theory, and social exchange theory. (2) The theories of group cohesiveness and social identity are clustered because they share very specific applicable mechanisms, being 'belonging' and 'community'.

Table 1 is a simplified representation of Figure 9 demonstrating the strength of the relation between the four clusters of theories (columns) and the four clusters of cooperation inducing mechanisms (rows). We labeled the four groups of theories as follows: (1) *group cohesion*,(2) *self-affirmation*, (3) *exchange*, and 4) *selection*. Moreover, we identified five groups of mechanisms of which four can be labeled: (1) *fairness*, (2) *esteem*, (3) *community*, and (4) *degrees of separation*. The fifth cluster has no strong relation with any of the four clusters of theories. We propose that any new cooperation inducing mechanism in P2P-TV systems should be evaluated along the four criteria of fairness, esteem, community, and DoS. In this way we believe that the contribution of that mechanism to social interactivity can be understood and fully exploited.

Cooperation Inducing Mechanisms and Tribler

To investigate the applicability of the four criteria for social interactivity we evaluated them along Tribler's current and future cooperation inducing mechanisms presented in Fokker et al (2007a):

1. **Make behavior publicly visible.** This refers to *fairness*, *esteem*, and *community*. In Tribler information is made as transparent as possible. However, the visibility of behavior for other users is yet limited to the download history ('common files' and 'also downloaded'), availability ('last seen'), and similarity ('similar taste').
2. **Show the similarity between users.** This refers to *fairness* and *community*. Two similarity features have been applied to Tribler (see Figure 10): (1) A heart representing the

Table 1. Overview and labeling of clusters derived from Figure 9. The number in a given cell denotes the proportion of positive links for that combination of clusters. The number between brackets before each label indicates the number of theories or mechanisms in that cluster.

I (3) 'group cohesion'	II (3) 'self affirmation'	III (4) 'exchange'	IV (3) 'selection'	
.00	.05	**.68**	.24	**A (7) 'fairness'**
.67	**1.00**	.50	.00	**B (1) 'esteem'**
.83	.00	.00	.17	**C (2) 'community'**
.08	.08	.00	**.75**	**D (4) 'DoS'**
.14	.14	.07	.02	**E (17) remainder**

Figure 9. Relation between theories and cooperation inducing mechanisms, based on a two-way hierarchical clustering (performed with JMP IN, release 5.1.2). A colored element denotes a positive link between a theory and a cooperation inducing mechanism. The dashed lines indicate the separation of clusters.

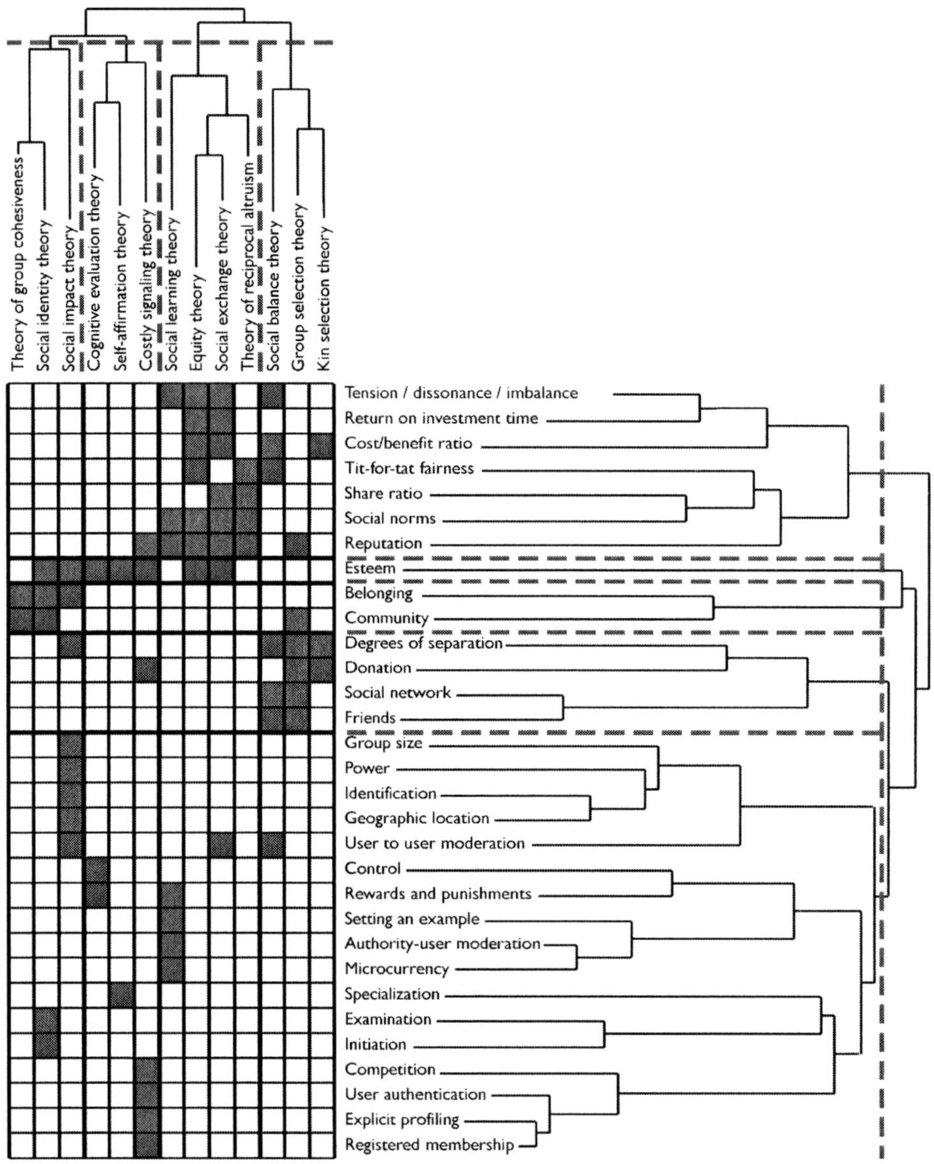

similarity in taste. Tribler calculates a top 20 of users who are most similar to the user; (2) The files a user has in common with another user. Both features are based on the same collaborative filtering algorithm in which users' download histories are compared in terms of overlap (Wang et al., 2008).

3. **Exploit the need to belong as well as to stand out.** This refers to *community* and *esteem*. The current version of Tribler does

not facilitate belonging to groups. As for standing out: users can reach a listing in the top 10 of users by uploading content.

4. **Show contributions in respect to a user's total, available resources.** This refers to *fairness*. This is not applied to Tribler yet, but it could help create a better understanding between users about fairness of the exchange. For instance, when user A has little idle bandwidth available because he is downloading a movie himself, user B will understand and accept that user A can't help him much when he would ask for donation of bandwidth.
5. **Make user-to-user moderation possible.** This refers to *community*, *fairness*, and *DoS*. User-to-user moderation has not been built yet. It would be a good mechanism for building user reputations.
6. **Make use of the degrees of separation.** This refers to *DoS* and *community*. Social distance has not been exploited yet, but could be a valuable mechanism in combination with delayed reciprocity particularly in communal relationships.
7. **Use different return-on-investment times for different relationship types.** This refers to *fairness* and *esteem*. This is a mechanism that is being investigated, because it is believed to be a perfect mechanism to induce cooperation that is flexible over time and resources.

Figure 10. Similarity features in Tribler (detail of main screen, May 2008)

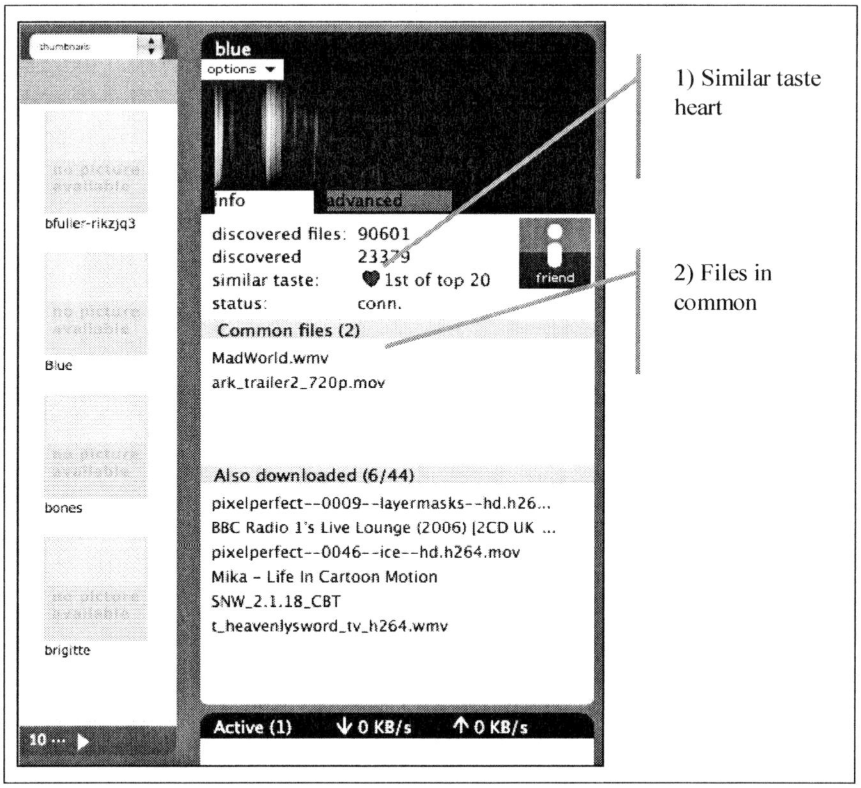

CONCLUSION AND FUTURE WORK

P2P technology may be vital to the television of the future, with cooperation being a challenging problem in P2P systems. We listed 13 theories, which provide fertile ground for coming up with solutions addressing that challenge. It resulted in four criteria along which cooperation inducing mechanisms can be evaluated on their contribution to social interactivity. Hence, a recipe for inducing cooperation in P2P systems is provided. This recipe is based on profound theoretical backgrounds, thus seeking to bridge theory and practice by applying social psychological theories to the relevant context of use, i.e. P2P-TV systems.

Social psychology literature on human cooperative behavior turns out to be a useful source for deriving motives for cooperation in P2P systems. The search resulted in a great number of proposed applications that could be clustered into the four criteria of *fairness, esteem, community,* and *DoS.* The examples given suggest that cooperation inducing mechanisms can be incorporated into P2P systems.

The present study suggests that a promising way to enhance the exploitation of available resources within a P2P network, is by expanding file exchange with flexibility over time and exchangeable resources. We are currently working on a study on the 'acceptability of delayed reciprocity' in a P2P-TV community among approximately 300 participants. In this, we investigate what delays of reciprocity users accept from other users under varying relationship types and exchangeable resources. The results will contribute to the development of new cooperation inducing mechanisms in P2P-TV.

REFERENCES

Adar, E., & Huberman, B. A. (2000). Free riding on Gnutella. *First Monday, 5*(10).

Answer.com. (2008). Peer-to-Peer: Definition and Much More from Answer.com. Retrieved May 27, from http://www.answers.com/p2p&r=67

Bandura, A. (1971). *Social Learning Theory.* New York: General Learning Press.

Baron, R. A., Byrne, D., & Johnson, B. T. (1998). *Exploring Social Psychology* (4th ed.). Needham Heights, MA: Allyn and Bacon.

Beenen, G., Ling, K., Wang, X., Chang, K., Frankowski, D., Resnick, P., et al. (2004, November 6-11). *Using Social Psychology to Motivate Contributions to Online Communities.* Paper presented at the CSCW'04, Chicago, Illinois, USA.

Bradner, E., & Mark, G. (2002, November 16-20). *Why Distance Matters: Effects on Cooperation, Persuasion and Deception.* Paper presented at the CSCW'02, New Orleans, Louisiana, USA.

Brembs, B. (2001). Hamilton's Theory. In S. Brenner & J. Miller (Eds.), *Encyclopedia of Genetics* (pp. 906-910). London, New York: Academic Press.

Cheng, R., & Vassileva, J. (2005). *User Motivation and Persuasion Strategy for Peer-to-Peer Communities.* Paper presented at the 38th Hawaii International Conference on System Sciences, Hawaii.

Chorianopoulos, K., & Lekakos, G. (2008). Introduction to Social TV: Enhancing the Shared Experience with Interactive TV. *International Journal of Human Computer Interaction, 24*(2), 113-120.

Cisco. (2008). *The Exabyte Era* (White Paper): Cisco Systems, Inc.

Clark, M. S., & Mills, J. (1993). The difference between communal and exchange relationships: What it is and is not. *Personality and Social Psychology Bulletin, 19,* 684-691.

De Valck, K. (2005). *Virtual Communities of Consumption: Networks of Consumer Knowledge and Companionship.* Erasmus University Rotterdam, Rotterdam.

Festinger, L., Schachter, S., & Back, K. (1950). *Social pressures in informal groups: A study of human factors in housing.* New York: Harper & Brothers.

Fokker, J., Buntine, W., Pouwelse, J., De Ridder, H., & Westendorp, P. (2006, August 6-11). *Tagging in Peer-to-Peer Wikipedia: A method to induce cooperation.* Paper presented at the OSIR'06, Seattle, WA.

Fokker, J., De Ridder, H., Westendorp, P., & Pouwelse, J. (2007a). *Inducing Cooperation in Peer-to-Peer Television Systems.* Paper presented at the EuroITV'07.

Fokker, J., De Ridder, H., Westendorp, P., & Pouwelse, J. (2007b). *Psychological Backgrounds for Inducing Cooperation in Peer-to-Peer Television.* Paper presented at the EuroITV'07.

Fokker, J., De Ridder, H., Westendorp, P., & Pouwelse, J. (2008, June 4-6). *Exploring the Acceptability of Delayed Reciprocity in Peer-to-Peer Networks.* Paper presented at the Persuasive'08, Oulu, Finland.

Fokker, J., Pouwelse, J., & Buntine, W. (2007c). Tag-Based Navigation for Peer-to-Peer Wikipedia. In R. K. J. Bandamutha (Ed.), *P2P Networks: An Introduction* (pp. 82-94). Hyderabad, India: Icfai University Press.

Frey, B., & Jegen, R. (2001). Motivation crowding theory: A survey of empirical evidence. *Journal of Economic Surveys, 5*(5), 589-611.

Goren, H., Kurzban, R., & Rapoport, A. (2003). Social loafing vs. social enhancement: Public goods provisioning in real-time with irrevocable commitments. *Organizational Behavior and Human Decision Processes, 90*, 277-290.

Grafen, A. (1990). Biological Signals as Handicaps. *Journal of Theoretical Biology, 144*(4), 517-546.

Greenberg, J., & Scott, K. S. (1996). *Why do workers bite the hands that feed them? Employee theft as a social exchange process.* (Vol. 18). Greenwich, CT: JAI Press.

Hamilton, W. D. (1964). The genetical evolution of social behaviour. I. *Journal of Theoretical Biology, 7*(1), 1-16.

Hardy, C. L., & Van Vugt, M. (2006). Nice guys finish first: The competitive altruism hypothesis. *Personality and Social Psychology Bulletin, 32*(10), 1402-1413.

Harkins, S., & Szymanski, K. (1989). Social loafing and group evaluation. *Journal of Personality and Social Psychology, 56*, 934-941.

Heider, F. (1958). *The Psychology of Interpersonal Relations*: John Wiley & Sons.

Homans, G. C. (1958). Social Behavior as Exchange. *American Journal of Sociology, 63*(6), 597-606.

Latané, B. (1981). The psychology of social impact. *American Psychologist, 36*, 343-356.

Mauss, M. (2000). *The gift: the form and reason for exchange in Archaic societies* (W. D. Halls, Trans.). New York: Norton.

McAndrew, F. T. (2002). New Evolutionary Perspectives on Altruism: Multilevel-Selection and Costly-Signaling Theories. *Current Directions in Psychological Science, 11*(2), 79-82.

Milgram, S. (1967). The small world problem. *Psychology Today, 1*, 61-67.

Milojicic, D. S., Kalogeraki, V., Lukose, R., Nagaraja, K., Pruyne, J., Richard, B., et al. (2002). *Peer-to-Peer Computing* (Technical report No. HPL-2002-57 (R.1)): HP Laboratories Palo Alto.

Nardi, B., & Harris, J. (2006, November 4-8). *Strangers and Friends: Collaborative Play in World of Warcraft.* Paper presented at the CSCW'06, Banff, Alberta, Canada.

Nielsen, J. (2006). Participation Inequality: Encouraging More Users to Contribute. Retrieved October 10, 2006, from http://www.useit.com/alertbox/participation_inequality.html

Pouwelse, J. A., Garbacki, P., Epema, D. H. J., & Sips, H. J. (2005, February 24-25). *The BitTorrent p2p file-sharing system: Measurements and analysis.* Paper presented at the 4th International Workshop on Peer-to-Peer Systems (IPTPS'05), Ithaca, New York, USA.

Rehn, A. (2001). *Electronic Potlatch: A study on new technologies and primitive economic behaviors.* Stockholm.

Ripeanu, M., Mowbray, M., Andrade, N., & Lima, A. (2006). Gifting technologies: A BitTorrent case study. *First Monday, 11.*

Ryan, R. M., & Deci, E. L. (2000). Self-Determination Theory and the Facilitation of Intrinsic Motivation, Social Development, and Well-Being. *American Psychologist, 55*(1), 68-78.

Skinner, B. F. (1953). *Science and Human Behaviour* (1st ed.). New York: The Macmillan Company.

Steele, C. M. (1988). The psychology of self-affirmation: Sustaining the integrity of the self. In L. Berkowitz (Ed.), *Advances in experimental social psychology, 21,* 261-301. New York: Academic Press.

Tajfel, H., & Turner, J. C. (1986). The social identity theory of intergroup behavior. In S. Worchel & W. Austin (Eds.), *Psychology of Intergroup Relations* (pp. 7-24). Chicago: Nelson Hall.

Trivers, R. (1971). The Evolution of Reciprocal Altruism. *Quarterly Review of Biology, 46,* 35-57.

Walster, E., Walster, G. W., & Berscheid, E. (1978). *Equity: Theory and research.* Boston: Allyn and Bacon.

Wang, J., Pouwelse, J., Fokker, J., De Vries, A. P., & Reinders, M. J. T. (2008). Personalization on a peer-to-peer television system. *Multimedia Tools and Applications, 36*(1-2), 89-113.

Wilson, D. S., & Sober, E. (1994). Reintroducing group selection to the human behavioral sciences. *Behavioral and Brain Sciences, 17*(4), 585-654.

Section III
Making it Work:
Social Television Systems

Chapter X
Getting to Know Social Television:
One Team's Discoveries from Library to Living Room

Gunnar Harboe
Motorola, USA

Elaine Huang
Motorola, USA

Noel Massey
Motorola, USA

Crysta Metcalf
Motorola, USA

Ashley Novak
Motorola, USA

Guy Romano
Motorola, USA

Joe Tullio
Motorola, USA

ABSTRACT

This chapter presents results from an ongoing social television project, in the context of other research in the field. The authors give a detailed description of the STV prototype used in their research, and summarize their studies, which provide the findings explained in the rest of the chapter. Three major research focuses are identified, namely evaluation and validation of Social TV systems, communication modality comparison, and detailed observation of user behaviors. Based on the findings in these areas, the authors list three major open questions and challenges for the field: multi-user support, new equipment requirements, and the creation of distinct and unique social television experiences. Finally, the chapter suggests that the emphasis within social television may be moving from research to design, implementation and deployment.

INTRODUCTION

Over the course of this decade, social television has gone from a marginal idea to a major focus area within the field of interactive TV (ITV). As members of an ongoing Social TV project in Motorola's applied research division, we have studied the issues around the subject extensively. In this chapter, we aim to discuss some of the things we have learned about social television, point out a few major unanswered questions and unsolved problems with social television, and offer a perspective on where social television is headed. Throughout, our own research and experiences are used as a jumping-off point, complemented by findings from the growing literature on the topic.

The main part of this chapter is devoted to three big questions that have attracted much attention. We look at attempts to evaluate the effectiveness and appeal of social television experiences, and find that their potential has been largely validated, although some important misgivings remain around privacy, disruptions, and systems that do not offer full freeform communication. We next examine comparisons of different communication modalities, primarily text, voice and video. Our findings indicate that contrary to previous, inconclusive studies, text is a better communication option than voice in the context of in-home social television, while video does not appear to be a suitable option. Finally, we describe the observed usage patterns of Social TV systems. We find that in naturalistic use, conversations are not as closely tied to the TV content as has previously been thought. On the other hand, television presence provides an important link between viewing behaviors and social interactions.

Because each topic draws widely on results from all our different studies as well as from the literature, the chapter is not broken up by individual study. Instead, to provide the necessary background for the reader to assess the findings we present, and explain where we are coming from, we first describe the prototype system used in our research, followed by a summary of the research we have conducted, before we start talking about what we have learned.

STV: A SOCIAL TELEVISION SYSTEM

At Motorola, we have explored several different kinds of social experiences around the TV. One of the experiences that has received most attention is to allow a small group of friends or relatives to share a feeling of contact or togetherness while watching TV; a "virtual couch." As part of this research, we have developed a series of prototypes, collectively known as STV, for use in lab and field trials (Table 1). The first iteration, STV1, consisted of a simple, single-session audio link between households, which allowed users to communicate via open room-microphones and hear their conversation partners through their television speakers, mixed with the TV audio (Harboe, Massey, Metcalf, Wheatley, & Romano, 2008a). There was no visual user interface, and the only control given to users was the ability to adjust the relative volumes of the TV audio and the voice audio using a remote control.

Later prototypes have been more elaborate, providing fully integrated Social TV systems that could be deployed in the field for extended periods of time. Since their features largely overlap (the biggest difference being an evolution in communication capabilities), we will describe them together in more detail. The main features of these prototypes are television presence (provided in the form of a buddy list and an ambient display), program suggestions, communication (including at various times graphical emoticons, pre-defined text messages, text chat and voice communication) and historical information such as viewing habits.

Table 1. Versions of the STV social television prototype.

Version	Presence	Suggestions	Communication
STV1	No	No	Voice
STV2a	Yes	Live, Future	Emoticons
STV2b	Yes	Live, Future	Emoticons, pre-defined text-messages
STV3	Yes	Live	Emoticons, text chat, voice

Presence

Presence information provides users with awareness of other users' current state, particularly their availability or whether they are currently active on the system. It can also include other information about their context. Television presence information, for example, lets users know about their buddies' current TV viewing. In STV, the two most important presence features are the buddy list and the ambient display.

Buddy List

The buddy list displays the user's contacts ("buddies") on the STV network, and shows their current status (Figure 1). Buddies who are not currently watching TV, or who have logged off in order to watch in private, show up in the list as unavailable. The STV buddy list is similar to buddy lists in instant messaging (IM) applications such as Windows Live Messenger[1] or AIM[2],but provides additional television-specific context, showing the station and program each buddy is watching. The user can tune directly to those stations by highlighting the buddy and pressing Select on the remote, joining whatever channel the buddy is watching.

Ambient Display

The ambient display provides a way for users to be aware of their buddies' presence status even when the TV is turned off. An Ambient Orb[3] (Figure 2) displays aggregate presence information by glowing in different colors depending on how many buddies are currently active watching TV on the system, with separate states for no buddies, one buddy, and multiple buddies active. The ambient display is described in more detail in Harboe et al. (2008b).

Figure 1. Screenshot of the STV buddy list.

Figure 2. Ambient "orb" display sitting next to the television.

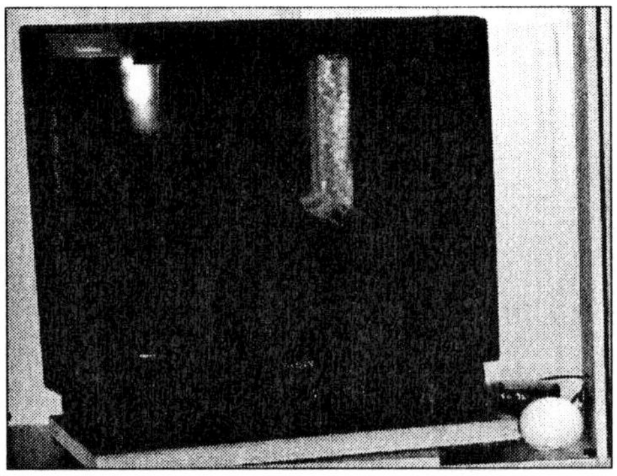

Program Suggestions

In addition to the ability to join what a buddy is watching through the buddy list, it is also possible to actively ask buddies to watch something. Program suggestions help users let each other know about specific shows, and arrange to watch them together. Upon receiving such a program suggestion, a user can accept and automatically tune to the program, or close the suggestion without accepting (Figure 3).

STV2a and STV2b also let users send suggestions for upcoming shows. If the recipient accepted the suggestion, the show was scheduled so that a notification would remind them when it was about to start. STV has no DVR capability, so there was no option to record the show. Perhaps because of this limitation, suggestions for upcoming shows were rarely used, and the feature was removed in STV3 to simplify the user interface.

Communication

Different iterations of STV have included different forms of communication. Although STV1 supported voice, this function was not included in the STV2 prototypes, which instead tested more limited, closed-form communication options: emoticons and pre-defined text messages. When these features proved unsatisfactory, STV3 restored voice support, and also added text chat, providing a choice of freeform communication options.

Figure 3. Screenshot of incoming program suggestion.

Closed-Form Communication

Two forms of simple, closed-form communication have been tested in STV, both of them restricted to users watching the same program. Graphical emoticons, present in all versions since STV2a, are meant as a quick way to comment on the program or exchange greetings without necessarily getting engaged in further conversation. Currently, this takes the form of "thumbs up" and "thumbs down" smileys (Figure 4b), which can be sent using dedicated remote control buttons. A third smiley, called "shoutout" and representing a generic greeting or exclamation, was included in STV2a, but proved redundant and was removed from later versions.

In STV2b, the shoutout was replaced with a set of around 20 fixed, pre-written text messages and responses, with phrases such as "How is this show?", "This sucks!", and "Call me" (Figure 4a). The content of the messages was picked based on input from study participants so as to express the most important things users might want to say to each other. After poor participant response in the field trial of this version, the pre-written messages were scrapped, and replaced in STV3 by full freeform communication in the form of voice and text chat.

Text Chat

The text chat feature of STV3 is similar to IM, in that users select a buddy from their list (the buddy must be active, but can be watching a different channel), type them a message, and that message then immediately appears on the buddy's screen, initiating a conversation. The chat window can easily be hidden so as not to obscure the TV program while waiting for a response. Users can be in multiple text chats at the same time,

Figure 4. Closed-form communication in STV. (a) Pre-defined text messages. (b) Emoticons.

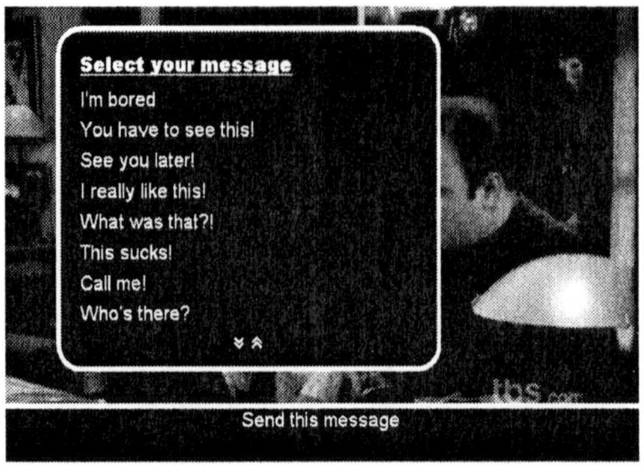

(a)

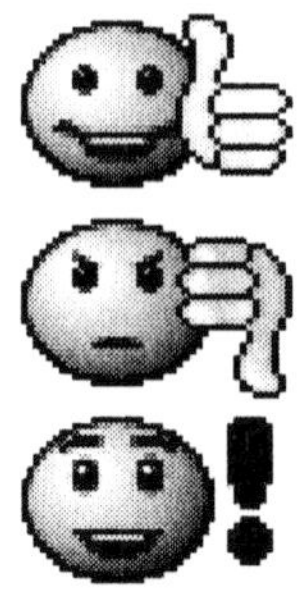

(b)

switching between them (Figure 5), but in our current system each conversation is one-to-one; we do not support group chat for text. Text entry is through a wireless Bluetooth keyboard, for ease of typing.

Voice Chat

In STV3, as in STV1, voice communication is made possible by establishing an audio link with one or more buddies. This is done in STV3 by selecting them from the buddy list and calling them. If any of the buddies accepts the incoming call, the audio link is set up and the call starts directly. Group calls are possible; a call can be set up with multiple people from the start, or more can be invited to an ongoing call (Figure 6). There is no particular ownership of a call; any participant can invite more people to join, and the call lasts for as long as any two participants remain. It is only possible to join a call by invitation, and each user can only be in one call at a time. While in a voice conversation, or when invited to a voice call, users can see a list of all the current participants in the call.

For voice communication, users have an echo-cancelling room microphone placed somewhere central in the room, for example on the coffee

Figure 5. Tabbed text chat window with two ongoing conversations.

Figure 6. Screenshot of the calling process. Conversation presence in the top right lists participants in the call.

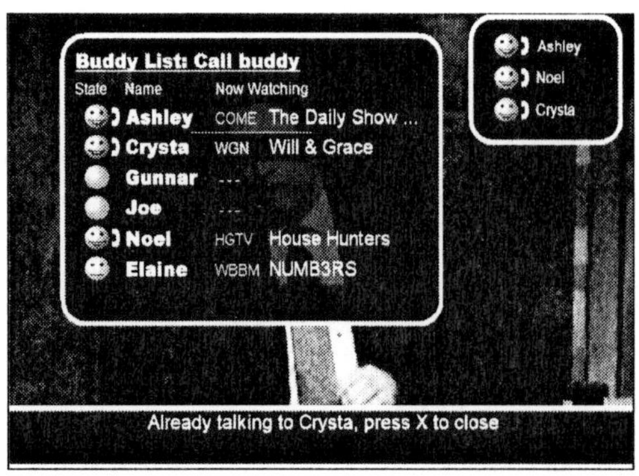

table. The voices of the remote participants are transmitted through the television speakers, mixed in with the audio from the TV program. The microphone sensitivity and the volumes of the voices and the television program audio can be muted and adjusted independently.

Users can start and maintain voice conversations with their buddies whether or not they are watching the same TV station or program. It is also possible to take part in both a voice and text conversations simultaneously, either with the same participants (e.g. using text as a backchannel in a voice conversation), or in different conversations with different people.

Historical Information

In order to make the TV content more socially meaningful, STV provides users with some awareness of their buddies' viewing habits. There is a viewing history which lists the programs each buddy has watched. Users can also see a list of favorite (most frequently watched) shows for each user, and for their buddy list as a whole (Figure 7a). These shows are also marked in the Program Guide grid with a little buddy icon (Figure 7b). Finally, another option allows users to view a list of shows they have in common with another buddy, meaning shows they have both watched.

A program only counts as having been watched if the user watched at least ten minutes in total of a single broadcast of that program. Programs that a user watches while logged off are not counted, and users can remove programs from their own viewing history.

OUR RESEARCH

The social television research in Motorola originated in several studies that we conducted on family communication and use of media. These found that "people who want to build/maintain relationships but live far apart can use mediated communication technologies to develop 'close' relationships through a virtual experience of being 'physically' close," and that sharing the experience of commercially created content is an important way to maintain a personal connection (Bentley, Metcalf, & Harboe, 2006). In one study in particular, we noticed people calling each other during a TV show they were both watching, so that they could watch it "together." This led us to explore the concept of social television.

Figure 7. Historical information in STV. (a) Most popular shows. (b) Program guide with popular shows marked with a smiley face.

(a) (b)

Our research to date has included paper prototyping tests, lab tests, usability studies, participatory design sessions and technical studies. However, the results presented here are drawn from four main studies (Table 2). Initial concepts were evaluated in a focus group study. The different STV prototypes were then tested in a series of in-home field trials. We will describe the methods used in each study in detail, followed by a summary of our data analysis methods, before going into our actual findings.

Focus Group Concept Study

We ran a concept study to test the appeal of communication and interaction through the TV, as well as four other, more advanced, concepts (which will not be addressed in this chapter). The primary purpose was to guide the future course of the project. The study consisted of seven focus groups (and a pilot group) followed by a data analysis workshop. In the focus groups, which lasted for two hours, participants were shown a number of storyboards depicting possible Social TV scenarios and were asked to discuss them with respect to their own needs and lifestyles. They also individually filled in worksheets with quantitative ratings (using Likert scales) and other information.

In all, 53 people participated in the focus groups. We had 6 groups of 8 participants, representing teens (17 to 19); young adults (25 to 35); and baby boomers (42 to 61), split into separate male and female groups. Finally, a group of 5 male Xbox Live gamers (18 to 25) was included. The groups were selected to include people with and without children, and for other demographic characteristics considered relevant. The concept study was run by an independent consultancy, with researchers from Motorola involved at each step of the design and execution.

STV1 Prototype Field Trials

For these field trials, we used the STV1 voice-only prototype (described above). The prototype was set up in the participants' own homes for a live TV show of their own choosing. We ran four trials with STV1. Each trial lasted for one hour, and the participants watched live programming that had been agreed upon in advance. Participants were recruited from our social networks using the friend-of-a-friend method. In all, there were 11 males and 8 females participating in 9 different households, and all were non-engineers (Table 3). The programs consisted of a basketball game, an American football game, an animated comedy followed by a sitcom, and a home decorating show. The participating households were connected to

Table 2. Overview of Motorola Social TV studies. Publications referenced: (1) Harboe., Massey, Metcalf, Wheatley, and Romano, 2007; (2) Harboe et al., 2008a; (3) Harboe et al., 2008b; (4) Metcalf et al., 2009; (5) Tullio, Harboe, and Massey, 2008; (6) Huang et al., 2009.

Study	System	Method	Participants	Publication
Concept Study	[N/A]	2-hour focus groups	7 groups, 53 participants	(1), (2)
STV1 trial	STV1	1-hour field trials	4 groups, 8 (+1) households	(1), (2)
STV2 trial	STV2a STV2b	2-week field trials	2 groups, 10 households	(3), (4)
STV3 trial	STV3	1,3-week field trials	2 groups, 9 households	(5), (6)

friends and their friends' families, except in the second trial which brought together a family by connecting a woman with her son, daughter-in-law and grandchildren. In the first trial, one of the 3 participating households was that of a Motorola employee familiar with the project, and data from this household was not used in the analysis. The participants were video-taped during the trial (the researchers were not present), and interviewed immediately afterwards. We also recorded the programs they watched.

STV2 Prototype Field Trials

We ran two separate in-home trials of STV2, using slightly different versions of the prototype (STV2a and STV2b). The focus in this iteration was on the presence awareness, with limited capabilities for communication through suggestions, emoticons, and (in the second half of the study) pre-defined text messages. Five households were recruited for each trial, and each trial was conducted over a two-week span. Participants were recruited using an independent recruiting agency. Its instructions were to find social groups in which the various household members were mutual friends, and all had strong ties with one another. A number of different kinds of relationships were represented in the two social groups, but neither group was completely tight-knit (Figure 8a, b). All the recruited participants were women. However, other family members such as husbands, fiancés, and children also used the system, and were secondary participants of the study.

We used multiple methods for data collection, including semi-structured interviews, automated usage logs, and voice mail diaries. Each household was interviewed three times. The initial interview lasted about half an hour and was used to collect background information about viewing habits, communication routines, and social relations between the participants. In a phone interview after the first week, lasting between 15 and 30 minutes, we gathered data about the participants'

Table 3. STV1 field trial sessions

Trial	Households	Participants	Programming
1	2 (+1)	5-10 (varying)	Basketball
2	2	5	American football
3	2	2	*Family Guy, Full House*
4	2	2	*Trading Spaces*

Figure 8. Social network graphs of relationships between households in our field trials. The solid lines represent friendships and close kinships (siblings), and the dashed lines represent more casual acquaintances.

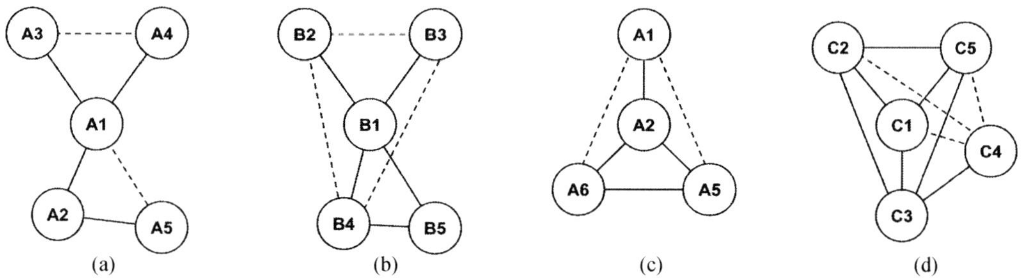

use of and reactions to the prototype in the first week. The final interview, lasting from an hour to an hour and half, was structured to collect more detailed information on their use of and reactions to the prototype.

We logged all interactions with the system, including when the TV turned on and off, and all button presses during system use. Voice mail diaries were used to collect information on behaviors surrounding the ambient presence features and other communication with other people in the study, since these did not appear in our logs.

STV3 Prototype Field Trials

This study, which used the STV3 prototype, was run in two stages: first a preliminary one-week trial with four households participating, then a more elaborate three-week trial with five households.

For the preliminary trial, we contacted three participants from the first group of the STV2 study (A1, A2, and A5) who formed a tight-knit group. A fourth participant, A6 (the sister of A2 and A5) was also added (Figure 8c). The participants were between the ages of 25 and 35. The system was deployed in their households for a period of one week. As in the STV2 study, an initial interview was used to gather background data, a phone interview was conducted mid-week to investigate participants' experiences with the system during the first half of the week, and a final interview was conducted at the end of the week to examine their experiences in-depth. We also logged all interactions and all text chats, and recorded all voice conversations held through the system.

For the longer trial, a recruiting agency recruited five participants, all of whom were males between 30 and 36. All were sports fans, and most were old friends (Figure 8d). The study was run during "March Madness," the American college basketball playoff series. For one week, the participants used the system without being able to connect to use the social features, in order for us to gather baseline data and to populate their viewing histories. After that, the social features were turned on, and they could use the system to communicate for two weeks, until the end of the study. The data gathering was as in the preliminary trial.

Analysis

In each study, we used a form of grounded theory analysis to interpret the data, applied as an affinity diagram technique (Bernard, 1998; Beyer & Holtzblatt, 1998). We reviewed the interview data, and (when available) the observed or recorded data from the usage sessions. We pulled out direct quotes and observations from the sessions and interviews that addressed the research questions. These items were grouped using the affinity method. Insights were drawn from the patterns that emerged, and the resulting organization of the data surveyed to answer the research questions.

For the focus groups and the STV2 and STV3 trials, we also performed statistical analyses of the data (ratings and usage logs, respectively). This quantitative data is mainly descriptive, especially for the field trials, since the small scale, non-experimental setup and numerous confounding variables don't permit us to draw statistically sound conclusions about any effects.

The results produced by our analysis, based on data from all the separate studies, form the material for the following section.

WHAT DO WE KNOW ABOUT SOCIAL TELEVISION?

In addition to the primary research just described, we have been able to draw on a growing social television literature. Research around social television has mainly clustered around a few big questions. We can sum these up as: validating the effectiveness and appeal of social features around video; exploring and comparing specific features and communication modalities to produce more optimal designs; and studying the behaviors around use of social television systems. Through the studies detailed above, we have made important contributions to answering each of these questions, which we will deal with in order.

Effectiveness and Appeal

As a concept that has never been widely deployed, social television requires some validation to test whether it is actually something people want or would enjoy, and whether it is useful or has other positive effects. The flipside of this question is the need to identify the concerns it causes and any potential negative effects it could have. This section will describe attempts to quantify the social and psychological effects of social television use, as well as people's response to it. We will then look at qualitative studies that describe people's first reactions to the idea without having used it, and compare these to studies of people's actual experiences with different Social TV systems. Finally, we examine the issue of privacy, which is crucial to the acceptability of social television.

Quantitative Measures: Ambivalence

A fundamental question to ask about social television is whether it really does provide a social experience. A number of researchers have studied the problem using quantitative methods.

Baillie, Fröhlich, and Schatz (2007) compared the "joint TV experience" and "social presence" of watching TV in the same room to watching together through the AmigoTV system. They found that using audio chat did not differ significantly on these measures from actually being in the same room, but communicating only through graphical symbols was significantly lower on both. Similarly, de Ruyter, Huijnen, Markopoulos, and IJsselstein (2003) found that group attraction was significantly higher for participants watching TV when they had a live video view of their friends watching the same thing than when they only had a sketch visualization or no view at all. Along the same lines, Weisz et al. (2007) found that the ability to text chat while watching video on the computer significantly increased feelings of closeness with and liking of other participants. Finally, participants in a 9-week trial of ConnecTV felt significantly more connected when using the ConnecTV system than when watching regular TV in a control condition (Boertjes, Klok, & Schultz, 2007). Plainly, there are social television features that increase television sociability, at least when compared to solitary viewing.

As reassuring as it is that social television clears this fairly low bar, the last study, the ConnecTV trial, deserves a second look. In addition to measuring connectedness, the researchers collected four other ratings from participants, on the scales pleasant–unpleasant, activated–calm, satisfied–irritated, and inspired–bored. On all scales, ConnecTV scored worse than just watching television. Participants were less pleased, less calm, more irritated and more bored. After the trial, only slightly over half of respondents (17, N = 31) regarded the system as an interesting addition to watching TV (Boertjes et al., 2007). Frustration over technical difficulties may have depressed ratings in this particular study, but the question remains: is Social TV something people even want?

In our own concept study, focus group participants rated their response to the ideas presented in storyboard scenarios on a 5-point Likert scale. The ratings showed a response that can best be described as lukewarm, falling mostly near the middle of the scale (M = 3.0, SD = 1.04, N = 53) (Figure 9). Weisz et al. (2007) report a similar result for watching films while chatting on the computer; enjoyment was moderate (M = 4.3, SD = 1.6, N = 15, on a 7-point scale).

In contrast, Regan and Todd (2004a; 2004b) report that 93% of participants in their study (N = 32) answered that they would like to IM with others while watching TV, and Baillie et al. (2007)

Figure 9. Focus group participant ratings of their interest in a generic social television concept. Likert scale from 1 (least interested) to 5 (most interested).

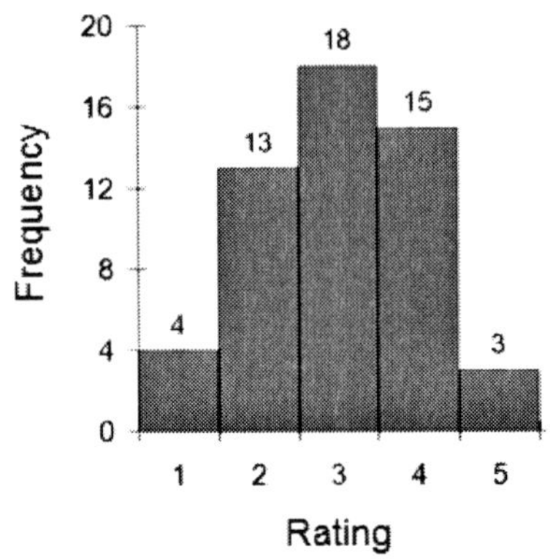

also found that 93% (N = 30) of their participants would like to use a Social TV application. Harrison and Amento (2007) got a somewhat weaker, but still positive, response to their CollaboraTV system, with 69% of participants saying they would use the system, and most agreeing that it made watching TV more enjoyable (M = 3.56, N = 32, on a 5-point scale). Results from a later CollaboraTV study by Nathan et al. (2008) were similar (M = 3.57, SD = 0.65, N = 16).

A lot of this variation can be attributed to differences between systems and study conditions, especially the recruiting. Both Regan and Todd and Baillie et al. specifically recruited participants believed to be receptive to buying this kind of product, while Harrison and Amento recruited employees of, and Nathan et al. former interns at, a corporate research lab. The participants in the other studies, on the other hand, were recruited from the general public, or from particular communities without a predictable bias. This may indicate that social television appeals primarily to certain demographics, and does not currently inspire widespread interest.

First Impressions: Wariness

In addition to responses measured quantitatively on Likert scales, there is a wealth of qualitative data on people's reactions. This information helps to paint a richer picture of the aspects of social television that excite people, and the things that concern them about the idea. Three similar studies all point towards the same concerns.

In our focus groups, the mixed reactions to the concept were expressed in words such as "intriguing," "interesting" and even "wonderful," on the one hand, but, "weird," "unnecessary" and "pointless" on the other. Matching descriptions were elicited by Ali-Hasan (2008) in a participatory design exercise and by Hess (2008) in a user forum. In each case, the concerns of skeptical participants soon turned to privacy. The focus group participants worried about others being able to see what they were watching at any given time. Most argued that this social transparency overstepped the boundaries of acceptability, and several invoked "Big Brother," perhaps recognizing in STV the telescreen from *Nineteen Eighty-Four* (Orwell, 1949/1990). As in the studies by Regan and Todd (2004b) and Ali-Hasan (2008), participants articulated their discomfort in variations of, "I don't watch porn, but if I was…"; except for a few male participants who were more worried about getting caught watching programs targeted at women. The more general point expressed in these examples is that people worry about being embarrassed about their TV viewing. After all, many popular TV shows, as well as television viewing in general, are held in low regard critically and culturally. These "guilty pleasures" may therefore not reflect well on the viewer: "If someone was watching stupid shows all the time, I would think, 'God, what a moron are *they*!' to be watching these dumb shows."

Some of our focus group participants could see no value in linking personal communication to broadcast content. They felt that a television program would be an irrelevant distraction from their interaction with other people, for which existing communication technologies (telephone, email, IM, etc.) or face-to-face meetings were sufficient. The lack of perceived utility was captured in the words "toy" or "novelty," terms also heard in other studies (Metcalf et al., 2009; Ali-Hasan, 2008).

Experiences: Interest and Appetite

While user panels and focus groups are cautious in their response to social television, field studies where people get a chance to use it reveal a different and far more positive reception. The enthusiasm is apparent in the testimonials about Zync, the video sharing IM plugin, collected by Shamma et al. (2008) "Let me start by saying, I absolutely love Zync, currently myself and my wife are about 2000 miles apart but we love to watch movies together and it allows us to talk and watch together and its the closest thing we have to actually being together."

The participants in the STV1 trials were similarly very positive, on the whole, towards the concept and their experience with the prototype. We consistently found evidence that social television added value over and above watching alone, in a number of different ways. It helped to relieve boredom, allowed the participants to share their interests, gave them someone to ask questions of and show off their knowledge to, relieved loneliness, and heightened the intensity of exciting moments (cf. Weisz et al., 2007). In some cases it came close to the experience of actually being together: "I kind of forgot I was talking through the television at some point. I just was talking and could hear Mom." "It felt like she was in the room."

Participants in the trial of the STV2 presence-oriented prototype expressed frustration about the lack of communication options (a reaction shared by participants in the Boertjes et al. (2007) pilot study of the similarly restricted ConnecTV system), and their overall reaction was more negative. When some of them had a chance to try STV3, which incorporated voice and text communication, they all agreed that it was a significant improvement, describing it as "a lot more fun." Other participants in the STV3 study said the same thing: "I certainly liked it. It was a nice means to communicate and interact with my friends who share some of the same interests." "It opens and creates new avenues and communication experiences that I didn't have with a normal TV."

Other questions find less clear-cut answers. Some focus group members were concerned being in communication with others would interrupt and disrupt their viewing, causing them to miss parts of the show they were watching. They argued that television was their "down time," the only chance they had to relax and unwind. In the tests of STV1, however, voice conversations generally did not prove to be greatly disruptive. This agrees with the findings by PARC researchers (Oehlberg et al., 2006; Ducheneaut et al., 2008) in a similar experimental setup. In our later prototypes, some participants expressed impatience with the amount of on-screen notifications and messages, calling it "a chore" to deal with, while others enjoyed the fact that they could use their downtime to stay in touch with friends. Weisz et al. (2007) found that text chat during video viewing is distracting, and their participants reported this as their main dislike, and Weisz and Kiesler (2008) again demonstrated a distraction effect. On the other hand, when Regan and Todd (2004b) asked participants to rate how much incoming IMs interfered with the enjoyment of the TV content, average ratings were uniformly low.

Just like the response of participants in our field trials was generally more positive than that of members of the focus groups, several of those people were more enthusiastic after having had hands-on experience with it than they were beforehand: "Before going into this I thought 'What would I ever use this for?' But it was a totally different experience actually doing it. Because I totally changed my mind: [...] I would totally use it!" Supporting this observation, Abreu and Almeida (2008) found that the proportion of subjects expressing high interest in a Social TV application rose from 27% before use to 67% after having tried it. (Baillie et al. (2007), on the other hand, found that participant reactions after having used the system were consistent with advance expectations. However, in that study, initial attitudes were already overwhelmingly positive.) To us, this underlines the value of practical testing in the study of new experiences. It is hard for people to predict their reactions to a hypothetical scenario. That is not to say that prototype tests are without their own biases; in all our field trials, as well as in the ConnecTV pilot study, technical problems (including degradation of TV image quality) were among the top complaints, and appear to have affected participants' overall reactions negatively.

People seem to respond more positively to social television when they try it than when it is only explained to them, and those who like it are a lot more enthusiastic. Some of the concerns people have when the concept is presented are not as serious as they imagine.

Effects on Privacy

As in the various user panels, aspects of privacy were also an important issue in our field studies. Although some participants expressed worry about government surveillance ("Can you imagine, like, back in the fifties, if they were Big Brother watching what we were watching? I would hope the government now doesn't want to come and… I mean, we don't have anything on our TV that's so…"), spammers, hackers, the possibility of strangers stalking them through the system, and the safety implications for their children, the gravest privacy concerns were in relation to their friends and family, both those on the system and those with whom they shared the home.

In practice the concerns were somewhat different from those anticipated by the focus group participants. The live television presence, which participants without direct experience worried about, did not prove to be a problem, perhaps because participants were able to log off if they wished to watch something in private. However, many felt that the viewing history feature, which logs all the TV programs watched publicly by a buddy, was too intrusive: "It's not really personal information, but it's not information required for anyone else to know. […] I don't think people would feel comfortable knowing what other people have watched." When the information was presented in summarized and aggregated form in STV3, participants had fewer objections.

STV3 also exposed another problem: the confidentiality of—and the confidence participants could have in—communications via the system. In the trials, there were numerous instances of participants being impersonated by other members of their households when using text chat. In some cases they were quickly found out ("it was the way [her husband] said 'Hey dude,' and my sister doesn't talk to me that way"), while in other instances the other person never realized the deception. Once people recognized this potential, reactions were strong: "Oh wow, that's scary... Heavens no! I'd be, like, angry if I knew that that happened. […] That's just, like, a violation of just, like, normal goodness!" "If she just pries a little bit, […] [my girlfriend] might find out something she doesn't necessarily need to know."

Everyone did not share this reaction. One participant was generally unconcerned about divulging information to the wrong person, because he felt that the conversations generally centered on topics that were not particularly sensitive: "What I was saying, even if it wasn't the person I [thought I] was talking to, I didn't feel like it was going to hurt anything. It wasn't anything that was sensitive information… so I wasn't too concerned about that." "I wasn't giving out my social security number."

Nevertheless, the overall trend is clear. When it comes to Social TV the primary threat to privacy is not Big Brother. It is people's actual big and little brothers, as well as their sisters, significant others, and other close kin.

Effectiveness and Appeal: Discussion

Evaluations of social television concepts and systems turn again and again to questions of sociability. The ability of the technology to provide a social experience is well established. When users talk about the benefits of having social television, nearly everything they say is about how it makes them feel closer to the other users: "It kept us more in touch with each other." Observations of their usage show that they take advantage of the social nature of the experience to engage in a rich set of behaviors.

Despite this alignment of potential capabilities with desired outcomes, and enthusiastic reception in certain studies, reactions to Social TV have not been uniformly positive. While some participant reservations could be caused by unfamiliarity with the concept (in focus groups, participatory design sessions and user panels) or technical problems (in prototype tests), two misgivings are not so easily dismissed: People worry about their privacy, although precisely what worries them is refined after experience with the system; and they are unhappy about disruptions that distract them from the TV viewing. These are problems that need to be addressed in the design of Social TV systems.

A final cause of dissatisfaction with some social television systems is that they are not social *enough*. Prototypes such as STV2 and ConnecTV, which do not allow freeform communication between users, cause much frustration. This frustration is satisfied in systems like STV3, which offers full text and voice chat support. It is therefore necessary to look at which features do satisfy people's social needs around television, and how they compare to each other.

Communication Modality Alternatives

The social television systems in the literature differ in the capabilities they offer for sharing a viewing experience. As pointed out above, merely offering presence awareness, or limited communication in the form of emoticons or pre-defined text messages to choose from, is not a satisfactory solution for users (Metcalf et al., 2009; Tullio et al., 2008; Boertjes et al., 2008; Baillie et al., 2007; However, see Schatz, Baillie, Fröhlich, & Egger, 2008). There is a need for more expressive communication. In existing systems, solutions to this have mainly taken the form of text chat (e.g. CollaboraTV) (Harrison & Amento, 2007), audio links (e.g. AmigoTV) (Coppens, Trappeniers, & Godon, 2004; Coppens, Vanparijs, & Handekyen, 2005), and occasionally video links (e.g. Reflexion) (Agamanolis, 2008), generally according to the designers' beliefs as to what will provide the most satisfying experience for users. A number of studies have sought to compare these different modalities.

Text vs. Voice

Geerts (2006) looked at voice chat and text chat in two separate systems, AmigoTV and Windows Media Center with Window Messenger. After using both systems in the lab, nine out of seventeen participants preferred the voice chat of AmigoTV, five preferred the text chat of Windows Messenger, and three had no preference. While voice was the more popular choice and was considered more natural, younger users and those with experience with IM text chatting tended to prefer text.

In a 144-person between-subjects lab experiment, Weisz and Kiesler (2008) compared communicating using text chat, voice chat, and both while watching streaming videos. Their results show that all options were enjoyable, and do not point to any of them as clearly preferable. Participants tended to prefer the option they personally experienced to other alternatives presented. Overall, text chat did slightly better both on enjoyment ratings and preference rankings, but on the other hand, users who had both voice and text available used voice almost three times as much. On the whole, the authors interpret the findings as support for adding voice.

In our focus group, participants were uniform in their opinion that communication should be by voice, and that talking should be made as natural as possible. They rejected out of hand the idea of texting for communication. Participants in the field tests of the voice-only STV1 system agreed, while those in trials of STV2, which had no freeform communication, were split on which they would prefer. Some talked about the benefits of voice: "A microphone or something to turn on would be cool. I wouldn't take the time to be typing in stuff back and forth; I'd pick up the phone and call them." Others strongly preferred text: "There's no way I want somebody talking to me during a TV show. Let the damn letters pop up. If I want to ignore them I'll close the window. I would never want someone talking to me during a TV show."

Text and Voice in Practice

In an attempt to resolve these conflicting views and inconclusive studies, we included both voice and text communication in the STV3 field trials. The results were unambiguous, based both on people's subjective opinions and on their actual use of the system. In the first, one-week trial, we counted 6 separate conversations, made up of combinations of 4 voice calls and 12 text chats. Of the 6 conversations, 3 involved voice, and all involved text. In the second trial, where participants had two weeks to use the communication tools, we counted 43 conversations, 6 of which involved voice, and all of them involving text.

Text was clearly the more frequent communication choice, and all the participants except one stated that they preferred it to voice. Their reasons for this choice varied. Some simply gravitated to what they were most familiar and comfortable with: "I guess that's a technology I'm used to, because I text-message a lot, and that's all I do at work." Other explanations were exactly the opposite: "[At work] I'm constantly on the phone, and you get burned out and it's just easier to text."

A number of participants felt it was easier to use and interfered less with watching TV: "Texting is what I think I'm more comfortable with, and you can pay attention more with the program, doing that. And when the voice worked, it was still fine, 'cause I could hear the program and hear the person. But it seemed like it took a little bit more of your attention or focus to make sure that you're hearing them and responding."

There was also a widespread feeling that contacting people over text was less intrusive and disruptive than trying to contact them over voice. "I don't want to inconvenience somebody by calling them or vice versa." Text was also seen as less intrusive to the rest of the household. "There's no sound or anything, so it's not disruptive," especially at night after children's bed time. The reluctance to use voice was closely tied to a sense of a social obligation being imposed. "I think the calling feature can be kind of tough to determine whether or not they're up for chatting. [...] Let's say you want to [text] first instead of just popping into a call, and so they don't have to turn you down and kind of feel bad about it or whatever." Considering this problem of initiating the conversation, it is interesting to note that all but one of the voice conversations started out with a text chat first. As one participant put it: "Texting sets up the conversation, 'Hey, I'll give you a call' or whatever... then I'll know to call."

Text may have been seen as less intrusive than voice because of the semi-asynchronous flow of a text conversation. While the back-and-forth in voice sessions was essentially instantaneous, text chats often included long pauses—many minutes, sometimes even hours—between a statement and the response. Participants felt that this offered more control over the rhythm of the conversation. "You don't feel like you have to be waiting on the other person's response." "I think you're more in a control in a conversation where you can text, where[as] when you talk, you're all in, you're in that conversation, and you can't just at the end of a sentence... you can't just walk away and then get back to them ten minutes later, like in a text." "So the same reason why you would send a friend a text, because they can answer it when they get to it, is the same reason you would text somebody on the TV. When you're on a voice call with them, you can't do anything else, you can't talk to anyone else, you don't feel like you can just get up and leave."

Video Communication

While there is now a substantial amount of data on text vs. voice, social television with video link communication has been less thoroughly studied. Agamanolis (2008) found that watching TV programs or movies together was "one of the most satisfying ways" to use the "magic mirror"-style video conferencing system Reflexion. And as already mentioned, de Ruyter et al. (2003) found that group attraction was significantly higher while watching TV with a video visualization of a remote group of users than when watching with a sketch visualization or without any visualization.

In our focus groups, many participants said that they would like for social television to include a video link, so they could see each other while talking. However, the field trials indicated that adding video to the voice connection would be problematic. When our participants were relaxing in their living rooms, they were very unselfconscious, which didn't always mean visually presentable. In the STV1 tests, all of our participants were barefoot, and most put their legs on a table or on the couch. In a more extreme example, a child walked naked into the living room during one of the sessions. "Good thing it's not live video! [My kid] has got no bottoms on." They joke about it—"That's just sick! (Laughter)"—but it is clear that it would be a genuine concern to them. One participant told us that "the best part is [that] it's not picture in picture, where you're talking to someone and they see you." "[People] don't want anybody seeing you." Mostly, having video just seemed unnecessary. "Visually you don't see them but it did feel like there was this big sense of camaraderie. But outside of the optical sensation of [not] seeing them it did seem like they were here. It was strange."

Communication Modality Alternatives: Discussion

In none of our studies did participants actually have a chance to try a video-mediated Social TV experience, so our findings on this issue cannot be conclusive. The difference between the opinions solicited in focus groups, the results of lab experiments, and the observations made in field trials demonstrate that we should be cautious in drawing firm conclusions from limited data. In addition, our findings on voice and text chat are themselves based only on two small groups of users, and may not be widely generalizable. However, if the results hold up, it would seem that text is the preferred mode of freeform communication for social television, followed by voice, followed by video (which surely faces all the same problems as voice, and others in addition).

Even this conclusion is not the full answer, however. Each communication modality can be provided in a number of different ways. For example, text chat can be enabled with a QWERTY keyboard or using other text entry methods and input devices, and voice communication can be open-microphone/open-speaker, or using personal headsets. The effects of such design variations remain mostly unexplored. Schatz et al. (2008) have found that they have substantial impact on the user experience of Social TV, observing that the method of text entry often determined participants' feelings about the text messaging features. The same authors also compared four ways to control voice conversations, and found significant differences in the experience of watching TV together and in the overall preference (Baillie et al., 2007). Additionally, they report that 50% of their participants preferred an open table microphone, 30% the headsets, and 20% had no preference.

The Uses of Social Television

Some of the most fascinating insights about social television have come from the detailed observation and description of the behaviors of people using it. Two main approaches have been used: observations in a laboratory, and observations in system field trials. Lab studies can be run on a large scale and allow close monitoring of participants, but sessions are typically short in duration, and the lab setting and experimental conditions can affect people's behaviors. Field trials can be much longer in duration and observe participants in their natural environment, performing natural tasks, but the ecological validity comes at a price: because they are time-consuming and difficult to set up they are typically smaller-scale, and rely more on interview data and records of actions than on high-fidelity monitoring. There are also studies with aspects of both approaches. We will first present an overview of the findings of a number of lab studies, then discuss two main themes identified in field trials: the content and style of conversations and the pattern of behaviors leading up to conversations. Finally, we touch on some of the ways social television can be harmful to social relationships.

Like Being There (In the Lab)

Lab studies are the most common method used to observe groups using social television. The key exemplar of this line of research is PARC's Social TV study (Oehlberg et al., 2006; Ducheneaut et al., 2008). In that study, the researchers compared the speech acts of a group of people watching TV in the same room to those of groups watching in two different rooms, connected via an audio link. Analysis showed that the nature and structure of the conversations were surprisingly similar in the two conditions. Participant interactions were tightly interwoven with the structure of the show, with conversations taking on a rhythm to minimize disruption of the programming.

In our field trials of STV1, we observed many of the same things. The flow of the conversations was remarkably natural, much more like a face-to-face conversation than a phone call. People would leave and come back, chime in from another room, fall silent when they were paying attention to the show, and chat with each other when nothing interesting was happening. In the bigger groups we often saw several conversations going on in parallel, and collocated groups frequently had side-conversations among themselves while still taking part in the larger, connected group.

Adding support to these results, Baillie et al. (2007) found that conversations over AmigoTV audio chat had characteristics similar to face-to-face conversations, and Geerts (2006) concluded that Social TV voice chat is more natural than text chat. Weisz et al. (2007) coded the content of text chats during video watching in the lab, using categories similar to ones identified by Oehlberg et al. (2006). In their data set, 55.3% of all statements were related to the content in some way, with 22.8% consisting of personal conversation. In a later lab study, Weisz and Kiesler (2008) found that 48.1% of the conversation concerned the video and 9.9% was personal. In a field study by Nathan et al. (2008), the numbers were 42.3% and 18.9%, respectively. When seen alongside the observational studies, these results point to social television experiences that are natural, unforced, but adapted to the rhythm of the concurrent content, which also provides the primary topic of conversation.

Just Talk

Although the above studies point in the same direction, it is important to recognize that none of them observed behaviors under natural conditions. In each, researchers arranged the viewing sessions, set up the communication link ahead of time, and, in several cases, selected the content. (With one slight exception: In the Nathan et al. study the researchers selected the content and gave participants viewing assignments, but did not otherwise interfere.) Because the initiative to watch and interact did not come from the participants, the way they do these things may be different from how they would act if the decision to engage was spontaneous.

In fact, when we examined these behaviors under less artificial conditions in the STV3 field trials, the results were very different. We found that personal conversation was far more common than conversation about television content. In the second group, for example, only eleven of the 43 text chats and three of the six voice chats contained any references to the current TV content. As one participant put it, "A lot of the text chats I had were more just general, 'Hey. How's your day? I haven't talked to you in a while. What are you up to?' It wasn't necessarily focused actually on the program. Which I thought was interesting, myself, 'cause I was thinking at the start that that's what we're gonna be focused on, we'll be talking about the programs." The following excerpt from the beginning of a voice conversation serves as an example:

C3: [C3].
C1: Hello. What's up?
C3: [C1], what's up buddy?
C1: How are you? You sound funny.
C3: Ha-ha, I'm just exhausted dude, Monday just eats me up.
C1: It does, huh? It's the first day of the week. You should be at your strongest.
C3: Na, I think it's just from really enjoying the weekend and come Sunday at night you're finally feeling decompressed. You know? And then it's like you go back to the grind.

In cases where conversations did touch on the concurrent programming, the comments were in passing. The conversation structure did not seem to be tightly interwoven with the structure of the programs. "We touched on watching it, but [our conversations] were separate from what we were watching." (Figure 10)

In fact, conversations were on a wide variety of topics. Participants reported that the topics and tone of their conversations bore resemblance to their face-to-face conversations. Participants often engaged in small talk and banter, as in this text chat:

C5: does [C1] always text about food?
C3: he must..i heard he sleeps with a bag of doritos under his pillow
C3: if [C1] were president, free zingers 4 everyone
C5: for sure
C5: time to take the dogto dumpout

Figure 10. C1 in a text chat with C3, commenting in passing on the team's green St. Patrick's Day uniforms.

C3: yeah, you have a good night bro
C3: sweet dreams
C3: sweet dreams on chocolate creams

However, participants also used the system to engage in more serious conversation. In one text chat, one participant inquires about how the other is doing after a recent move:

C2: How are you enjoying the burbs?
C5: we love it
C5: nice and quiet
C2: It is a change from the city. Like the quiet too.
C2: Only thing is a good night out is going to Bennigans.haha

In another example from a text chat, participants used STV3 to provide live context updates about themselves and their families, and to set up face-to-face meetings:

A2: i'm back
A2: whats [your son] doing?
A1: being a pain in my butt. I am going o
A1: to put him to sleep soon
A2: but hes so cute
A2: sounds good
A1: yea he is
A2: i miss him
A2: we should get together really soon
A1: he gets his looks from his dad...

Just like the content of the STV3 conversations was largely independent of the TV programming, participants did not fit the rhythm of their speech in the audio conversations to the TV content. Unlike the long silences between statements that we observed in the STV1 trials whenever participants were paying attention to the program, here conversation was constant, with the longest pauses between utterances less than 5 seconds in length (except at times when users had technical difficulties with the system), and most no longer than a second or two (Figure 11). As a result, each audio session generally corresponded to a single, unbroken conversation. The data shows no indication of participants ever pausing to listen to the TV, and they generally kept the program volume low (but not off) during their conversations.

Overall, despite the coupling of communication with television and the fact that participants could only communicate with others who were also watching television, television content was generally not the main focus of conversation. Participants in the STV3 trials did enjoy the opportunity to communicate through the television. However, the voice and text chat features were not primarily used to create a social television experience of "watching together," but rather as

Figure 11. The waveform of the first five minutes of a typical audio session. Participants are talking continuously, and pauses are short (<3 s).

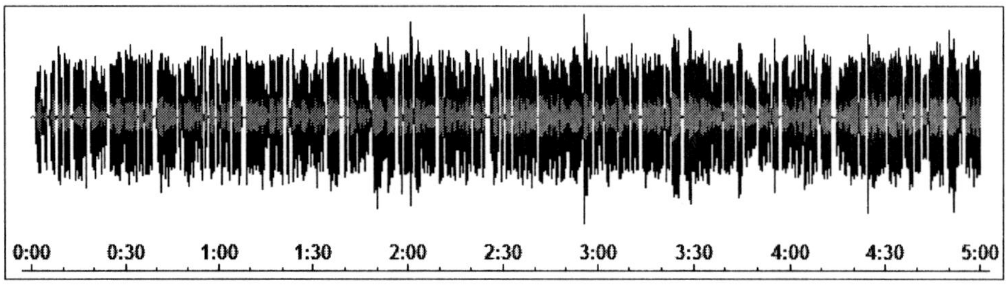

another communication medium that had little to do with the television content.

Usage Patterns

As touched on above, one explanation for the different patterns of use found with STV3 compared to previous studies is that the social dynamics of communication are to a large extent determined by when and how conversations are initiated, as well as how they are terminated. If social television designers wish to encourage different types of social experiences, they need to understand the steps that lead up to the actual conversations. We investigated this issue in our studies of STV2 and STV3.

STV2 and STV3 both incorporated an ambient display, intended to help participants establish contact with each other. We saw evidence that the ambient information did help to engage them in the social television experience. Many of our participants told us of occasions when they turned on the TV because the orb indicated that others were watching. For example, one participant told us "as soon as I come into the house or I wake up or come into the room, that's the first thing. It draws my attention, and the first thing I do is turn on the TV." Another said that the orb "has made me a little bit more aware, makes me want to, when it does change colors, to see which of my buddies are on."

Once their TV was on and participants could see not only who was watching but what was being watched, their curiosity was stimulated further: they would often take a peek at the shows their buddies were watching. Said one participant, "If they're watching it, then maybe it's a good episode or something." Some participants used what their friends were watching as a sort of menu of program options to choose from. No one needed to suggest a show; they would simply be curious and join them. One participant described that she would turn on the TV "just to see who's on, and then most of the time we would all end up watching the same thing."

In the STV2 trials, participants often used lightweight messages like thumbs-up as greetings and to acknowledge others watching the same thing. For example, one said she sent messages to express "'Hey! You're watching the same show. Good show!' Kind of... acknowledging." In other instances, suggestions were used in a similar way, as a way to greet someone who just came on. Lightweight messages were also taken as a signal of availability to communicate. Sometimes this would culminate in a phone call:

"I noticed the orb was blue, so I knew somebody had their television on, and sure enough, it was [B2's household]. And I knew that her husband's at work and her kids were at school, so I deduced it was [B2]. And so without even saying 'Who's there?' I immediately went to her channel, which was Oprah, and I sent her a thumbs-up. And then she thumbs-upped me, and then two seconds later I said [to myself], 'This is dumb!' And then [I called her, and] we had a whole conversation." (B1)

When freeform communication was integrated in the system with STV3, we saw a similar pattern. The orb would alert participants that a buddy was available, and they would then test the buddy's interest in communicating:

"I sit down, and then I see the light go blue or whatever color, that means someone else is on there; and I go look and see, and then if I'm in the mood... I think if people are in the mood they'll type a line or two, and if they're not interested and just want to have a quiet night without having the keyboard in front of them in the family room, [they'll] just ignore it." (C5)

As an indication of how important this gauging of interest is, we can observe that with only one exception, all voice conversations were preceded by a text chat.

To generalize the steps reported: the participants become aware of their friends' availability through the ambient display, turn on their TV (thereby themselves showing up as available to the other participants), look at their buddy list, and often either join what one of their friends were watching or suggest that their friend join them. Once they were viewing together, they would usually send messages or emoticons. From this idealized flow, we can define different stages of interaction (Table 4).

Moving through the stages, the user becomes progressively more engaged with the experience, going from peripheral presence awareness to immersive participation through a number of intermediate levels. Part of the reason for this is that at each stage, the user has access to more detailed social information. However, the most interesting thing to note may be the two-way dynamic at work. As users become more deeply engaged, they also become increasingly present to and connected to their buddies, which tends to nudge their buddies to become more deeply engaged. In this way, the system encourages more social interaction and communication at every level.

Anti-Social Television

Although the purpose of STV is to connect people and strengthen social bonds, we have always been on the lookout for possible negative effects social television could have on sociability. In our focus groups, some participants were concerned about the risk that easy communication through Social TV could be a substitute for face-to-face socializing. Similarly, there have been concerns that it might increase TV viewing, making people less active and less social.

In our field trials, we saw little evidence of either of these effects. Although STV3 participants reported that the system substituted for some phone calls or text messages, they did not get together any less frequently in person. In fact, our results suggest that STV3 may have encouraged increased face-to-face interaction. Participants reported extending and receiving opportunistic invitations or suggestions to meet in person through STV3 that would otherwise not have occurred:

"There was one time that we started texting on the system and he's like, 'Hey, we're going out for dinner. Do you guys wanna join us?' First of all, we probably wouldn't have talked to him that night [without STV3] but since we did it was pretty neat. It was something that opens communication between everybody." (C1)

In the STV2 trials, several participants felt that the system made them watch more TV, while

Table 4. Different stages of interaction with the system, along with some notable characteristics.

Interaction	Presence specificity	Context detail	What others experience	Degree of participation
Noticing ambient display	Aggregate buddy list (+inferences)	TV is on/off	No presence info	Peripheral awareness
Viewing buddy list		The program(s) being watched	TV is on/watching program	↓
Co-viewing program	Aggregate household (+inferences)	Shared viewing context (common ground)	Sharing the viewing experience	↓
Message probes		Reactions to show, responses to messages	In contact, communicating	↓
Full two-way communication	Individual	User-provided context (+ background audio)	Rich conversation	Full involvement

with STV3 all agreed that they did not watch more television. Our usage logs do not indicate an increase in viewing.

Another concern about a system like STV is that the awareness it provides will not lead to feelings of closeness, but instead emphasize differences and cause alienation. The ability to observe friends' viewing habits may lead to the realization that you have less in common with them than you thought. In fact, while many participants told us in the initial interviews that they thought they had similar tastes to the others in their group, they in fact learned that their interests differed significantly. As one participant expressed it, "I have noticed that for the most part we don't watch a lot of the same shows." Several people commented that one particular household watched what they variously called "weird shows" or "crazy movies." That participant in turn observed, "they watch something totally different from what I watch."

Similarly, a presence indicator can also be an absence indicator. There were many occasions on which only one participant was watching TV, so that there was no one to talk to over the system: "The oddest thing, it's a nice night to be in… and not one person is on Social TV tonight!" Some participants would repeatedly check their buddy list, and sometimes try to contact people who were offline before giving up.

Even when other buddies showed up as being on the system, attempts to contact them were not always answered. "We were sending suggestions and nobody was answering us back, and it was like, 'Are these people just leaving their TV's on?'" Some participants felt like they were being ignored: "I thought, 'OK, maybe she just doesn't feel like responding.'" One participant speculated that her friends "just wanted to watch their show, and they were ignoring you."

Participants never explicitly stated that they felt rejected or were hurt by such non-responses, or that being the only person on the system made them feel lonely. However, turning the question around, they *were* concerned about how others would take it if *they* didn't answer. Accordingly, there was a social pressure to respond. For example, one participant felt "obliged to send something back" when she received a message, because "it just felt like they're doing something, so should I do something back…"

Another potentially disturbing problem which we saw scattered evidence of in the field trials was a displacement of sociability. In the STV1 sessions, one of our participants confided that because he was interacting with his friends, he was paying less attention to his son than he would have done otherwise. By bringing outsiders into the family room, social television can intrude on important "quality time," and weaken the closest ties even as it builds relationships with more distant people.

As stated by a participant in the STV3 trials, evenings are traditionally "our time with our significant others that we live with." By effectively bringing other people into the living room, social television challenges the home sphere. In some cases, this can literally drive away the ones who are closest: "My wife would probably just look at it, see something ridiculous [my friend] would write and say, 'Oh, you guys are stupid,' and she'd go out and leave the room." Taken to its ultimate consequence, this could lead to a situation like that reported by Quico (2003): "I left my wife because I am in love with the TV set."

The Uses of Social Television: Discussion

Earlier studies have tended to portray the conversation experience of social television as very much an aspect of the viewing experience, with both the structure and the content of the communications tightly linked to the program content. In contrast, the STV3 field trial reveals a communication system that goes beyond simply being a tool for communicating about television.

That is not to say that integrating voice and text communication with the TV is irrelevant. Through the presence mechanism, TV viewing influences the set-up of conversations in a number of ways, most of them positive. Although there is always a risk of causing harm to existing social relationships, so far the overall effect has been to strengthen social ties and even cause new ones to form.

WHAT DON'T WE KNOW ABOUT SOCIAL TELEVISION?

Many of the important unanswered questions about social television arise from what we have learned so far. Research into the appeal of social television has uncovered challenges around privacy and disruption that need to be resolved. Evaluations of different types of communication modalities have indicated high-level preferences, but left details unanswered. And finally, studies of social television behaviors have found that the process of initiating contact, and the way remote communication interacts with collocated relationships, require further investigation. Here, we will look at these questions from a slightly different angle, organizing them under three big questions: how the challenges raised by the fact that TV is often a shared device and a shared activity can be resolved; which missing technical pieces are necessary to realize social television; and which direction social television should go in to provide a distinct and particularized social experience.

How Do We Support Multiple Users within the Home?

In households with more than one person, the main television is usually a shared device, what we might call "public within the home." It is used by multiple different people, both separately at different times and in groups at the same time. This fact has a major impact on social television.

Of all the privacy concerns surrounding social television, the ones most manifest in the field are those over how to protect communication from other people in the household. Let us call this the "Little Brother" problem. At different times and in different situations, Little Brother could be a sibling, a girlfriend, boyfriend or spouse, a parent, a child, or a roommate. Little Brother complicates social television in a number of ways:

- Little Brother is nosy, and particularly interested in the private matters and personal relationships of his family. He listens in on family members' conversations unannounced. He impersonates them on text chat with their friends. He sometimes conspires with the real person to fool others, sometimes with that person's buddies, and sometimes he acts by himself.
- Little Brother is (at least sometimes) a child, and it is important to his parents that he is protected from inappropriate messages like swearing and adult conversations, kept safe from strangers, and that they can restrict and control how he uses the system.
- However, Little Brother does not want to be monitored.

In addition to these privacy issues, Little Brother faces another two problems of television sociability:

- Little Brother has his own friends. Most of the time he wants to talk only to them and see only what they are doing, and doesn't care about his other household members' friends.
- Little Brother is (usually) a family member, and watching TV is traditionally his quality time with the others in the household. Social television makes it easier for outsiders to intrude on these family moments.

In order to address these problems, a way to manage the multiple identities within the household is needed. This may include separate accounts for each user, with a logon process to verify identity. Since multiple users may be watching at the same time, the presence states needs to be flexible enough to represent this, and the interface needs to bridge or merge separate buddy lists and allow interaction from different user accounts. To date, Media Center Buddies is the only documented social television system with explicit support for multiple users (Regan & Todd, 2004b). For the most part, the design requirements for such capabilities remain unexplored, although ideas from the general interactive TV field may come into play (e.g. Farré, 2006).

Several approaches could be taken to limit the increase of intrusion and reduction of intimacy caused by social television. One avenue could be social television applications that support co-located sociability, socializing within the home, as suggested by comments by Chorianopolous (2007). In another direction, the shared nature of the TV makes it a potentially more inclusive social device. We have seen several instances where STV helped make other family members part of a relationship, such as when the girlfriend of one participant for the first time had a real conversation with one of his friends, or as the husband of another participant said about his wife's relationship with her sisters: "Usually I wouldn't know what they talk about. Here I get more involved, probably more than I wanted to." It remains to be seen if this involvement can be maintained once each household member can contact *their own* friends.

If, then, multiple users want to use social television individually at the same time, there are some scarce resources that need to be managed. Two of the most important ones are screen real-estate and system control. TV's have traditionally only had one remote control, which limits direct input to a single user at any one time. Similarly, only a limited amount of information can be displayed on the television screen, and will always interfere with the TV programming. Potential solutions to these problems include appropriating other devices, such as cell phones, for use as control and input mechanisms (Lin & Chen, 2005; Cesar, Bulterman, & Jansen, 2006), and employing secondary screens to display supplementary and personalized information (Fink, Covell, & Baluja, 2006; Cesar et al., 2008). This brings us to the issue of equipment.

What Equipment Do We Need?

Even if social television proves itself viable in field trials and pilot deployments, significant barriers to mainstream adoption remain. One of the main bottlenecks for deployment is the requirement of specific hardware. For this pragmatic reason, determining the necessary hardware components of a social television system is important. From the point of view of social television, current hardware is limited in terms of processing power, middleware support and networking ability, in the availability of input devices for communication (including text, voice, video and other forms), and in the use of peripheral and secondary displays.

We found an ambient display to be a popular addition to the STV system, and an important part of the experience. Because such displays are rare in the home, the question becomes whether the functionality can be provided through other equipment. One candidate are the secondary screens mentioned above. As already pointed out, the more important function of secondary screens is to move the display of (and often interactions with) advanced and personalized features off the TV. The effects of such a separation between TV content and social application on how social television is perceived and used are not well understood.

The most important hardware requirement, however, is very likely the choice of input device, or input devices. This choice is primarily driven by the communication modality. For text chat, a text input device is needed; for voice chat, a microphone; and for video communication, a video camera. A pointing device is desirable for on-screen drawing; closed-form communication (like emoticons and multiple-choice messages) can make do with a regular remote; while a game controller is ideal for rich avatar interactions.

Current findings point to text as the preferred communication modality over voice, and seem to show that video communication is not suited for social television. Nevertheless, efforts are likely to continue with all of these means of communication, and others. In any case, the demonstrated inadequacy of closed-form communication indicates that regular remote controls are insufficient for input.

We can predict, then, that a flood of different input devices is heading for the living room, if people are willing to let them in. However, the details remain open. For voice, what are the merits and drawbacks of a room microphone compared to headsets? For text, are any entry methods other than a full QWERTY keyboard acceptable? We know that these choices affect the experience, but the exact effects are unknown.

The numerous complicating factors with deploying social television for the TV, and the restricted capabilities of most current set-top boxes, raise the possibility that social experiences can more easily make a breakthrough as part of the burgeoning TV and video offerings on the Internet and on mobile platforms (Schatz, Wagner, Egger, & Jordan, 2007). The chance that widespread deployment may happen through systems and devices already primarily used for communication only makes the question more acute: in what way can Social TV differentiate itself, so that it offers a communication experience that is not interchangeable with existing offerings?

Where Should Social Television Be Going?

From the original definitions of social television (whether narrow or broad), both research results and system developments have tended to blur the line between Social TV and other social network and communication technologies. As online video becomes increasingly integrated with chat (e.g. Zync) (Shamma, Bastea-Forte, Joubert, & Liu, 2008), the distance between text chat over social television and over IM becomes smaller. Findings that show that most communication over social television is not content related make such systems seem less distinct. And with the prospect of using a cell phone or computer for the interaction, it becomes hard to separate social television interaction from any other form of communication.

For social television to become merely an extension of existing communication systems would be mildly disappointing, given some of stated ambitions of the field. We believe that the potential for social television to contribute unique and powerful user experiences exists. In order to achieve it, two goals should be pursued: close integration with the activity of television viewing, and close integration with the television content.

Our studies have shown that television presence information influences communication initiation. Presence awareness as part of a frequent and extended laid-back, recreational activity provides a spur to striking up conversation. At the same time, the degree of contact and transparency imposes social pressure, not just to communicate, but also around the patterns of TV viewing. Social television systems should seek to harness these forces positively (cf. Consolvo, Everitt, Smith, & Landay, 2006), for example by improving the mechanism of word-of-mouth recommendations of high-quality programming, and be designed to minimize the risk of alienation (Sokoler & Svensson, 2008). Future research should continue to explore the social effects of exposing TV viewing to public scrutiny.

Alongside these efforts, communication modes that are more directly focused on the content deserve more attention. These include shared program tags, exchanging clips, and social bookmarks that point to moments of interest within a program. Previous studies have shown that commercial content such as music is appropriated by people for their personal and social uses (Bentley et al., 2006), and one could take the perspective that social television is about using television content as a resource for personal and social interaction. Even more promising is the creation of social television experiences tailored to specific programs, such as multiplayer competition within quiz shows (Luyten, Thys, Huypens, & Coninx, 2006), fact checking or commentary during political debates, or a gossip channel as part of entertainment news. Enabling third-party creation of these custom experiences requires standardized protocols and a social television API. The ability to define interaction formats could allow a wide variety of structured social experiences for the television (van den Bergh et al., 2007).

Social TV systems can certainly provide generic text and voice communication experiences over the television. Much of the research and development to date has been explicitly of implicitly aimed at reaching and validating this capability, and most of the major questions have now been answered. Moving on from this early achievement, social television should set its sights on making a difference and creating an independent identity for itself.

CONCLUSION

Several years ago we embarked on a project with a simple idea: to create a "virtual couch" that would let people watch TV together even when apart. We quickly found ourselves on a lengthy exploration of a much bigger topic; an exploration that has involved extensive prototyping, reading, and primary research. We have made some surprising discoveries, and on at least one occasion had to change our course.

In that time, the uncharted waters of the topic have been mapped, at least to some extent, by ourselves and others. On three big questions about social television we have at least partial, tentative answers. The exploration is not yet complete, and there may be major discoveries yet to be made in the further reaches of the subject area. However, according to our bearings, some of the emphasis is moving from research to more practical concerns of design, implementation and deployment. Future advancements in social television are likely to come from new products as much as from dedicated research.

ACKNOWLEDGMENT

We would like to thank all the other people who have worked on or been involved with the Social TV project, including Vernell Chapman, Ambiga Dhiraj, Frank Bentley, Dave Wheatley, Josh Hurwitz, Larry Marturano, Seonyoung Park, Timothy Sohn, and many others. Thank you.

REFERENCES

Abreu, J. F., & Almeida, P. (2008). From Scratch to User Evaluation: Validating a Social iTV Platform. In A. Lugmayr, S. Kemper, M. Obrist, T. Mirlacher, & M. Tscheligi (Eds.), *Adjunct Proceedings of EuroITV 2008* (pp. 31–32). Tampere, Finland: Tampere University of Technology.

Agamanolis, S. (2008). At the Intersection of Broadband and Broadcasting: How Interactive TV Technologies can Support Human Connectedness. *International Journal of Human-Computer Interaction, 24*(2), 121–135.

Ali-Hasan, N. F. (2008). Exploring Social Media Scenarios for the Television. In *Proceedings of the 2008 International Conference on Weblogs and Social Media.*

Baillie, L., Fröhlich, P., & Schatz, R. (2007). Exploring Social TV. In *Proceedings of the 29th International Conference on Information Technology Interfaces* (pp. 215–220). IEEE.

Bentley, F., Metcalf, C., & Harboe, G. (2006). Personal vs. Commercial Content: The Similarities Between Consumer Use of Photos and Music. In R. Grinter, T. Rodden, P. Aoki, E. Cutrell, R. Jeffries, & G. Olson (Eds.), *Proceeding of the Twenty-Fourth Annual SIGCHI Conference on Human Factors in Computing Systems* (pp. 667–676). New York: ACM Press.

Bergh, J. van den, Bruynooghe, B., Moons, J., Huypens, S., Handekyn, K., & Coninx, K. (2007). Model-Driven Creation of Staged Participatory Multimedia Events on TV. In P. Cesar, K. Chorianopoulos, & J. F. Jensen (Eds.), *Proceedings of the 5th European Conference, EuroITV 2007* (pp. 21–30). Berlin: Springer.

Bernard, H. R. (Ed.) (1998). *Handbook of Methods in Cultural Anthropology.* Walnut Creek, California: AltaMira.

Beyer, H., & Holtzblatt, K. (1998). *Contextual Design: Defining Customer-Oriented Systems.* San Francisco, California: Morgan Kaufman.

Boertjes, E., Klok, J., & Schultz, S. (2008). ConnecTV: Results of the Field Trial. In A. Lugmayr, S. Kemper, M. Obrist, T. Mirlacher, & M. Tscheligi (Eds.), *Adjunct Proceedings of EuroITV 2008* (pp. 21–22). Tampere, Finland: Tampere University of Technology.

Cesar, P., Bulterman, D. C. A., & Jansen, A. J. (2006). An Architecture for End-User TV Content Enrichment. In G. Doukidis, K. Chorianopoulos, & G. Lekakos (Eds.), *Proceedings of the 4th European Conference, EuroITV 2006* (pp. 39–47). Athens, Greece: Athens University of Economics and Business.

Cesar, P., Bulterman, D. C. A., Geerts, D., Jansen, J., Knoche, H., & Seager, W. (2008). Enhancing Social Sharing of Videos: Fragment, Annotate, Enrich, and Share. In *Proceedings of the 16th international Conference on Multimedia*, in press. New York: ACM Press.

Chorianopoulos, K. (2007). Content-Enriched Communication: Supporting the Social Uses of TV. *The Journal of The Communications Network, 6*(1), 23–30.

Consolvo, S., Everitt, K., Smith, I., & Landay, J. A. (2006). Design requirements for technologies that encourage physical activity. In R. Grinter, T. Rodden, P. Aoki, E. Cutrell, R. Jeffries, & G. Olson (Eds.), *Proceeding of the Twenty-Fourth Annual SIGCHI Conference on Human Factors in Computing Systems* (pp. 457–466). New York: ACM Press.

Coppens, T., Trappeniers, L., & Godon, M. (2004). AmigoTV: Towards a Social TV Experience. In *Proceedings of EuroITV 2004.*

Coppens, T., Vanparijs, F., & Handekyen, K. (2005). *AmigoTV: A Social TV Experience Through Triple-Play Convergence* (Technology White Paper). Retrieved September 5, 2008, from http://www1.alcatel-lucent.com/com/en/appcontent/apl/T0205-Amigo_TV-EN_tcm172-195461635.pdf

Ducheneaut, N., Moore, R. J., Oehlberg, L., Thornton, J. D., & Nickell, E. (2008). Social TV: Designing for Distributed, Sociable Television Viewing. *International Journal of Human-Computer Interaction, 24*(2), 136–154.

Farré, D. G. (2006). Presence-Induced Implicit Interaction and Audience Measurement in Interactive Television. In G. Doukidis, K. Chorianopoulos, & G. Lekakos (Eds.), *Proceedings of the 4th European Conference, EuroITV 2006* (pp. 105–109). Athens, Greece: Athens University of Economics and Business.

Fink, M, Covell, M., & Baluja, S. (2006). Social – and Interactive – Television: Applications Based on Real-Time Ambient Audio Identification. In G. Doukidis, K. Chorianopoulos, & G. Lekakos (Eds.), *Proceedings of the 4th European Conference, EuroITV 2006* (pp. 138–146). Athens,

Greece: Athens University of Economics and Business.

Geerts, D. (2006). Comparing voice chat and text chat in a communication tool for interactive television. In A. Mørch, K. Morgan, T. Bratteteig, G. Ghosh, & D. Svanaes (Eds.), *Proceedings of the 4th Nordic Conference on Human-Computer interaction* (pp. 461–464). New York: ACM Press.

Harboe, G., Massey, N., Metcalf, C., Wheatley, D., & Romano, G. (2007). Perceptions of Value: The Uses of Social Television. In P. Cesar, K. Chorianopoulos, & J. F. Jensen (Eds.), *Proceedings of the 5th European Conference, EuroITV 2007* (pp. 116–125). Berlin: Springer.

Harboe, G., Massey, N., Metcalf, C., Wheatley, D., & Romano, G. (2008a, May). The Uses of Social Television. *Computers in Entertainment, 6*(1), 1–15.

Harboe, G., Metcalf, C. J., Bentley, F., Tullio, J., Massey, N., & Romano, G. (2008b). Ambient Social TV: Drawing People Into a Shared Experience. In *Proceeding of the Twenty-Sixth Annual SIGCHI Conference on Human Factors in Computing Systems* (pp. 1–10). New York: ACM Press.

Harrison, C., & Amento, A. (2007). CollaboraTV: Using Asynchronous Communication to Make TV Social Again. In A. Lugmayr & P. Gołębiowski (Eds.), *Adjunct Proceedings of EuroITV 2007* (pp. 218–222). Tampere, Finland: Tampere International Center for Signal Processing.

Hess, J. (2008). Community Awareness in Social TV Environments. In A. Lugmayr, S. Kemper, M. Obrist, T. Mirlacher, & M. Tscheligi (Eds.) *Adjunct Proceedings of EuroITV 2008* (pp. 33–34). Tampere, Finland: Tampere University of Technology.

Huang, E. M., Harboe, G., Tullio, J., Novak, A., Massey, N., Metcalf, C. J., & Romano, G. (2009). Social Television Comes Home: A Field Study of Communication Choices and Practices in TV-Based Text and Voice Chat. In *Proceeding of the Twenty-Seventh Annual SIGCHI Conference on Human Factors in Computing Systems*, in press.

Lin, C. C. & Chen, M. S. (2005). On controlling digital TV set-top-box by mobile devices via IP network. In *Proceedings of the Seventh IEEE International Symposium on Multimedia* (pp. 8–16). IEEE.

Luyten, K., Thys, K., Huypens, S., & Coninx, K. (2006). Telebuddies: Social Stitching with Interactive Television. In *CHI '06 Extended Abstracts on Human Factors in Computing Systems* (pp. 1049–1054). New York: ACM Press.

Metcalf, C., Harboe, G., Massey, N., Tullio, J., Romano, G., Huang, E. M., & Bentley, F. (2009). Examining Presence and Lightweight Messaging in a Social Television Experience. *Trans. on Multimedia Computing, Communications and Applications* (TOMCCAP), in press.

Nathan, M, Terveen, L., Harrison, C., Yarosh, S., Stead, L., & Amento, B. (2008). CollaboraTV: Making Television Viewing Social Again. In *Proceedings of UXTV 2008*, in press.

Oehlberg, L., Ducheneaut, N., Thornton, J. D., Moore, R. J., & Nickell, E. (2006). Social TV: Designing for Distributed, Sociable Television Viewing. In G. Doukidis, K. Chorianopoulos, & G. Lekakos (Eds.) *Proceedings of the 4th European Conference, EuroITV 2006* (pp. 251–259). Athens, Greece: Athens University of Economics and Business.

Orwell, G. (1990). *Nineteen Eighty-Four*. London: Penguin Books.

Quico, C. (2003). Are communication services the killer app for Interactive TV? Or: I left my wife because I am in love with the TV set. In J.

Masthoff, R. Griffiths & L. Pemberton (Eds.) *Proceedings of the 1st European Conference, EuroITV 2003* (pp. 99–107). Brighton, UK: University of Brighton.

Regan, T., & Todd, I. (2004a). *Media Center Buddies: Instant Messaging Around a Media Center* (Technical Report MSR-TR-2004-47). Redmond, WA: Microsoft Research. Retrieved September 5, 2008, from ftp://ftp.research.microsoft.com/pub/tr/TR-2004-47.pdf

Regan, T., & Todd, I. (2004b). Media Center Buddies: Instant Messaging Around a Media Center. In *Proceedings of the Third Nordic Conference on Human-Computer interaction* (pp. 141–144). New York: ACM Press.

Ruyter, B. de, Huijnen, C., Markopoulos, P., & IJsselstein, W. (2003). Creating social presence through peripheral awareness. In *Proceedings of the Tenth International Conference on Human-Computer Interaction* (pp. 889–893).

Schatz, R. Baillie, L., Fröhlich, P., & Egger, S. (2008). Getting the Couch Potato to Engage in Conversation: Social TV in a Converging Media Environment. In A. Lugmayr, S. Kemper, M. Obrist, T. Mirlacher, & M. Tscheligi (Eds.), *Adjunct Proceedings of EuroITV 2008* (pp. 25–26). Tampere, Finland: Tampere University of Technology.

Schatz, R., Wagner, S., Egger, S., & Jordan, N. (2007). Mobile TV Becomes Social: Integrating Content with Communications. In V. Luzar-Stiffler & V. H. Dobric (Eds.), *Proceedings of the ITI 2007 29th International Conference on Information Technology Interfaces* (pp. 263–270). Zagreb, Croatia: SRCE University of Zagreb.

Shamma, D.A., Bastea-Forte, M., Joubert, N., & Liu, Y. (2008). Enhancing Online Personal Connections Through the Synchronized Sharing of Online Video. In *CHI '08 Extended Abstracts on Human Factors in Computing Systems* (pp. 2931–2936). New York: ACM Press.

Sokoler, T., & Svensson, M. S. (2008). PresenceRemote: Embracing Ambiguity in the Design of Social TV for Senior Citizens. In M. Tscheligi, M. Obrist, & A. Lugmayr (Eds.), *Proceedings of the 6th European Conference, EuroITV 2008* (pp. 158–162). Berlin: Springer.

Tullio, J., Harboe, G., & Massey, N. (2008). Investigating the Use of Voice and Text Chat in a Social Television System. In M. Tscheligi, M. Obrist, & A. Lugmayr (Eds.), *Proceedings of the 6th European Conference, EuroITV 2008* (pp. 163–167). Berlin: Springer.

Weisz, J. D., Kiesler, S., Zhang, H., Ren, Y., Kraut, R.E., & Konstan, J. A. (2007). Watching Together: Integrating Text Chat with Video. In *Proceeding of the Twenty-Fifth Annual SIGCHI Conference on Human Factors in Computing Systems* (pp. 877–886). New York: ACM Press.

Weisz, J. D., & Kiesler, S. (2008). How Text and Audio Chat Change the Online Video Experience. In *Proceedings of UXTV 2008*, in press.

ENDNOTES

1 http://get.live.com/messenger
2 http://www.aim.com/
3 http://www.ambientdevices.com/cat/orb/

Chapter XI
ConnecTV:
Share the Experience

Erik Boertjes
TNO Information and Communication Technology, The Netherlands

Jente Klok
TNO Information and Communication Technology, The Netherlands

Omar Niamut
TNO Information and Communication Technology, The Netherlands

Martijn Staal
TNO Information and Communication Technology, The Netherlands

ABSTRACT

The combination of content and communication has proven to be a powerful and successful concept. Many online services not only allow for consumption of content, but also give their users the possibility to exchange views on the content among each other. YouTube for instance, not only allows its visitors to watch movie clips, but also to discuss them, to review them, and to send recommendations about them. Electronic Program Guides on the web more and more offer social functionality in addition to listing television programs; thus integrating content and communication. Users can discuss their favorite TV shows in forums that are organized around TV programs. ConnecTV is a social interactive TV service that combines communication with watching television. It makes watching TV a social activity, and aims to give its users the feeling of 'watching together'. ConnecTV was developed in the B@Home research project (B@Home). Among the project's goals was an investigation of the type of services that will become feasible when 'fiber to the home' is widely introduced, giving households broadband internet connections with significantly more capacity than today. Although the functional design of ConnecTV was carried out in close cooperation with experts from the media industry, the real test for end-user acceptance would be a field trial with the service. In addition, a trial would reveal the effects on the users' viewing behavior, and would give insight in how to make a positive business case around ConnecTV. In 2007 ConnecTV was implemented and a field trial was held in about 50 households in the town of Enschede, in The Netherlands. In literature, many systems have been proposed that combine social networks with consuming content in general, or with watching TV specifically. Examples are AmigoTV (Coppens, Trappeniers, & Godon, 2006), 2BeOn (Abreu, Almeida, & Branco, 2001), SocialTV (Harboe, Massey, & Metcalf, 2006), ChaT.V. (Fink, Covell, & Baluja, 2006) and CollaboraTV (Harrison & Amento, 2007). Although some of ConnecTV's functionality can be found in these services as well, other functions (like following a buddy, or switching to the most popular channel) are new. The main contribution of the research described in this chapter though lays in the field trial of ConnecTV. With some of the above-mentioned social TV services field trials have been performed (e.g. (Harboe, Massey, & Metcalf, 2006)), but they were limited in size (typically two or three groups), and did not use extensive logging of user activities. To our knowledge, a field trial of a social TV service at the scale of the ConnecTV field trial with such extensive logging, surveys and experience sampling has not been performed before. This chapter starts with explaining the functionality of ConnecTV and its implementation. The main focus of this chapter is the field trial: the set-up, the research questions, the research methodology and the results of the field trial are discussed extensively. The chapter concludes with a description of the most viable approaches for a positive business case, and gives an outline for future work.

THE CONNECTV SERVICE

Developing ConnecTV was an iterative process. Early in the project, an interactive mock-up was realized and presented to experts from the media industry (cable operators, interactive TV application developers, etc.) and potential end-users. After several iterations, a prototype was built that functioned in a lab environment. The step towards a system that could be field tested was still significant. A field trial version to be used by 'real' people is fundamentally different from a prototype that is merely meant for demonstration purposes.

We put a lot of effort in redesigning the user interface and user interaction and tested it with a dozen people. While the prototype had many functions, we carefully selected those that we thought were essential for the concept behind ConnecTV and could be implemented within the available time constraints. We ended up with the functions described below.

The Buddy List

The buddy list lists each of the user's buddies together with the TV program that they are currently watching (Figure 1). The list is divided in two categories: online (with buddies that are watching TV) and offline (with buddies that are not watching). In addition to the traditional human buddies, we introduced so-called 'theme-buddies'. Each of them represent a certain genre, such as Comedy, Sports, etc. If a program in one of those genres is broadcasted, the corresponding theme buddy gets online, and indicates the program name.

Change to the Channel that a Buddy is Watching

This function is one of the four interactive functions that ConnecTV provides from the buddy list (Figure 2). When selecting a buddy, users can change their channel directly to the same channel as their buddy is watching.

Figure 1. Buddy list in ConnecTV

Figure 2. Available functions related to an online buddy. From top to bottom: 'Change channel to Nederland 1 (which is the channel Leon is watching)', 'Send a recommendation', 'Invite a friend', 'Follow a friend'.

TV Program Recommendation

Users can recommend the program that they are watching to one or more of their buddies. If a user watches a program that he thinks might be interesting to a buddy, he selects the buddy and sends a message which is displayed on the target user's screen saying '[User X] recommends to switch to [channel C] to watch [program P]'. The target user can then decide to accept or deny this invitation with one touch of a button (Figure 3).

Figure 3. Receiving a recommendation from a buddy. "I am watching RTL4, would you like to watch this channel as well?"

Follow a Friend

This function allows someone to replicate the viewing behavior of one of his buddies. A user chooses from the buddy list a buddy to follow. When this buddy switches channels, the TV of the user does the same. Until one of the two ends this function, the user will constantly watch the same program as the buddy.

Invite a Friend

This function is the complement of the previous one: Invite Friend allows users to invite one of their buddies to follow them in the way described under 'Follow a Friend'.

Buddy Status Messages

When one of the user's buddies is watching the same channel as the user, a corresponding message is displayed on screen: '[Buddy X] is currently watching this program as well' (Figure 4). In addition, a small buddy icon appears in the lower right corner of the screen, indicating the 'presence' of another buddy on this channel. Other changes in a buddy's status are displayed as well:

Figure 4. Status message: "Leon is watching this channel as well"

- '[Buddy X] stopped watching this program'
- '[Buddy X] is watching TV now'
- '[Buddy X] stopped watching TV'
- '[Buddy X] stopped following you'

Most Popular Channel

ConnecTV allows users to switch to the channel that is watched by most of their buddies. When pressing the corresponding button, the system checks what channel is most popular among the user's buddies and switches to that channel. If multiple channels are equally popular, the user can switch between them in a round-robin fashion, pressing the same button multiple times.

All messages are displayed on-screen; all interaction is done with the remote control of the media center PC on which ConnecTV was installed (see next section).

IMPLEMENTATION

The following list of requirements served as the starting point for the implementation:

- The service should operate on the basis of a client-server model. All client-side functionality should be operated by a remote control, in order to not deviate from the traditional TV operation. Also, the service should extend, rather than replace, the existing TV service.
- The functionality offered by the service should be enticing and should appeal to the user, so users would be invited to use the functionality. The functionality should be sufficient to provide information for our research purposes.
- The service should allow for logging all relevant user actions and provide means for in-service experience and behavior measurements.
- The client equipment should be adjusted to the trial environment with regards to the available TV and data signals and the general home situation. The client implementation should run on the available hardware and allow for dynamic service configuration, maintenance and upgrade. Given that the service would impact such a basic service as TV, the service should be robust to prevent any service break-down.
- The server-side solution should provide for flexible user management. The server platform itself should be manageable so that administrators can check and/or change user states and information.

The technical feasibility of the social TV service had been demonstrated in earlier stages of the project, on both PC-based and IPTV set-top box platforms. Each of these prototypes operated in a client-server architecture. Given the field trial environment, we used an analogue TV source signal because of its widespread availability. A media center PC was selected for the client-side equipment to foster rapid development and upgrade of the service and to easily allow for the analogue TV source.

Figure 5 displays the functional architecture and protocol flows that constituted the service. On the server side, two dedicated servers were

Figure 5. Technological design of the service

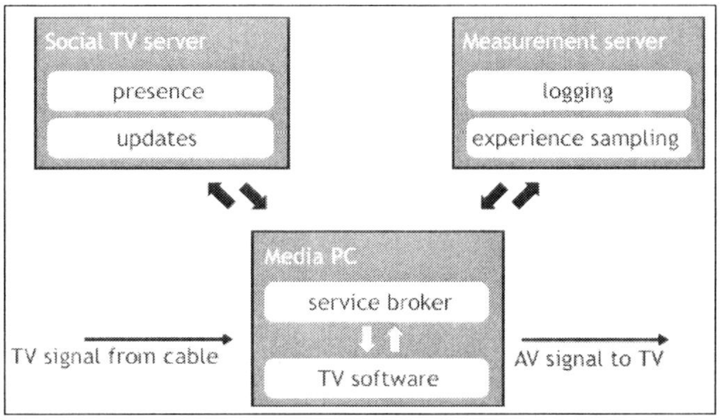

installed, one hosting the presence and service update functionality and a second server was responsible for the user action logging and in-service response measurement.

On the media center PC, front-end TV software was deployed based on the commercially available SageTV media center software (SageTV). Since this software offered limited messaging and state awareness functionality, a dedicated Java-based service broker operated on the client PC as a message proxy between the TV software and the servers. The service broker further handled the sorting of the buddy list in categories and alphabetically, user identification, and determination of the most popular channel. A dedicated XML-based protocol was developed for communication between the service broker and the TV front-end software. On the Social TV server, the open-source software Openfire was installed. This server handled the routing of messages exchanged between service brokers from different users, as well as maintaining and updating the presence information of the users. The messages between service brokers were implemented using the Jabber/XMPP protocol (Jabber, XMPP).

The measurement server was developed using TUMCAT technology (TUMCAT). This allowed for logging all user behavior as well as presenting the user with questionnaires while they were watching TV. A dedicated TUMCAT log message protocol was employed between the service brokers and the measurement server.

SETTING UP THE FIELD TRIAL

Finding participants was not an easy task. The social nature of ConnecTV required that participants were organized in groups of people that know each other well enough to use the ConnecTV functionality among each other. We put posters and flyers at places that were likely to be visited by groups of people that know each other, like sports clubs, schools, churches and community centers. People could sign up on a website as a group: people that would like to join needed to bring in at least three of their friends or acquaintances as well.

After two months, we found 52 participants, divided over 8 groups of people ('buddy groups'), varying in size from 4 to 10 people. We had groups of people that were colleagues; others joined the same sports club, did the same studies, or were family. The group of participants consisted mainly of men (91%). Most of the participants were highly educated (73%). Their ages ranged from 21-25 (23%), 26-35 (45%), and 36-45 (23%). Clearly, the participants were not representative for the Dutch population. We feel however that they do fit a typical early adopter's profile of people that are eager to test a new service, are very critical towards it, but at the same time very willing to help think about improvements.

The field trial was divided into three phases (see Table 1); a baseline measurement phase in which the users could only watch 'normal' TV without being able to use any of the ConnecTV's functions and two phases in which all ConnecTV functionalities where available. The field trial was split into two phases for practical reasons: despite intensive testing of the ConnecTV services in the lab, due to technical problems only about half of the participants could use ConnecTV in phase 1. In phase 2 these problems had been solved.

RESEARCH METHODOLOGY

The main research questions our field trial was to answer were:

- To which extent does ConnecTV affect people's TV-viewing habits?
- To which extent does ConnecTV change the TV-viewing experience?
- What are promising ways of commercial exploitation of ConnecTV?

Besides, we were interested in participants' use of the new social TV functions, and in their opinions about them:

- How do participants use the functions of ConnecTV?
- What do participants think about ConnecTV?

Table 1. Field trial setup

Fase	Duration	# users
Baseline measurement	2 weeks	50
Field trial phase 1	4 weeks	24
Field trial phase 2	5 weeks	38

To answer all these questions, we used three different data gathering methods: online surveys, logging of all system and user actions, and so-called experience samples. Logging and experience samples were provided by TUMCAT, a set of experimental tools of TNO ICT dedicated to measure user experiences at the moment, location and in the context where they occur (TUMCAT).

Online Surveys

Participants were asked twice to answer an extensive online questionnaire. The first survey was sent out at the start of the trial, before the participants had been able to use ConnecTV, asking for their general data, attitudes and expectations. The second survey was sent out at the very end of the trial, asking about the usage of the ConnecTV service, user experiences and commercialization.

Logging

During the trial, all activities of the users were logged by TUMCAT, e.g. when the TV was turned on, when a participant switched to watch another channel, when a participant called up the buddy list, when he or she sent out a recommendation (and to which buddy), etc.

Experience Samples

Throughout the pilot, participants received experience samples: short questions dedicated to inquire about experiences and intentions related to activities of the participant at that moment. In this trial, multiple choice questions were projected on their TV-screen. The experience samples could be triggered by:

- Actions of the user, e.g. a number of times after receiving a recommendation, the invited buddy was sent an experience sample asking him how the invitation affected his experience. The sender of the recommendation would receive an experience sample asking him if he was checking whether the invited buddy was following up the invitation.
- Time, e.g. at predefined points of time during the trial, an experience sample was sent out to find out how participants were feeling at that moment.

The above data gathering methods where used to measure the TV-viewing behavior, TV-viewing experience, and people's opinion about ConnecTV.

Measurement of Usage of ConnecTV

The usage of ConnecTV functions was mainly recorded by logging: every time a participant turned on the TV, switched channel, or used a ConnecTV function like sending out a recommendation, it was logged by TUMCAT. After a number of specific actions, participants received an experience sample connected to their behavior. Also in the final survey, people were asked about their use of ConnecTV.

Figure 6. Circumplex model, from Desmet & Hekkert (2007)

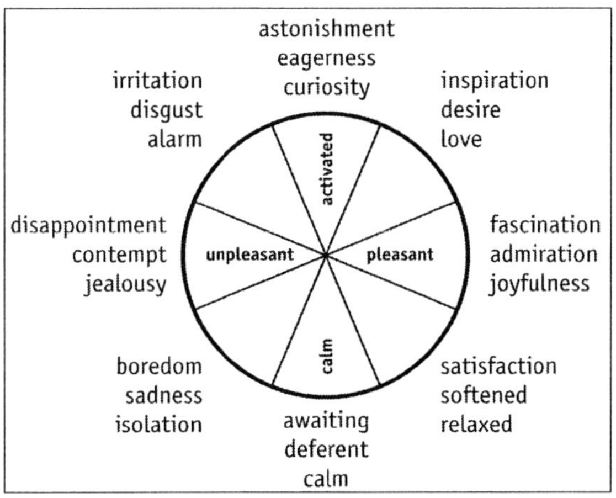

Measurement of TV-Viewing Behavior

The participants' TV-viewing behavior was mainly measured by logging when a participant turned on the TV and when they switched to another TV-channel. In both surveys more qualitative data was measured: e.g. the type of programs people mainly watched.

Measurement of TV-Viewing Experience

To measure the TV-viewing experience, we first needed to model this experience. As a base, we took the Circumplex model described by Desmet & Hekkert (2007). This model describes the experience of users while interacting with a product, along two central axes: unpleasant – pleasant and calm – activated (see Figure 6).

We adapted this model for two reasons (see Figure 7):

1. The social aspect is a very important part of ConnecTV. To measure the experience connected with this, we added the dimension "Isolated – Connected"
2. We wanted to use experience samples to question participants about their experiences while using ConnecTV. This implied our questions to be asked in simple terms, which meant that we needed to simplify the Circumplex model.

To measure the experience of participants while watching TV, the following five dimensions were used:

1. Unpleasant – Pleasant
2. Calm – Activated

3. Bored – Inspired
4. Satisfied – Irritated
5. Isolated - Connected

The actual experience of participants along these dimensions was measured in the following ways:

- In both surveys, various scenarios were described (e.g. watching TV along or in the company of others) and participants were asked how they would feel in those scenarios
- On arbitrary moments during the pilot, experience samples were sent out to participants, asking them how they felt.

Figure 7. Measurement of TV-experience in five dimensions – simplification and modification of the Circumplex model.

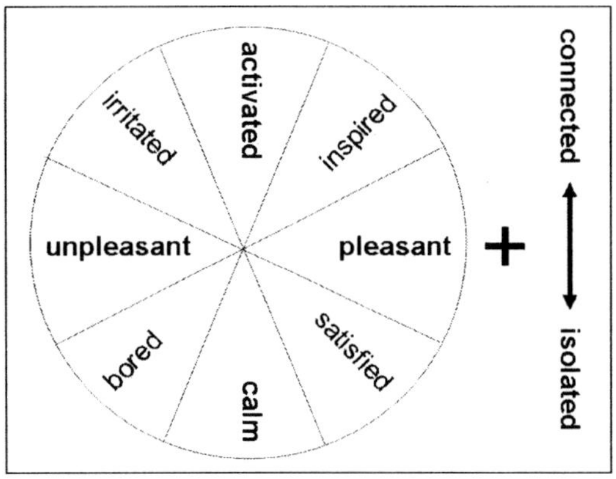

Measurement of People's Opinion about ConnecTV

What participants thought of ConnecTV and its functions, was measured in two ways: through the surveys and through experience samples, which were sent a number of times right after a participant used a specific function of ConnecTV.

RESULTS OF THE FIELD TRIAL

Usage of ConnecTV

The buddy list is the central function of ConnecTV: it provides access to all other ConnecTV functions. Therefore, calling up the buddy list provides a good indication of the general usage of ConnecTV. Table 2 shows that in phase 1, on average each participant used the buddy list slightly over twice a day, in phase 2 this dropped to once a day. The main reason participants gave for calling up the buddy list was curiosity about which buddies are watching TV. The buddy list was typically called up at the start of a TV watching session. Participants indicated that seeing what their buddies were watching, often guided their own choice in what program to watch.

Changing to the channel a buddy is watching at that moment was the ConnecTV-function that was used most (see Figure 8). In the final survey, 10% of the respondents claim to have done this at least once a day. In the end survey, a similar amount of respondents say they sent out recommendations at least once a day. However,

Table 2. Number of times the buddy list was called up

Phase	# buddy list was called up	Average per person per week
Baseline measurement	35	≈0
Field trial phase 1	1664	17
Field trial phase 2	1266	7

this number is overrated since the logging data showed that all in all these functions were used only around 40 times in each phase. As Table 3 shows, a little over 20% of these recommendations were accepted, while 24% - 36% were refused by their addressees. The main reason for this was that they already felt they were watching something interesting at that moment.

Effect on Viewing Behavior

During the baseline measurement, logging showed that respondents were watching on average 16.1 hours per week TV (Table 4). This is a little more than the average 14.8 hours that they estimated in the survey. Both logging data and survey confirm that after a slight increase in viewing hours in phase 1 (average 17.9 hrs), the average hours spent watching TV decreased (average 12.3 hrs). Participants gave the following reasons for this decrease:

- Some people had less free time available
- A few felt that the image quality of ConnecTV was not good enough
- One respondent felt discouraged in watching TV when ConnecTV showed him that he was the only one of his buddies watching TV.
- Another respondent had been shocked to find out how much time a week he was spending watching TV when asked in the baseline measurement, so he decided to dedicate less time to watching TV.

In addition, technical problems during the trial caused some people to watch TV without using the ConnecTV PC. We could not measure the viewing behavior of those users, which may have added to the measured decrease in viewing hours.

In the final survey, 20% of the participants said that ConnecTV caused a change in the channels they were watching. The information of the viewing behavior of their buddies was pointed out as the most important reason for this. People used the buddy list as an inspiration to choose the cannel to watch.

In the final survey, about one third of participants stated that they surfed channels less

Figure 8. Usage of ConnecTV functions

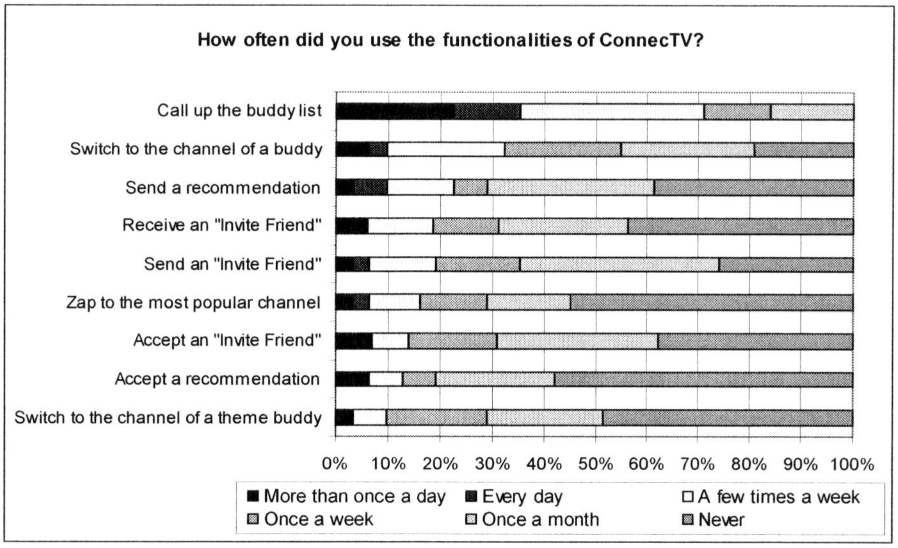

because of ConnecTV. Logging data confirm this: the number of channel changes in phase 2 was a third less than the number of channel changes in phase 1. The substantial channel change delay of ConnecTV was pointed as an important reason for this decrease. However, 13% of the participants state that through ConnecTV, they started to surf channels even more, mainly out of curiosity.

Effect on Viewing Experience

As described, the viewing experience of participants was measured on five dimensions.

On the dimension Unpleasant – Pleasant, all measurements were between neutral and half pleasant.

On the dimension Calm – Activated the scenario of watching alone was experienced as quite calm. The other measurements were between neutral and half activated.

The measurements on the dimension Bored – Inspired were mainly neutral. However, during the pilot the participants showed being bored a little.

On the dimension Irritated – Satisfied, the participants mainly felt between neutral and half satisfied. Only the experience samples during the pilot showed more irritation.

The dimension Isolated – Connected showed most differentiation. Participants expressed feeling most on their own (halfway between neutral and isolated) in the scenario in which they watched TV alone; in both scenario's of watching TV with other people (either in real life or through ConnecTV), they expressed to feel a little connected, but in the experience samples during the pilot they express to feel quite connected (average 5.3, with 0 = isolated and 6 = connected).

Through the experience samples, the actual experience at the moment of watching TV was measured. Figure 9 combines the average scores on the experience samples during the baseline measurement and phase 1 of the field trial on each of the five dimensions.

The experience samples show that during the baseline measurement, participants felt in general quite neutral: a little active, a little pleasant, a

Table 3. Number of recommendations and follow-up actions

	Total #	Follow ups		
		Refused	Time-out	Accepted
Field trial phase 1	41	24%	54%	22%
Field trial phase 2	39	36%	41%	23%

Table 4. Viewing hours per phase

TV hours per week	Baseline measurement (16,1 hrs)	Field trial phase 1 (17,9 hrs)	Field trial phase 2 (12,3 hrs)
0-5 hrs	27%	26%	47%
6-10 hrs	27%	15%	12%
11-15 hrs	9%	15%	12%
16-20 hrs	9%	9%	6%
> 20 hrs	27%	35%	24%

little isolated, a little bored and a little irritated (see the dark line in Figure 9).

In phase 1, while using ConnecTV, the following shifts occur (see the light grey line in Figure 9):

- Participants feel a little less pleasant
- They feel more activated
- They are more irritated, and more bored
- They feel a lot more connected.

In the final survey, participants were asked whether specific functions of ConnecTV changed their feeling of watching TV together. On average, about 40% of the participants stated that ConnecTV has increased their experience of "watching TV together", another 40% of participants say that their experience has not changed, while 7% report a decrease in the feeling of watching TV together (the remaining 13% declared 'not applicable'). The increase in 'feeling together' was distributed evenly among quite a few functions of ConnecTV: sending and receiving recommendations, information about which buddy is watching TV, and which channel they are watching.

People's Opinion about ConnecTV

In the final survey, participants could state in a number of ways what they thought about ConnecTV. More than half of the participants said to regard ConnecTV as an interesting addition to watching TV. The functions "[buddy X] is watching this channel now" and calling up the buddy list were deemed to be most handy; see Figure 10.

Most participants agreed that ConnecTV needs a number of technical improvements, like image quality and channel surfing speed. Besides, participants felt that ConnecTV would be best enhanced by adding more communication functions. 60% of the participants is satisfied with the simplicity of use, the lay-out of the screen and the lay-out of the buddy list.

Figure 9. Experience while watching TV (source: experience samples)

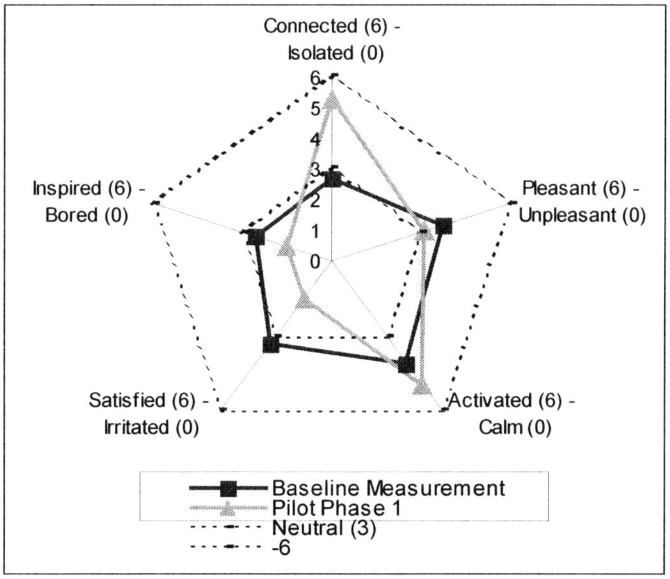

Figure 10. Appreciation of the different functions of ConnecTV (source: final survey)

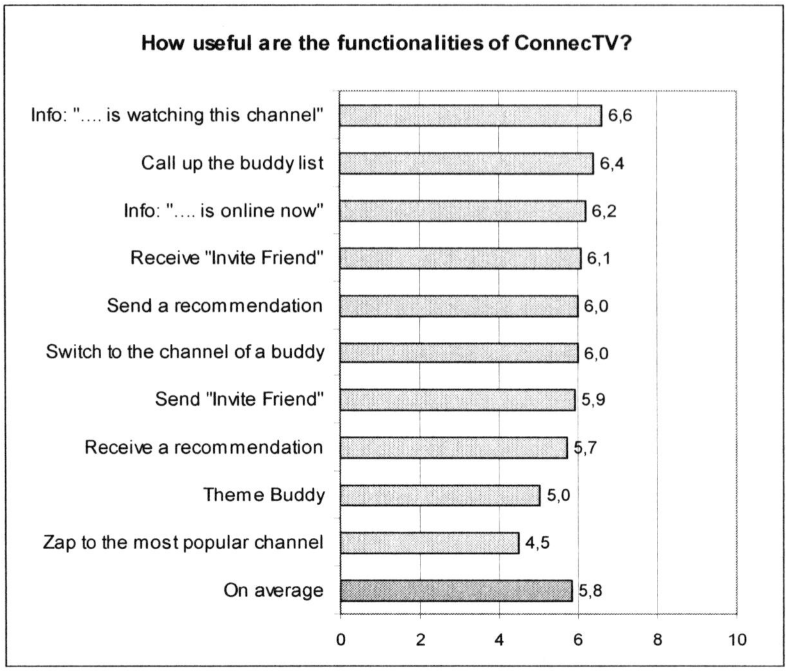

BUSINESS ASPECTS OF SOCIAL TV SERVICES

Next to user satisfaction, business aspects are equally relevant for a successful introduction of a social TV service. That is why we tried to gain insight in the business perspective of social TV services and tried to define the steps to be taken for a commercial introduction and exploitation of these services. When looking at the commercial aspects, our focus was not limited to ConnecTV but to Social TV services in general.

The ConnecTV field trial gave important input for these questions. It showed the popularity of each of the functions and showed the perceived value to a service. In the survey, we asked users for what functionality they would be willing to pay for, and how much. This gave an indication of the revenue potential of each of the functions and contributed to a reliable starting point for choosing a set of functionality that is interesting from a commercial point of view. In addition to the field trial, interviews have been held with organizations from the area of TV services like TV providers, producers, STB manufacturers, consultancy companies, broadcasters etc. We discussed with them the commercial opportunities of social TV services.

The factors that determine the success of ConnecTV can be divided into service, technical, organizational and financial aspects.

Service Aspects

Our research shows that a social TV service can provide added value on top of other solutions for finding content and improve the TV viewing experience. From the interviews, we concluded that youth and males between 20-35 years are the most important target group for a social TV service. ConnecTV was actively used by the pilot members. However, the commercial version should guarantee quality of service and support more forms of content such as recorded content and internet video. Nevertheless, the commercial introduction of a social TV service may turn out to be complex because a lot of competition from similar services on the internet exists. Their familiarity with communities and digital contacts will lead to a quicker insight into the benefits of a social TV service. This in turn will lead to a quicker adoption.

Technical Aspects

Here we focus on the intellectual property issues. When comparing the technologies that are needed to realize ConnecTV's features with existing patent claims, we found that there are a small number of large companies that have much influence on the market in the areas of interactive TV, buddy list and messaging related technologies. Most of the essential features of a social TV service are potentially covered by one or more patents. Thus, the ConnecTV service provider needs to consider licensing costs when delivering the service to customers.

Organizational Aspects

On the organizational side we can conclude that social TV services are most likely to be offered by TV Service Providers. TV Service Providers can most easily make the needed link between a social TV service and their TV platform. For making such a service successful it is important that it is used by a large audience right from the start. This can be stimulated by cooperating with other TV Service Providers (e.g. cable operators). In addition, the social TV service's position would be strengthened if it is provided over the internet as well.

Financial Aspects

Although we found that many consumers see ConnecTV as a welcome addition to watching TV and the ConnecTV functionality has a high perceived value, few ConnecTV users are willing to pay for (part of) the service. Most of the revenues are therefore expected to come from advertising and increased sales of social TV-related services like paid content services and upgrades of existing services. For a successful business case sufficient mass of users and links with social networking sites such as Facebook or Hyves are essential. Because the scale of the user base is so important, social TV services seem especially viable for larger TV providers and cooperating TV providers.

In the field trial, many users liked to send recommendations, and about 40% of users accepted these recommendations. By using the social TV service there was a 20% change in the TV channels that were watched compared to the situation in which the service was not being used. Also 90% of the users switched regularly to the TV channel that a friend was watching. This way, a social TV service might open up the way to the long tail: content that users are not likely to find just by surfing channels. An extra source of revenue could therefore be generated by letting minor broadcasters pay for the extra viewers that a social TV service directs to their content and advertisements.

During and after the programs participants of the field trial talked a lot about their viewing behavior. Because of this social interaction around TV programs a social TV service can be a way to make consumers more loyal to TV programs and commercials, which again indicates that social TV services offer added value for broadcasters and advertisers.

On the financial side most revenues can be generated by consumers that upgrade their data or TV subscription and by revenues from advertisers. A very limited set of ConnecTV users is willing to pay for (part of) the ConnecTV service. In order to realize a social TV service like ConnecTV, an estimated total sum of 1.3 million is needed, assuming one of the bigger TV providers in The Netherlands offers the Social TV functionality as described in this chapter. This investment can be recovered within approximately 3,5 years. A large extra source of revenue could be generated by letting minor broadcasters pay for the extra viewers that ConnecTV directs to their content (and advertisements).

CONCLUSION AND FUTURE WORK

Conclusion

The most basic functionality of ConnecTV was also the most popular. Users valued the buddy list and being able to see what their buddies were watching, as the most valuable feature. Logging data showed that it was also used significantly more often than the other functions. Switching to the channel that a buddy is watching was the runner up in the list of most popular functions. Least popular functionalities were ‘changing to the most popular channel’ and the functionality related to the ‘theme buddies’.

About 70% of the participants appreciated being able to send recommendations. In phase 1 of the trial, participants sent recommendations mainly out of curiosity. In phase 2 the main reason was the expectation that the recipient would appreciate the recommendation. Only half of the recommendations were acted upon.

ConnecTV changed the viewing behavior of the participants in two ways. One third of the participants indicated that they did less channel surfing than before. This was confirmed by the logging data: the number of times the participants changed a channel was reduced by 30 percent. In addition, participants indicated that they watched other channels than before. Some participants indicated that seeing what their buddies were watching, caused them to discover new programs.

ConnecTV changed the viewing experience by giving the users the feeling that they are more connected than before. Participants indicated that the contribution to this feeling was evenly divided among the functions of ConnecTV. Interestingly, ConnecTV turned out to initiate other forms of communication between the participants. People started to call each other, to send text messages, and to talk face-to-face, in response to what they noticed from the buddy list about each other’s viewing behavior.

ConnecTV should offer more functionality. The functionality that was lacking according to most of the participants was the ability to chat with other users. Also a voice or video connection would be a valuable addition to ConnecTV.

A social TV service like ConnecTV is most likely to be introduced as a commercial service by TV Service Providers. However, the commercial introduction can be complex. On the service side, it faces a lot of competition and substantial growth of digital TV households is needed.

On the internet, a lot of substitute services are available that can fulfill part of the same need as ConnecTV does. Hardly any of such services can be found on TV, but because of the substitutes on the internet, the ConnecTV added value can be hard to explain.

Although ConnecTV can generate multiple small sources of revenue, it can be complex to realize a steady and large stream of revenue that significantly exceeds the investments for introducing ConnecTV.

Further Work

We plan to extend the functionality of ConnecTV by giving the user more means of communication, such as live video and an audio connection with their buddies. Users indicated that they would like to use the Media PC for recording TV programs.

Integrating this into ConnecTV would enable users to record a program for their buddies, or watch recorded content together. Another interesting extension would be to integrate an Electronic Program Guide with ConnecTV. It could show the programs that a user's buddy has recorded or is planning to record and it would enable users to recommend programs from the guide to their buddies, in addition to recommending the program that they are watching right now.

We plan to make ConnecTV a cross platform service by making part of its functionality available on multiple devices. This would allow users for instance to receive or send recommendations on their PC or mobile phone.

More research could be aimed towards measuring viewing behavior. Experience samples could be shown on other devices than the TV, for instance a mobile phone, to make the sampling less intrusive.

REFERENCES

Abreu, J., Almeida, P., & Branco, V. (2001). 2BeOn: Interactive Television Supporting Interpersonal Communication. In *Proc. Eurographics Workshop on Multimedia 2001.*

B@Home. The Future of Broadband, Multimedia Services in the Home. From http://bathome.freeband.nl

Coppens, T., Trappeniers, L., & Godon, M. (2006). AmigoTV: towards a social TV experience. In *Proc. European Conf. on Interactive Television 2006.*

Desmet, P. M. A., & Hekkert, P. (2007). Framework of product experience. *International Journal of Design, 1*(1), 57-66.

Fink, M., Covell, M., & Baluja, S. (2006). Social and Interactive-Television Applications Based on Real-Time Ambient-Audio Identification. In *Proc. of EuroITV Conf. 2006.*

Jabber. *Open instant messaging and presence.* From http://www.jabber.org

Chapter XII
Asynchronous Communication: Fostering Social Interaction with CollaboraTV

Brian Amento
AT&T Labs – Research, USA

Chris Harrison
Carnegie Melon University, USA

Mukesh Nathan
University of Minnesota, USA

Loren Terveen
University of Minnesota, USA

ABSTRACT

With the advent of digital video recorders and video-on-demand services, the way in which we consume media is undergoing a fundamental change. People today are less likely to watch shows at the same time, let alone the same place. As a result, television viewing which was once a social activity has been reduced to a passive, isolated experience. CollaboraTV was designed to address this new mode of television viewing by directly supporting asynchronous communication. We demonstrated its ability to support this communal viewing experience through a lab study and a month-long field study. Our studies show that users understand and appreciate the utility of asynchronous interaction, are enthusiastic about CollaboraTV's engaging social communication primitives and value implicit show recommendations from friends. Our results both provide a compelling demonstration of a social television system and raise new challenges for social television communication modalities.

INTRODUCTION

Television is undeniably a major component of modern society. In the United States, it is not only the dominant media activity, but is also considered the most exciting and influential media type (FCC 2006; Putnam 2000; TBA 2006). Despite increasing competition from the internet, television usage has been steadily increasing, and is now at it's highest level since viewing data was first collected, a 50% increase since the 1950s, and a 12% increase from 1996. The average person watches 4.5 hours of programming a day, with the average household tuned in for more than 8 hours (FCC 2006; Putnam 2000). This consumes almost half of people's total leisure time (BLS 2006; Putnam 2000).

Given the significant place that television holds, our research focuses on understanding the social aspects of television viewing – especially in today's age of social behavior-altering technological advances – and the utility of social television systems for meeting the new challenges that such advances bring about.

Declined Social Interactions Around Television

Television was once championed as the "electronic hearth" which would bring people together (Tichi 1991). Indeed, television shows provide a common experience, often affording even total strangers a social connection on which to initiate conversation. This effect blossomed in the 1950s, when two-thirds of all Americans tuned in to watch "I Love Lucy" (Putnam 2000). However, a fundamental shift in how we consume media is degrading such social interactions significantly – an increasing number of people are no longer watching television shows as they broadcast. Instead, users are favoring non-live media sources, such as Digital Video Recorders (DVRs), Video-On-Demand services (e.g. Apple's iTunes Video Store), and even rented physical media (e.g. DVDs via Netflix). To complicate matters further, televisions are outnumbering people in the average home; less than a fifth of households have a single television (Fairbank 1995; AP 2006). This is leading to a decline in ability for people to interact and is eroding once strong social ties. People are increasingly watching TV without their families, with some studies suggesting at least half of Americans usually watch alone (Putnam 2000). However, all indications point towards a lack of ability to communicate, not a lack of desire.

The "Water-Cooler Effect": A Thing of the Past?

Television shows often act as a conversation starter, enabling the "water-cooler effect" (Putnam 2000), where groups congregate and discuss a television show, automatically assuming everyone in the group has seen it. For example, co-workers could discuss a show from the previous night at work the following day. However, this effect is heavily dependent on a property of live television: shows have a fixed broadcast time. This means that after a show has aired, everyone who wanted to see it, must have watched it (or missed it). DVRs enable people to watch shows days, weeks, and even years after they first air. This trend towards asynchronous viewing, although not omnipresent today, is becoming a dominant media consumption mode. DVRs are already found in 20% of American homes (Leichtman 2007) and worldwide adoption is predicted to reach 250 million users by 2011 (ABI 2006). On-demand commercial video downloads are also booming, jumping 255% from 2005 to 2006 (NPD 2006). Similarly, Netflix, the most popular DVD rental company, has experienced nearly 50% growth in subscribers annually since 2002 (Netflix 2006).

What does asynchronous viewing mean for water cooler talk? Many people will not have watched the most recently aired episode by the following day. In fact, some people may be multiple episodes or even seasons behind. This makes conversing about a show considerably more problematic. If a group of friends meet and talk about the latest episode, those who have not seen it are left out. It is even possible that some may avoid the conversation entirely, fearing that yet-to-be-seen episodes will be spoiled. Equally likely is that people will moderate their conversations in order to prevent revealing spoilers to friends that are one or more episodes behind. However, this hampers the exchange of important and interesting details that were revealed in recent episodes. Moreover, by the time lagging people do catch up, and want to discuss the show, it is likely others will have forgotten important details or simply have lost interest, substantially degrading the quality of interaction.

Unprecedented Level of Program Choice

Television viewers today can easily be overwhelmed by the number of channels as well as programs. Gone are the days when one could flip through a 40 page TV guide and decide on what to watch. Given the plethora of content today, the task of finding something relevant to watch has become very difficult. As a result, viewers often resort to randomly scanning multiple channels ('channel surfing') to find a show of interest. Digital TV content providers have tried to respond to this information overload problem in a few ways. Electronic program guides, which are little more than digital forms of their paper predecessors, still require viewers to sift through a multitude of static choices. Vendors also provide basic search capabilities along with TV programming service, but again the task of searching for the illusive "good" show rests on the viewer. Despite the availability of technology aids, viewers still prefer channel surfing as a method to select what to watch, often expressing considerable dislike for on-screen program guides (Taylor 2003). DVR systems, like the TiVO, even automatically record shows that it thinks the user will like. But the quality of such system-recommended shows is questionable (Zaslow 2002).

To address these concerns, a new communication paradigm for television is needed, one that captures users' comments during and immediately after watching a show. This is when the material is fresh and engaging, and peoples' commentary rich and plentiful. This content then needs to be saved – the enthusiasm and detail preserved – and shared with users who watch the content at a later time to recreate the experience of watching together. Also, the social nature of such a system involving groups of people who often watch shows together can be exploited to tackle the problem of content overload by using the viewing habits of friends as recommendations. CollaboraTV achieves all this using a unique set of communication primitives and visualization techniques.

Our efforts to address the aforementioned issues are organized around 3 research questions (RQs):

- **RQ1.** Does CollaboraTV provide users a sense of social presence? Does it enrich the viewing experience?
- **RQ2.** How well does CollaboraTV support the task of asynchronous television viewing?
- **RQ3.** Can friend networks present in social television systems be leveraged to help users choose what to watch?

The remainder of the chapter is organized as follows. We first survey related work, illustrating how we build on or advance it. We then describe the CollaboraTV system, highlighting distinctive features that appreciably affect usage. The heart of the paper describes a lab study followed by a month-long deployment of CollaboraTV to support social interactions while consuming video content; we focus on how our results answer our three research questions. Finally, we discuss the implications of our results for future design and research.

RELATED WORK

Social Television Systems

Television-based communication has been the focus of substantial research. Many of the ideas in CollaboraTV are extensions of successful elements in previously developed systems. However, CollaboraTV distinguishes itself in several significant ways.

Xerox PARC's SocialTV (Oehlberg 2006) envisioned the use of a shared audio channel, where groups of users could interact verbally. The project also introduced the idea of a movie theater-themed visualization scheme for user presence. Indeed, CollaboraTV's virtual audience is a direct descendant of this concept. However, unlike avatars in SocialTV, CollaboraTV's are dynamic and used as a conduit for communication (comments, gestures and emoticons).

AmigoTV represents one of the earliest efforts in this domain (Coppens 2004). Like SocialTV and CollaboraTV, avatars are used to visualize user presence. The system offers a series of faces as avatars, and allows users to select a demeanor (e.g. happy face, angry face). This simple mechanism allows avatars to operate on an additional dimension: emotion. Users can also generate shared video effects, for example a flaming ball whizzing across the screen. Like SocialTV, AmigoTV allows users to communicate via speech.

ConnecTV is a tightly integrated instant messaging and television application (Boertjes 2007). From a user's perspective, friends are placed into one of three groups: "watching this channel", "watching another channel", and "not watching". In addition to being able to chat with friends that are watching the same show, the system also allows messages to be sent to friends watching different shows. The latter serves as an invitation for that user to switch channels and join the friend. These invitations have the ability to operate in an asynchronous way. If a user is not available when the invitation is sent, it will be saved until that user comes online. Once logged in, the invitation is shown, and the user can start watching the associated content asynchronously from the originating friend. Media Center Buddies, system developed by Microsoft, is similar (Regan 2004).

The Telebuddies system (Luyten 2006) promotes communication amongst synchronous viewers using a series of events incorporated into the media stream. The authors offer a quiz example, where users are formed into groups and compete. A text-based chat interface is provided to allow users to deliberate.

Unfortunately, evaluation of social television applications has been fairly limited (Geerts 2006, Harboe 2007, Boertjes 2007). There has been a significant amount of work evaluating the use of ambient displays to convey TV watching status (Harboe 2008). Although there has been considerable investigation of particular social television elements (Geerts 2006; Harboe 2007; Weisz 2007), comprehensive field studies that cover the usability, impact, and potential adoption of full systems are limited. Currently, these systems do not distinctly support asynchronous communication. With a clear trend towards on-demand media consumption, systems that do not support this form of communication have significantly diminished value. In these regards, CollaboraTV is unique.

Avatars

The benefit of avatars in communication systems has been widely researched (Diederiks 2003; Persson 2003; Salem 2000; Smith 2000). The use of animated characters in television has also been investigated (Chorianopoulos 2006). Television is traditionally watched in groups, which makes avatars an obvious technique to simulate presence. Avatars humanize remote users, or in the case of CollaboraTV, potentially teletemporal (literally meaning far in time) as well. Moreover, avatars have been shown to be ideal for non-textual communication; emotions, gestures and postures offer a universal dimension of expression that people find familiar and natural.

SYSTEM DESCRIPTION

CollaboraTV was designed from the ground up to support synchronous and asynchronous viewers in a unified interface. The resulting system allows communication in and between these two viewing modes, providing an unprecedented level of interaction potential. Additionally, CollaboraTV provides a suite of novel features that leverages existing user behavior to further enhance the social experience.

Temporally-Linked Annotations

Our initial prototype version of CollaboraTV allows users to create text comments while watching a show. The text content is attached to the media stream at the corresponding time index. When other users encounter this point in the show, the comment is displayed on screen for several seconds.

This annotation method inherently supports asynchronous communication. Previously generated comments will be shown to later viewers as they watch the show. Although the mechanism is simple, the effect is great – past users appear to make comments as if they were watching in parallel with you. In a synchronous situation, when one or more users are concurrent, comments are shown immediately to all group members because the comment is attached to the show at the point everyone is presently viewing. This allows people to chat (live) like they would with an instant messaging client. Additionally, because these (chat) comments are attached temporally to the media, subsequent viewers will see the conversation unfold in "real-time" even though the conversation took place in the past.

This communication scheme has several important qualities. Foremost, synchronous user groups can communicate without hindrance while simultaneously interacting with asynchronous commentary. Secondly, lone viewers who would otherwise see no communication are exposed to a wealth of previous interaction. This may motivate them to participate in the conversation knowing that subsequent viewers will see their remarks. Lastly, as more users watch and comment, the richer the dialog becomes for later viewers.

CollaboraTV supports a second type of temporally-linked annotation: the interest point. Like comments, interest points are attached to the media stream at the time index they were created. However, instead of containing text, interest points are used to indicate a positive or negative reaction to a show's content (e.g. a particularly funny joke or exciting action scene). Users are free to associate the polar nature of this feature however they see fit (e.g. thrilling/dull, witty/cheesy, and suspenseful/predictable).

Virtual Audience

The most prominent feature of the CollaboraTV user interface is the virtual audience. A series of avatars are used to visualize both current and past viewers (i.e. synchronous and asynchronous users respectively). The effect of a movie theater, where people are seated and shown as silhouettes against the content, is used. The simple black figures are unobtrusive, immediately recognizable, and work well at a variety of scales. Each avatar is named and has a static seating position throughout the show. A "watching live" label is displayed below avatars that represent synchronous viewers. Figure 1 shows a screenshot of the initial prototype version of CollaboraTV.

In addition to embodying concurrent and teletemporal viewers, the virtual audience is used as the primary communication conduit. Temporally-linked comments are shown in translucent, comic-book-like speech bubbles, which are rendered above the source user's avatar. In addition, avatars raise and lower their arms to make thumbs up/down gestures. These correspond to user-generated positive and negative interest points.

The virtual audience is a critical component of the CollaboraTV communication strategy. Because user actions are coupled with a personal avatar, one that is seen by friends and family, providing commentary and interest feedback becomes an outlet for expression and not just data entry.

Buddies, Sharing and Privacy

A user may consider their comments and interest points personal, and thus only wish to share them with friends and family. To achieve this, CollaboraTV employs a simple social network in order to determine how user-content is shared. Specifically, a unidirectional subscription model is used. If comments, emoticons and interest points are to be exchanged between two users, they must explicitly befriend each other by adding the other to their buddy list. This would be most typical among friends and families, who want bi-directional communication. However, there are some instances where unidirectional communication is valuable. For example, it would be possible to watch a television program with one of the show's actors. This would be achieved by subscribing to that user. However, because the subscription is only one-way, the user would not appear in the actor's virtual audience.

Interest Profiles

Although temporally-linked comments, emoticons and interest points are primarily used for communication purposes, their location, frequency and content can be leveraged for other purposes. Interest profiles are a notable example. These are created by interpolating a continuous interest level from a series of interest points. Data from a single user or group of users can be used. Because positive and negative interest is assumed to be transient, the interest level slowly decays to a nominal level. (Figure 2)

Figure 1. The CollaboraTV prototype interface with all panes active. A) Information about current show. B) Alternative, chat-room-like visualization (inset, used only in lab study). C) Interest profile visualization. D) Progress bar that displays the position in the media stream. E) Virtual Audience. Panels A, C and D fade away after a preset period of inactivity.

Figure 2. A set of discrete interest points is used to form a continuous interest profile.

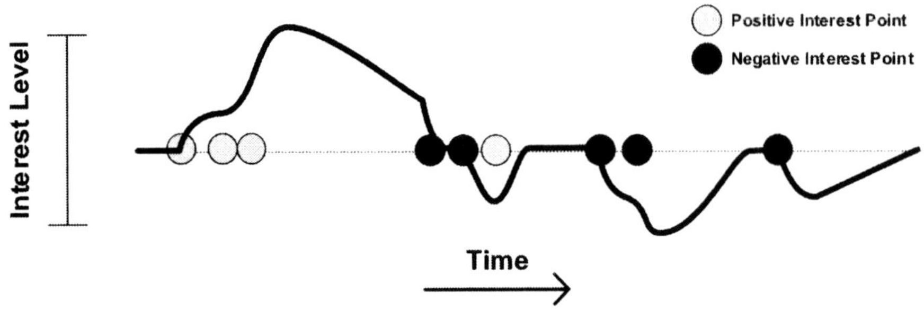

These data of how a viewer's interest level swings between positive and negative over the time period of a TV show may be interesting to viewers. In response, the feature is exposed graphically in the CollaboraTV user interface. While a show is watched, two interest profiles are calculated: one for all previous viewers and one for the current user. The two interest profiles are shown together in Figure 1. This helps visualize the difference between these two profiles, displaying the magnitude of disagreement between the user and the group. It was thought that highlighting sections of agreement and disagreement may spark additional discussion, increasing social interaction.

Interface Design

The CTV GUI consists of the following 5 elements:

1. **Buddy Information (View/Join):** A list of buddies of a user along with an indication of whether or not the buddy is currently viewing CollaboraTV. A user has the ability to 'join in' on a buddy, who may already be watching a show, and start watching at the same point in the show as the buddy. Thus a user can catch-up with a buddy who may not be too far along into a show.
2. **Program Guide:** A listing of all available shows in the CollaboraTV system, along with show descriptions. Shows viewed by buddies of a user are marked in the program guide along with any collected social usage data. The program guide also has a list of shows that have been viewed by a user's buddies, as well as a list of popular shows across all users of the CollaboraTV system.
3. **Virtual Audience:** Discussed previously.
4. **Progress Bar:** Displays the current position in the media stream. The temporal locations of annotations – comments, interest points and emoticons – are denoted using icons on the progress bar (Figure 1). When the show progresses to an annotation as seen on the progress bar, the avatar corresponding to the user who made the annotation is animated to display a comment bubble, thumbs up/ thumbs down gesture or emoticon on its face.
5. **Interest Profile Visualization:** Discussed previously. Users have the option to move to particular points on the interest profile timeline and begin watching the show at the chosen point of interest.

System Architecture

CollaboraTV uses a client-server architecture. The prototype client was built using Java and provides the GUI functionalities described previously to access available content, launch programming, and control playback (Harrison 2007). The prototype uses MythTV, an open source media center application with DVR capability, for recording and playing shows and accessing show information for its program guide.

EXPERIMENT: LAB STUDY

Our first study was a lab based study of the initial CollaboraTV system (Figure 3). In addition to the research questions listed above, we also wanted to see how users could communicate asynchronously and what types of interactions would arise while using the CollaboraTV system. We were also interested in seeing if our virtual audience interface could provide a better sense of presence and help to elicit more interactions between users.

Experimental Setup

In this study, users were seated in a room with a large rear-projection screen. A wireless (Bluetooth) keyboard was used for entering text annotations. For all other functionally, a wireless (radio frequency) remote control was provided. A labeled color printout of the remote control was provided for reference. Twenty-three televisions shows were available for users to watch. These varied in content and genre. Advertisements were removed, and clips shortened to approximately seven minutes in length. We kept the amount of content low to help ensure overlap between users.

Figure 3. A partial virtual audience from a live user's perspective. Other avatars represent past viewers. Note the speech bubbles (comments), thumbs up and thumbs down avatar gestures (interest points) and the avatar emoticons (emotions). These annotations are also temporally displayed on the progress bar seen above the virtual avatars.

Before the study began, the authors watched each show and generated some preliminary comments and interest points. Thus, no less than two avatars (three including participant's avatar) were present in the virtual audience. This was deemed necessary to engage people in the asynchronous technology and help them better evaluate if comments left in the past were interesting and whether they would respond to comments knowing previous users may never see them.

In order to investigate whether the virtual audience increased user participation, an alternative commentary interface was created. This was modeled after a generic chat-room-like application. Comments were listed sequentially in a vertical orientation, prefixed with user names with no audience display.

Comments and interest points created by users, persisted between sessions. This meant that avatars in the virtual audience represented real people (in this case, colleagues). Additionally, users understood that future study participants would see their avatar's comments and interest points. This provided the necessary social context to engage users. It was thought that if users knew their comments were to be deleted once they finished, there would be no motivation to communicate. Also, this exposed participants to a variety of people, commentary, commentary styles, and audience sizes.

Participants

Thirty-two participants (24 male and 8 female, with a mean age of approximately 42) volunteered for a small food reward. Participants were recruited from a large corporate research lab using an email campaign. All but one participant watched television on a regular basis.

Procedure

Experimental conditions were balanced across participants to remove any order effects. Users were first shown a short video providing an overview of CollaboraTV. This was followed by a 10-minute, hands-on walkthrough, which familiarized participants with the system and its features.

Following the tutorial, participants were asked to select and start a show of their choosing. Based on the participant's experimental condition, either the virtual audience or the chat-room-like visualization was enabled. Users then watched the show, generating comments and interest points as they saw fit. After the first show had finished, users were asked to select a new program to watch. This time, the other comment visualization method was used. For the third (and final) clip, users watched a common show, ensuring overlap with other participants.

After the final show concluded, users answered an online survey eliciting feedback about the system and their perceptions about asynchronous communication. The survey used a Likert-style one to five scale to determine user's level of agreement to questions about CollboraTV. Each trial took approximately one hour to complete.

RESULTS: LAB STUDY

Overall reaction was positive. Users understood and appreciated the value of asynchronous communication immediately. When asked if CollaboraTV made watching television more engaging and enjoyable, over half of participants agreed or strongly agreed (m=3.58, sd=1.09). The ability to create comments was popular, with 76% of participants agreeing or strongly agreeing the ability made television viewing more fun (m=3.87, sd=1.05).

Asynchronous Communication and the Virtual Audience

The virtual audience appears to be an effective and engaging visualization. User participation was significantly greater when compared against the alternative, chat-room-like visualization ($p<.05$). On average, users generated 21% more interest points and 17% more comments when the virtual audience was enabled. This was further reinforced by survey results, which indicated that two-thirds of participants agreed or strongly agreed that they preferred the virtual audience (m=3.97, sd=0.87). A follow up question asked users to consider an alternative interface where the virtual audience was present, but comments were decoupled from (i.e. not associated with) particular avatars. Responses were overwhelmingly negative (m=2.15, sd=1.06).

The virtual audience's ability to support both synchronous and asynchronous viewers was also popular. Users had overall negative option when asked to consider a virtual audience that only displayed synchronous users (m=2.12, sd=0.89). Equally negative was the reaction to a virtual audience that only showed past viewers (m=2.09, sd=.91). Also, a pair of questions asked participants to consider two systems, one where their comments and interest points were only shown to synchronous viewers, and the other where their comments and interest points were shown to both synchronous and later (asynchronous) viewers. Users strongly favored participation in the system that supported both communication modalities (m=3.09, sd=1.07 vs. m=3.94, sd=0.83).

System Redesign

The lab study was useful in helping us evaluate the initial system. However, it is difficult, in a lab setting, to see how users might integrate this technology into their daily viewing habits and determine the true value of asynchronous communication while watching television. Since our prototype system was not very portable, because of numerous hardware constraints, we decided to re-implement CollaboraTV to run within a web browser. This would allow users access to the system from their homes and help us to reach a wider audience.

The new version of CollaboraTV uses a client built in Adobe Flex. A central JBoss server in coordination with a Flash Media Server serves out video content to the clients, and coordinates the communication between multiple clients. A central MySQL server provides a common data store for user generated data (annotations, buddy lists, show ratings) as well as show related data.

Based on user comments from the lab study, we made a few changes to the client interface. Most of these changes were required to make the interface more mouse friendly. The prototype was controlled through a remote device, but the new web version, needed to support the mouse. Besides text comments and interest points, CollaboraTV now allows users to temporally link emotion icons or emoticons to the media stream. Users can express their feelings at the time of watching a show by selecting from a set of emoticons that represent common expressions (happy, sad, angry etc.) Along with text comments and interests points, emoticons provide CollaboraTV users with a rich set of primitives that help digitally recreate the communal viewing experience. Emoticons are seen as corresponding animations of the avatars. For example, when a user selects the 'happy' icon, the user's avatar turns around and displays a smile on its face. All actions are smoothly animated. These subtle effects enhance the perceived interactivity. Figure 1 provides an example of each of these items.

EXPERIMENT: FIELD TRIAL

In order to gauge the true utility of CollaboraTV, we conducted a field study to investigate the usefulness of CollaboraTV to support the communal viewing experience. Participants accessed and used CollaboraTV from their personal computers, interacting with their television buddies freely and naturally over the duration of the study. Such an open and natural design preempted the need to simulate the experience of watching programming remotely with other users of the system, and also gave users sufficient time to form an opinion about the system.

Participants

The participants were recruited using a mailing campaign directed at students who had interned at a large corporate research. The summer interns presented an interesting group to recruit from as they had interacted with each during their internship period. They often stayed in the same summer housing, traveled to work together, had lunch together as well as socialized during many events for summer interns. Thus, new friendships were made and new cliques formed. After the summer internship, the interns returned to their home towns, within and outside the US. This group had an important property that made it ideal for our study – cliques of friends who were now geographically separated. The presence of cliques within the potential participant pool was vital, as CollaboraTV depends on the notion of buddies. We used a survey to find participants amongst this group of interns and their friends who regularly watched television. 16 qualified subjects participated (14 male and 2 female, with a mean age of approximately 25). Their backgrounds were mainly in IT.

Procedure

Participants used CollaboraTV for 4 weeks. They completed a pre-study survey about their TV viewing habits and selected a group of buddies with whom they would like to watch shows on CollaboraTV. Participants could either choose buddies from the list of interns or invite other friends. Following the initial survey, we invited each participant to view a short online video providing an overview of CollaboraTV. At the end of the study period, we conducted another online survey where participants informed us of their experiences with CollaboraTV.

Participants viewed streaming video from the CollaboraTV server using a web browser with the Adobe Flash plug-in installed. Once logged in to the system, they could select from a list of 37 shows (600 episodes total) across 17 genres. The list of shows to be recorded was based on a listing of the highest rated shows across different genres on a popular TV show rating website. During the course of the study, participants could request shows to be added to the system. They could also modify their buddy list, and make annotations while they watched a show.

We told participants to use CollaboraTV just as they wanted: they could watch any show of their choice, add or remove anybody to their buddy list and use any system feature. To ensure minimal usage levels, we asked participants to view at least 2 shows. We offered a modest incentive: a randomly selected user who participated at the minimum level received a $50 gift certificate. 70% of the subjects met the requirement.

Experimental Design

To test the effect of CollaboraTV's social features in providing users' a sense of social presence (RQ1), we asked users to create a buddy list and observed the interactions that took place between buddies. These buddies were in most cases other interns with whom participants had fostered friendships over the last year as a result of a joint summer internship experience. In other cases, participants invited a family member or a friend to be their CollaboraTV buddy. Annotations made by participants while watching a show were persisted and shown to their buddies. Users understood that their buddies would see their avatar's annotations and this provided the necessary social context to engage users. Besides observing the logs of annotations made by users, we asked about their experiences of using the social features of CollaboraTV while watching shows with their buddies.

Numerous existing social television systems have explored the dynamics of synchronous communication. To better understand the potential of asynchronous communication, one of CollaboraTV's most distinctive features, our study focused on this mode of interaction (RQ2). We asked users to watch specific shows with a buddy(s) in a time-shifted manner. A new episode of a show found in the viewing history of a user or buddy was chosen. Participants were asked to view the chosen show, or alternately a show of their choice, and were given a week to complete this task.

Digital television services today pose a significant content selection problem. Rather than taking the algorithmic route such as recommender systems, CollaboraTV employs a simple yet powerful mechanism. Social Influence Theory tells us that when we do not know what to do, we often times do as others do. CollaboraTV aids such behavior via certain interface features. Users can view a list of shows that their buddies have been watching called PopularShows(Buddies). Also available is another list called PopularShows(All) that anonymously displays names of shows that have been recently viewed by any system user, not only buddies. Shows recently viewed by buddies are indicated in the program guide. To learn if friend networks present in social television systems could be leveraged to help users choose what to watch (RQ3), we observed usage logs for instances when users selected a show displayed under either PopularShows(Buddies) or PopularShows(All). We also asked survey questions to find out if participants found these lists useful.

RESULTS: FIELD TRIAL

At the start of the study, participants completed a survey that gave us sufficient background regarding their familiarity with different video viewing technologies and their attitudes towards television viewing. Given their IT background, almost all participants reported themselves as expert computer users, with high weekly computer usage. On the question of television expertise and usage, almost all of the participants had used televisions for more than 10 years, with 75% viewing between 1-3 hours per day.

We also asked questions about familiarity and usage of DVR technology, to which 56% of participants said they use it often, while 37.5% of users had either heard of it or used it a few times. We felt that the familiarity with DVR technology would lend itself nicely to the asynchronous viewing mode in CollaboraTV. Only 3 participants had used online social television systems before. Figure 4 shows usage levels of some popular video-related technologies.

With this background about participants, we now address the core of our results, which are organized around our 3 research questions. We begin with RQ1, describing how participants felt about the social features of CollaboraTV. We follow by identifying opportunities for improving the communication primitives based on participant feedback, and then address RQ2, which focuses on asynchronous viewing. At this point, we explore the related issue of privacy in social television systems, before finally looking at a possible solution to the content selection problem described in RQ3.

RQ1. Does CollaboraTV provide users a sense of social presence and enrich the viewing experience? We asked participants questions to understand if they even valued the experience of watching with others in the first place. 81% indicated that they watched television with family or friends, and said that watching with others was more fun than watching alone. 19% also added that watching programs/films that elicit strong emotions with friends or family helped bring them closer together. These responses encouraged us to believe that participants would use and value the social annotation primitives of CollaboraTV at the onset of the study.

Annotation Activity

During the period of the study, we maintained logs of all the annotations made by the participants with the intention of using annotation activity as

Figure 4. Amount of usage of different video-related technologies

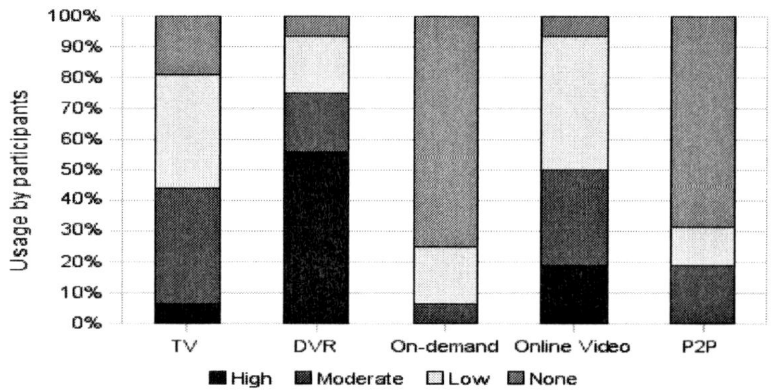

a measure of success of CollaboraTV in enriching the viewing experience. In all, participants created 213 annotations, an average of 14 per person (min: 0; max: 72; std: 22.5). 137 (64%) were text comments, 24 (11%) were interest points, and 52 (25%) were expressions. The difference in usage of the 3 types of annotations could be because of the ease of use and expressiveness of text chat. In many cases, viewers used chat-style emoticons embedded in their text comments.

Chat Content

After the study was complete, we assigned each of the participants' 137 messages to one of 6 categories. We adopted the classification scheme used by Weisz (2007). A reliability check was performed between two independent coders, with an inter-rater reliability of 83% after two iterations. Each line of chat was coded under one of these categories -- show content, evaluations of the show, personal topics, laughter, system-related comments and greetings/partings. Laughter, which was often embedded in text comments, was coded separately as either occurring by itself or co-occurring with text. A breakup of the chat content is shown in Table 1.

Our results were similar to those reported by Weisz (2007). A significant amount of chat was about the show content as well as personal interactions triggered by the content. For example, a participant said, "do you like Chandler? Would you date him?", to which a buddy said, "haha prob not, he's too much like my first bf" while watching an episode of a popular sitcom. Laughter in the form of "hehehe", "lol", ":)", ":D" etc. occurred very frequently, with users often mixing it in with chat, but more often occurring by itself. Close to 20% of chat consisted of solely laughter. This is roughly the same as the number of times the emoticon representing 'laugh' was used during the study. More over, the high usage of chat-style emoticons such as ":D", ":)" could explain why emoticons were not directly used as much. The

Table 1. Examples of chat in each coding category

Category	**Example chat (original form)**	**% chat**
Television Show	"does she leave him there or something?" "that's dakota fanning indeed" "Colbert took the book apart"	34.3
Evaluations	"oh man . . . bad driver" "if this wasn't already planned, this dude's the funniest astronaut out there" "nice choice of music"	8
Personal	"I want to sit next to you" "do you like Chandler?" "I'm hungry"	18.9
System	"it works fine even with limited bandwidth" "they should cut out all commercials" "test . . . "	16.8
Laughter	":D", "hehehe", "lol" and many variations thereof "I've seen this show too many times :)"	18.9 solo 5.1 mixed
Greetings & partings	"hi", "hey", "bye" and many variations thereof "I'm out of here"	2.2

chat-style emoticons are quicker to create and users are already familiar with them.

Attitude Towards Text Chat and Other Social Interaction Features

In the final survey, participants were asked if they liked the social interaction mechanisms that CollaboraTV offered such as text comments, emoticons and interest points. More than half the participants agreed or strongly agreed that CollaboraTV is fun to use (m=3.57, sd=0.65), while 71% agreed or strongly agreed that the system is easy to use (m=3.71, sd=0.73). Participants for the most part (57.1%) had no trouble learning to use the system (m=3.64, sd=0.84). When asked if the system was useful to them, 85.7% responded affirmatively (m=3, sd=0.55), saying in 78.5% of the cases that it supported what they wanted to do to a large extent (m=3.07, sd=0.73).

Participants liked the movie theater seating effect of avatars and expressed the desire for interactions between proximately seated avatars. They were not very enthusiastic about the possibility of customizing their avatars.

We asked participants to evaluate each of the social interaction features on the dimensions of how fun and easy they were to use. Given that viewing annotations made by others via their avatars may be considered as distracting, we also asked users if this was the case in their experiences with CollaboraTV. Figure 5 shows the responses of participants.

Note that viewing an interest graph or joining a buddy in viewing a show are activities with no direct external factors vying for the users immediate attention, unlike animated avatars. In the particular scenario of the distractions caused by chatting and viewing others comments, 64% of participants were ambivalent about creating text comments, responding with a 'neither agree or disagree' while close to half were undecided about their attitude towards viewing others comments while viewing television. Likewise, 50% of the

Figure 5. Evaluation of CollaboraTV's social features by participants, on a 1 to 5 Likert-scale

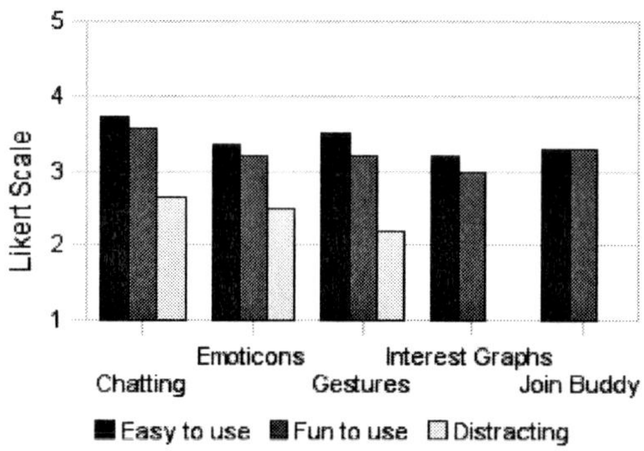

participants were undecided about how distracting the virtual audience as a whole was. This is despite the fact that more than half the users found chatting to be a fun activity. This confirms what researchers have reported in the past about chatting while consuming video being fun as well as distracting simultaneously (Weisz 2007). To accommodate such users, it would be worthwhile to investigate alternative visualization schemes, in particular ones with smaller visual footprints. These could be offered as alternative visualization modes in a future CollaboraTV version.

Overall Activity Level

CollaboraTV had a collection of close to 600 episodes from 37 shows in 17 popular genres during the study, which were updated based on requests from participants. Despite this large and flexible selection, the positive survey responses/ general feedback to the social features, CollaboraTV experienced only low to moderate levels of activity. On average, participants viewed 4 shows and created 14 annotations. Several factors could be used to explain this paradox. One obvious reason could be the lack of critical mass. As one characteristic participant said: "very few people [online]; the more people and friends there are the better the experience". We could solve this by providing users a robust invitation mechanism.

A reason for low number of annotations could be the common habit of viewers carrying out some other task in parallel to watching television. 68% (m=3.75, sd=0.93) admitted to normally doing something else along with watching television, explaining that they seldom solely watch television. Another reason could be the current input modalities of CollaboraTV, where users are expected to use their keyboard to make annotations. We discuss alternate unobtrusive temporal annotations that could remedy this in our 'Future Work'.

In the pre-study survey, 4 participants had indicated a preference to watch television alone, as they felt watching with others was distracting. At the end of study, we checked to see if their opinion had changed as a result of using a social television system first hand. 3 of these 4 users, who had earlier strongly agreed that watching programs with others is distracting, admitted that the virtual audience was not distracting while watching a show. All 4 were ambivalent when judging if the CollaboraTV experience with its virtual audience was more engaging and enjoyable compared to traditional TV. Finally, one of the 4 now strongly agreed that the ability to join shows and watch in parallel with other users with a virtual audience is compelling. These changes in opinion, though minor, serve as encouragement to us.

On the whole, over half of the participants agreed (m=3.21, sd=0.89) that when compared to traditional TV, the experience of viewing in the company of a virtual audience is more engaging and enjoyable. Also, participants said the ability to join shows and watch in parallel with other users was compelling (m=3.29, sd=0.83). In both cases, 35.7% agreed or strongly agreed.

RQ2. How well does CollaboraTV support the task of asynchronous television viewing? The asynchronous viewing capability of CollaboraTV is meant to help preserve interactions like the "water cooler conversations" that build social capital. But before studying the utility of this capability, we wanted to learn if participants valued "water cooler conversations". We verified this by asking users questions regarding the social role of television in their day-to-day lives. 75% (m=4, sd=0.73) strongly agreed that watching the same programs as friends or family provided them with common ground for conversations. Additionally in 19% (m=2.75, sd=0.86) of the cases they explained that watching programs that elicit strong emotions with friends or family brings them closer together.

Next we asked participants how frequently they (a) used conversations about programs/films as conversation starters (b) used examples from programs/films to illustrate points in conversations with others, (c) used conversations about programs/films as a way to clarify values and opinions to others. In response 37.5% (m=2.69, sd=1.2) admitted to (a), 37.5% (m=2.69, sd=1.2) said they often employed (b) and 18.8% (m=2.44, sd=1.03) said they clarified their values as in (c).

Sharing and Privacy

Any system that encourages social interaction in the form users sharing their annotations would have to address the privacy concerns of users. In CollaboraTV, the privacy of all annotation types - comments, interest points and emoticons - is handled in the same manner. At the onset, it was unclear if users placed different levels of sensitivity on these items. In the final survey, users were asked how they would prefer their different annotations to be shared. The results in Figure 6 indicate that users generally consider comments and emoticons to be more personal than interest points. When queried about the granularity of control desired in sharing comments, 57% preferred a global program setting that applies to all shows. Show or episode level granularity did not appeal to participants.

The most striking observation however, is that participants favor anonymously sharing their annotations globally, irrespective of the type of annotation, versus any other scheme of sharing. They even prefer to share annotations anonymously, over publicly sharing within their friend group. A strong bias towards anonymous sharing schemes is visible in Figure 6. However, this is not a blind bias, as participants also seem to care about sharing their comments with as many people as possible This can be seen in their preference of the 'anonymous global' scheme over the 'publicly within group, anonymously globally' scheme or even the 'anonymously within group only' which would seem the most conservative of the schemes. Participants seem to desire striking a balance between their privacy and benefit to others, by wanting to enrich the viewing experiences of others with their anonymous comments. The willingness of users to consider the social good and share bodes well for a system such as ours.

RQ3. Can friend networks present in social television systems be leveraged to help users choose what to watch? At the start of our study we had asked participants questions about how they plan (if at all) what they watched on television. 55% strongly disagreed when asked if they planned ahead on what to watch (m=2.63, sd=1.15). While users claimed to tune in to watch a specific show often (62%, m=3.63, sd=1.02), they also still explained that they watched programs/films when it is convenient, rather than to see something specific for the majority of the time (56%, m=3.38, sd=0.96).

Figure 6. User preferences for sharing of different types of annotations

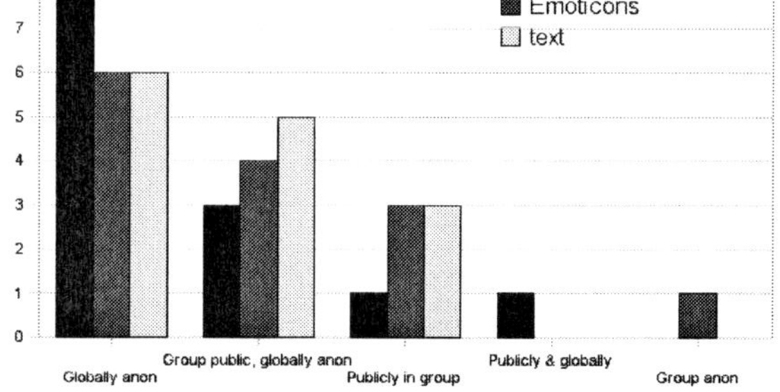

So when users do find a convenient time to watch some television in general, how do they decide on what to watch? Participants said that they mostly browse channels until they find something interesting to watch (51%, m=3.38, sd=0.96). This reiterates the popularity of the 'channel surfing' behavior as reported in previous research such as Taylor (2003) that studied how TV users select a show to watch. Based on these responses, and using the categorization described by Logan (1995), we found 25% of participants to be "watchers" or those who know exactly what they want to watch, and 50% to be "grazers" or those who generally channel surf.

Close to 90% of our participants watch only low to moderate amounts of television weekly. This minimal television viewing, coupled with the ad hoc "grazer" habits described earlier, could mean that users would likely not want to waste their precious viewing time channel surfing. CollaboraTV employs a simple, yet powerful method to help users select a show to watch quickly. Users can view two lists -- PopularShows(Buddies) and PopularShows(All) -- which are lists of shows that have been recently viewed by buddies or all system users respectively. The idea here is that when users are having trouble deciding which show to watch, they can turn to what their like-minded friends have been watching recently, or optionally pick a show that a lot of system users seem to be watching lately. This show recommendation technique is based on the theory of social influences.

We asked participants if their viewing choices were influenced by those around them. More than half the participants said that they watched video content recommended by family, friends or colleagues (m=3.19, sd=0.75). Further, a fair number of them admitted to having watched content that others had been talking about or were the current media buzz (m=2.9, sd=0.57). These responses give credence to our recommendation technique.

At the end of the study, we asked users if they found the PopularShow() lists useful. 57% (m=3.71, sd=0.73) strongly agreed that the ability to see what others in their buddy group have been watching i.e. PopularShows(Buddies), was useful. In contrast, only 42% found the PopularShows(All) list useful. Despite responding positively towards the utility of the PopularShow() lists, observing our usage logs revealed low usage in terms of users actually clicking on shows in these two lists. This could be because the lists are not very accessible in their current location inside the program guide. To make the recommendation method using these lists more effective, future versions of CollaboraTV will have a splash screen similar to those found in popular social networking site. Recent activity traces of buddies will be displayed, which in turn would encourage users to explore and interact more with the system. 62% of participants also said that they already share video content today(e.g. by sending a friend a link to a website). Such explicit behavior will also be supported allowing users to send and receive show recommendations.

As another approach to solving the show selection problem, we can use interest profile data for collaborative filtering based recommendation. Most contemporary systems rely on an overall rating (e.g. two thumbs up, five stars). Interest profiles, on the other hand, provide essentially a scene-level granularity. For some content, even finer grained feedback may be available, for example, the individual jokes and gags in a sitcom. With this level of detail and personalization (e.g. user liked this joke, but not this one), the accuracy of recommendations could be exceptional.

DISCUSSION AND FUTURE WORK

In our studies, we have taken a first step towards creating social television systems that support both synchronous and asynchronous viewing modes. Further, we have demonstrated the importance and utility of a rich set of communication primitives in recreating the social experience of viewing television with others, while also illustrating the value of receiving recommendations implicitly from other users of the system. We now discuss the future implications of our work.

Alternate Temporal Annotations

To provide us with future direction, we asked study participants to consider 3 alternative annotation categories (1) illustrations, where users could draw something on the screen, (2) audio, where users could record short voice annotations, and (3) video, where users could attach a video clip of themselves. 52% voted for voice-based annotations, while 32% showed interest in option (1). Video annotations did not seem to resonate with respondents, which could be due to the overhead of maintaining a presentable countenance while watching television and loss of privacy in general.

Integration with Social Networking Sites

In answering RQ3, we explored the potential of exploiting buddy networks within CollaboraTV for the purpose of making implicit show recommendations. The next logical step would be to tie CollaboraTV to popular social networking websites such as Facebook or MySpace that have open APIs for developers. We are currently working on a Social EPG which integrates a social network, recommender engine, and social usage data into an EPG. It can display a visualization of shows that a user has been watching recently, allowing their buddies to discuss the show or even follow a link back to CollaboraTV to view a show that everyone is talking about.

Show Highlights

Interest profiles can be used in several powerful ways. We are currently working on features that exploit this social usage data. One application for the data is an interest-based media seek feature. Most media navigation controls provide the ability to skip forward and backward using a fixed-time step. However, by taking advantage of interest profile data, it is possible to provide a skip feature that advanced users to the next section that was highly rated by previous viewers. Consider if this feature was available for a soccer game. The portions of high interest (as rated by previous users) are likely to be intense periods of game play, shots on goal, and actual goals. Using the interest-based seek feature, one could progressively advance through the game, seeing only the most highly rated sections.

Another example is a feature that collapses programming down to a fixed time. For example, a user could ask the system to collapse a baseball game down to only the most exciting (i.e. highly rated) hits. This would be achieved by identifying the highest rated sections of the game, removing these clips, and then reassembling them into a new (and shorter) program. This feature could also be applied to a genre of programs, for example, news, where the most highly rated stories and interviews of the day would be pulled out and synthesized into a custom news program (Figure 7).

Figure 7. Highly rated sections from three programs are identified and synthesized into a new program.

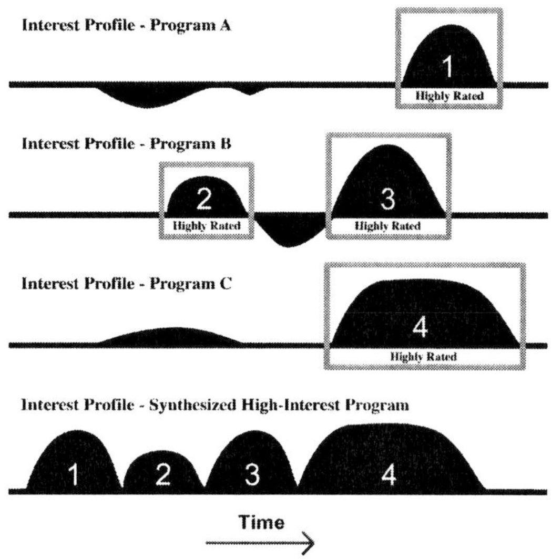

Show Recommenders

From our survey, 64% of users were willing to share their annotations with a show recommendation system (m=3.57, sd=1.28) to receive personal recommendations, while 57% were comfortable with sharing their annotations for the purpose of providing recommendations to other system users. Because this implicitly collected data provides a rich representation of a user's experience while watching video, we are exploring recommendation techniques based on user annotations. Implicit user feedback in the form of annotations can be used as a good substitute for explicit ratings. Such a method would do away with the issue of scarcity of ratings that many collaborative filtering-based recommender systems face and hopefully provide additional data to help in forming better user clusters for generating recommendations.

Limitations and Learnings

While having a set of participants who are geographically spread across the world provided an ideal setup for testing CollaboraTV, this was not without its shortcomings. Delivering instructions via email and following up with participants who we cannot meet face-to-face can be quite challenging. We believe that it would be more fruitful to have a set of users who we can meet in person at the beginning and the end of the period of study to learn first hand about their experiences with the system. Online surveys lack the ability to explore serendipitous questions and conversations that are possible with meetings.

We chose an open ended model of running our experiment to observe realistic interactions and use cases that are hard to capture in manufactured lab settings. However, we came to realize that in order to study interesting user behaviors (e.g. synchronous and asynchronous modes of communication), we would have to engage with participants with specific tasks geared towards these behaviors. It is important to balance the freedom of using the system as one wishes with specific tasks and events.

CONCLUSION

CollaboraTV allows people to interact in a synchronous, asynchronous, and mixed television-viewing situation. This unique ability offers users an unparalleled way to communicate, and ultimately reconnect with friends, family and colleagues. The lab study and comprehensive field study have yielded results that show the virtual audience was successful in engaging users and humanizing people who were remote or teletemporal. Users are enthusiastic about asynchronous communication and found value in receiving implicit show recommendations from their friends. Overall, we believe CollaboraTV achieves our goal of bringing people closer together, making the ever present "electronic hearth" social again.

REFERENCES

ABI Research. (2006). *Worldwide DVR market analysis*. Technical report. Oyster Bay, NY: ABI Research.

Associated Press. Average home has more TVs than people. Retrieved January 29, 2006, from USA Today: http://usatoday.com/life/television/news/2006-09-21-homes-tv x.htm, September 2006.

Boertjes, E. (2007). Connectv: Share the experience. *In 5th European Conference on Interactive Television. EuroITV '07* (pp. 139–140).

Bureau of Labor Statistics (2006). *American time use survey*. Technical report, U.S. Dept. of Labor, Washington, DC.

Chorianopoulos, K. (2006). Animated Character Likeability Revisited: The Case of Interactive TV. *Journal of Usability Studies, 1*(4), 171-184. UPA Press.

Coppens, T., Trappeniers, L., & Godon, M. (2004). AmigoTV: Towards a social TV experience. *In 2nd European Conference on Interactive Television. EuroITV '04*. Brighton, UK.

Diederiks, E. (2003). Buddies in a box: animated characters in consumer electronics. *In 8th international Conference on Intelligent User interfaces. IUI '03* (pp. 34–38). ACM Press.

Fairbank, M. (1995). Sending signals: Kids speak out about values in the media. Maullin & Associates (USA). A Children Now Poll, Los Angeles, CA.

Federal Communications Commission (2006). *12th annual assessment of the status of competition in the market for the delivery of video programming*. Technical report, FCC, Washington, DC.

Geerts, D. (2006). Comparing voice chat and text chat in a communication tool for interactive television. *In 4th Nordic Conference on Human-Computer Interaction 2006: Changing Roles*. NordiCHI '06 (pp. 461–464). ACM Press.

Harboe, G., Massey, N., Metalf, C., Wheatley, D., & Romano, G. (2007). Perceptions of value: The uses of social television. *In 5th European Conference on Interactive Television. EuroITV '07*, (pp. 116–125). ACM Press.

Harboe, G., Metcalf, C., Bentley, F., Tullio, J., Massey, N., & Romano, G. (2008). Ambient Social TV: Drawing People into a Shared Experience. *In CHI '08: Proceedings of the SIGCHI conference on Human factors in computing systems*, New York, NY, USA. ACM Press.

Harrison, C., & Amento, B. (2007). CollaboraTV: Using asynchronous communication to make TV social again. *In Proceedings of the EuroITV 2007 Conference (TICSP)*, (pp. 218-222).

Leichtman Research Group, Inc. *Press release: DVDs now in over one of every five U.S. households*. Retrieved September 18, 2007, from: http://www.leichtmanresearch.com /press/ 082107release.html, August 2007.

Logan, R., Augaitis, S., Miller, R., & Wehmeyer, K. (1995). Living room culture – an anthropological study of television usage behaviors. *In Human Factors and Ergonomics Society 39th Annual Meeting*, (pp. 326–330).

Luyten, K., Thys, K., Huypensm S., & Coninx, K. (2006). Telebuddies: social stitching with interactive television. *In CHI '06: CHI '06 extended abstracts on Human factors in computing systems* (pp. 1049–1054). New York, NY, USA. ACM Press.

NetFlix, Inc. *Corporate facts sheet*. Retrieved September 18, 2007, from: http://ir.net ix.com.

NPD Group, Inc. *Npd group, inc. tv-dvds soften video sales decline*. Retrieved September 18, 2007, from: http://npd.com/press /releases/press061024. html, October 2006.

Oehlberg, K., Ducheneaut, N., Thornton, J., Moore, R., & Nickell, E. (2006). Socialtv: Designing for distributed, sociable television viewing. *In 4th European Conference on Interactive Television. EuroITV '06*.

Persson, P. (2003). Exms: an animated and avatar-based messaging system for expressive peer communication. *In GROUP '03: Proceedings of the 2003 international ACM SIGGROUP conference on Supporting group work*, (pp. 31–39), New York, NY, USA: ACM Press.

Putnam, R. D. (2000). *Bowling Alone*. New York, NY: Simon & Schuster.

Regan, T. & Todd, I. (2004). Media center buddies: instant messaging around a media center. *In NordiCHI '04: Proceedings of the third Nordic conference on Human-computer interaction*, (pp. 141–144), New York, NY, USA. ACM Press.

Salem, B., & Earle, N. (2000). Designing a nonverbal language for expressive avatars. *In CVE '00: Proceedings of the third international conference on Collaborative virtual environments* (pp. 93–101), New York, NY, USA: ACM Press.

Smith, M. A., Farnham, S. D., & Drucker, S. M. (2000). The social life of small graphical chat spaces. *In CHI '00:Proceedings of the SIGCHI conference on Human factors in computing systems*, (pp. 462–469), New York, NY, USA: ACM Press.

Taylor, A., & Harper, R. (2003). Inside the Smart Home, chapter Switching on to Switch O (pp. 115–126). London: Springer-Verlag.

Taylor, D. (2007, April). *The digital water cooler*. Retrieved June 9, 2008, from http://www.fabricoffolly.com/2007/04/digital-water-cooler.html, April 2007.

Television Bureau of Advertising (2006). *Media comparisons study*. Technical report, TBA, New York, NY.

Tichi, C. (1991). Electronic Hearth: Creating an American Television Culture. New York: Oxford University Press.

Weisz, J. D., Kiesler, S., Zhang, H., Ren, Y., Kraut, R. E., & Konstan, J. A. (2007). Watching together: integrating text chat with video. *In CHI '07: Proceedings of the SIGCHI conference on Human factors in computing systems*, (pp. 877–886), New York, NY, USA: ACM Press.

Zaslow, J. (2002, November). *If tivo thinks you are gay, here's how to set it straight*. Retrieved June 5, 2008, from WSJ Online: http://online. wsj.com /article email/SB1038261936872356908. html, November 2002.

Chapter XIII
From 2BeOn Results to New Media Challenges for Social (i)TV

Jorge Ferraz Abreu
University of Aveiro, Portugal

Pedro Almeida
University of Aveiro, Portugal

ABSTRACT

This chapter focuses on traditional and emergent challenges for the Social (i)TV area focusing on explaining the development and evaluation of one of the first Social iTV prototypes and looking at the challenges new media is introducing to this research field. The authors begin by explaining the conceptualization, development and evaluation process of the 2BeOn system and continue with the most important results from it's evaluation with a particular focus on the results that can be important when developing any Social iTV platform. In the last part of the chapter recent developments in the broadcast of TV and Audiovisual content, namely considering the Internet as a medium, are addressed. In this scope authors propose a categorization of emergent online distribution platforms along with a set of social activities users perform on those platforms. Taking in consideration some of the challenges surrounding the presented scenario the chapter ends with the conceptualization of UMCA, a system that could increase social interaction activities performed during the consumption of online AV/TV content.

SOCIAL (I)TV - PROGRESSES AND CHALLENGES

Interactive television platforms that promote services aimed to support socialization practices around TV viewing, acting as a kind of "social glue" (Light, 2004), are at the core of Social (i)TV research (Harboe, 2007). Despite the introduction of the first iTV systems with real time communication services being considerably recent, it is interesting to notice that a considerable amount of research as been done in this area. Since the year

2000, with AOLTV being the first commercial iTV system integrating an Instant Messaging service (Kawamoto, 2000), there have been a number of iTV conceptual models, prototypes and systems that enriched this concept adding user awareness in relation to what others are watching. This type of presence/information, besides enabling a feeling of togetherness, opens up several interaction possibilities among users of an established social network: a user can alert his buddies to something interesting that is on air or simply peep at what others are watching and start a conversation based on a common TV content, no matter their geographical location. In this context, numerous developments have emerged, such as: 2BeOn—University of Aveiro (Abreu, 2002); Reality IM—Accenture Labs (Chuah, 2002); Amigo TV—Alcatel (Coppens, 2004); Social TV—PARC (Kharif, 2005), Social Software - described by Coates (2005); ConnecTV—TNO (Boertjes, 2007); CollaboraTV—AT&T (Harrison, 2007); Find-A-Friend - University of Siegen (Heß, 2007); Living@room—CSP (Andrea, 2007).

However, even with this considerable number of prototypes, applications and related research, there are still a lot of open questions, being Social (i)TV far from a fully explored area. In the following sections, and based on the experience gained with the 2BeOn prototype, some answers to the following topics are addressed:

1. Considering the different sociability practices promoted by television viewing, what are the communication, presence and content sharing services for a suitable mediation of those activities in an iTV system?
2. As the interface design of a Social iTV application is a central piece to achieve its' proposes, what are the fundamental issues in this domain?
3. How to evaluate a Social iTV application in order to find out if there is a correlation between users' digital literacy and: (a) learning curve; (b) the suitability of the application to support and promote interaction between viewers allowing an increase of the communication frequency; (c) the users' wish to make use of such an application.

Taking into account the ever-increasing sociability practices that arise from the current use of AV (audiovisual) distribution platforms (for both PC and mobile based devices), this chapter also focuses on the techno-social context in which users are currently sharing AV content and interacting through it.

THE 2BEON SYSTEM

The 2BeOn system was conceptualized, prototyped and evaluated with one main goal: to support and promote interaction between viewers, enabling an increase of communication and sociability practices (Abreu, 2007). Users involved in the evaluation of the prototype were not restricted to members of an existent social network (e.g. friends or family members), since viewers that did not know each other but that were in tune with the same TV content were also considered.

As stated by Gray (1992) *"A very important part of the pleasure of television serials is to gossip about them the following day."* (Gauntlett, 1999, p.128). In this context and in parallel with its main goal, 2BeOn aims to capitalize the TV ability to induce conversations, strengthen sociability practices and enable a new form of TV enjoyment in a virtual, nevertheless collective, viewing environment.

Conceptualization Process

The conceptualization process of this application was based on a theoretical framework[1] (mainly centred on sciences and technologies of communication) in order to define a set of functional requirements needed to support viewers' social practices. These include commenting, criticizing or gossiping about TV content, in real time or after the broadcast (Dahlgren, 1995); warning someone of a TV program being broadcasted; or sharing AV content. At this stage of the development process, a central issue was the identification of these typical communication practices around TV content and the evidence of emergent transmedia and multitasking behaviours, like using PC mediated communication services while watching TV (Lafrance, 2005; Oumard, 2008).

Design Guide Principles

The aforementioned theoretical contextualization also reflected about behavioural aspects of viewers (the main users of the application); visualization patterns; and television's technical constraints, in order to supply relevant inputs to establish a suitable set of Design Guide Principles (DGP) for iTV interfaces. These DGPs (21 in total), which were applied in the development of the prototype, can be arranged in the following 3 categories based on their impact at:

- Functional level (based on behavioural factors, such as attention levels to the TV set, predisposition for interaction and social contexts of visualization).
- Graphic design.
- Interaction design.

Main Features

Awareness about user status and channel viewed is of great importance to enable a feeling of togetherness and to promote interaction between known viewers. The role of such presence mechanism is even more important in the current scenario of TV audience fragmentation and ever-increasing individual TV consumption. As sustained by Chuah (2002), this kind of information can have the ability to change the motivation for the selection of TV programs, it being based not only on their content but also on the social context surrounding it.

One necessary feature of these awareness systems is the possibility of guaranteeing users' privacy. When the user does not want others to be informed about the current TV channel he is watching, he can simply "block" that particular channel. In this situation the users' status changes automatically to "busy"—the same status when he is performing other tasks, like instant messaging, reading messages or managing his social network. The portability of personal information was also guaranteed allowing users to login in the application wherever they were.

The support of informal interaction between known users was based on 3 services:

1. ClipEmail used for sending short alerts
2. IM designed to support longer real time conversations[2]
3. TVPR (TV Program Recommendation).

Communication between unknown viewers and identification of matching telly profiles was also considered, since occasional conversations about TV also occur between unfamiliar people and affinities might emerge among unknown viewers sharing the same type of TV consumption. For these purposes a TVChat service (whereas chat rooms are in sync with TV zapping) was implemented and a search tool for matched television profiles was conceptualized.

These features have been grouped into 4 sections of the system:

- FriendsOn - buddy list and access to IM, ClipEmail, TVPR and Email services;
- MsgOn - management of incoming messages while offline or busy;
- ChannelsOn - management of channels and access to the TVChat service;

- + FriendsOn - management of the social network.

The Development Process

Since the beginning of this research it became clear that it would be necessary to outline a strategy for demonstrating the sociability concept of the application to a group of pilot users, in order to try and test its functionality and interface approach and to determine its acceptability level. This is the global goal of a high-fidelity prototype (Chorianopoulos, 2004).

The prototype development was carried out based both on a functional and a technical dimension (Strauss, 1997), targeted at fulfilling the identified functional requirements.

The functional dimension comprised the design of the graphical and interaction user interface, which was assisted by the achieved set of DGPs (based on behavioural, technical and usability factors) and interaction storyboards. The main communication strategies, choice of colours, fonts, iconography, spatial distribution and other graphical elements of the interface were also defined at this stage.

Some of the solutions that were applied, based on a sample of DGPs, are shortly described here, since they can be of value for the development of other iTV applications (for details please see Figures 1 and 2).

Considering DGPs with impact at the functional level, it is necessary that the application is set to:

- *Different levels of users' attention to TV program*—it is important to consider the situations in which the user, consciously or not, reduces its level of attention to television programs.
  Adopted solutions:
  a. sound alerts were incorporated to catch the user's attention to special events.
  b. based on the user's typing frequency, when using the IM service, the interface (with an embedded layout) is dynamically dimensioned. This way, as the typing frequency increases, the text field expands and the TV area shrinks and vice-versa.
- *Predisposition for interaction*—the application must be designed so that users don't feel forced to interact and perform tasks (Peng, C., 2002) allowing the interaction to occur only when wanted.
  Adopted solutions:
  a. any communication by instant messaging is preceded by an invitation. Thus, the user can easily refuse the beginning of an interaction.
  b. if the user does not login he can watch regular TV, that is, without any interactivity. If logged on he can define himself as being offline, simply by pressing a dedicated key on the remote.
- *Different social contexts of visualization (individual or in group)*—in a situation where the user is not alone in the room there are several factors to consider. It is required to ensure privacy (for example, preventing that messages addressed to him are viewed by all) and it is also required that his interactions do not mix with other elements being watched. Regan (2004) also addressed this problem from another perspective.
  Adopted solutions:
  a. a family/group account was implemented so that the user can login with a contact list based on people known to all elements of the group.
  b. conceptualized, yet not fully implemented, was a feature that allowed the user to choose to have his messages directly displayed on the TV set or, for example, to be redirected to his

mobile phone. This solution fits in the secondary screen approach, which has been used in a number of different iTV prototypes, more than a decade ago, as described by Cesar (2008).

Regarding some of the DGPs that influence graphic design, the interface must be designed in order to:

- *Avoid irrelevant information*—given the limitation on the amount of information that can be displayed on a standard definition TV set, it is advisable to limit the maximum number of selectable icons that are available at the same time on the screen.
  Adopted solutions:
  a. limitation of the number of icons in each screen to a maximum of 4, which also allowed to trigger them through the chromatic shortcuts of the remote.
  b. use of colour codes to differentiate users status (online and offline) and privacy of channels (active and blocked channels).
- *Give primacy to television programs*—it is essential to preserve, as much as possible, the integrity of television content in order to guarantee traditional television experience. Thus, the interface must be designed and positioned to obstruct as less television images as possible.
  Adopted solutions:
  a. positioning of the logo and interaction menus in the right upper corner of the screen.
  b. vertical placement of menus.
  c. use of overlaid and embedded interface layouts.
  d. automatic scalability of the IM interface.

Finally, there are DGPs with impact at the interaction design level, pointing out that the application needs to fit to:

- *User interaction adapted to television*—it is not accurate to assume that users of iTV platforms have high digital literacy, so interaction with TV should avoid typical PC navigation/interaction patterns, like drop-down menus and vertical scroll of pages (Gawlinski, 2003).

Figure 1. 2BeOn main menu (left) / zoom of FriendsOn interface—start of an IM communication (right)

Figure 2. automatic dimensioning of the IM interface: normal size (left) / expanded field text (right).

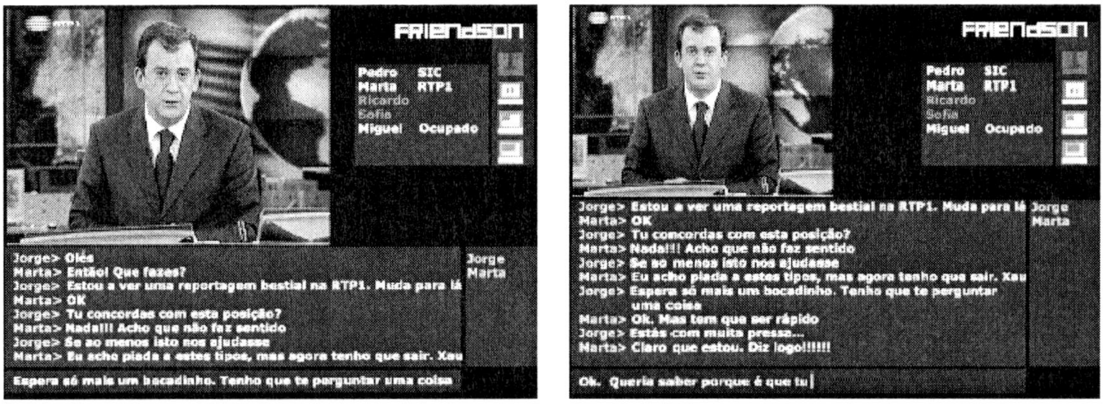

Adopted solutions:
The initial purpose of using a remote control acting like a mouse was quickly abandoned and replaced with:

a. chromatic keys—shortcut buttons to activate screen icons matching the same colour.
b. standard orthogonal cursor—the selection area activates, for a brief period of 2 seconds, on screen labels linked to each icon.
c. reserved keys for some special features.

- *Optimization of the navigation experience*—by reducing the number of steps needed to perform a particular action and keeping the response time below one second.
  Adopted solutions:
  a. prediction and anticipation of the next interaction.
  b. automatic sorting of contacts and channels lists, based on the frequency of communication with each of the contacts and viewed channels.

As far as the technical dimension is concerned (which addresses the design of the hardware and software architecture that supports the required features), its development was assisted by block diagrams of the architectural modules and flowcharts correlated with the interaction storyboards. The prototype was not based on a commercial middleware; instead, Macromedia Director was used with PC clients (running as Set Top Boxes) connected to regular TV sets via VGA to PAL adaptors. Two components or plug-ins were essential: BuddyAPI (which allowed the control of the TV tuner application) and WinShapper (used for the definition of transparent areas supporting both overlaid and embedded interfaces). For more details on other technical solutions please refer to Abreu (2002).

The Evaluation Process

At the time the 2BeOn prototype was developed, published iTV evaluation methodologies were very scarce and mainly based on traditional usability engineering concepts and methods (Chorianopoulos, 2004). With the growing number of available Social iTV solutions, there is an increasing need to understand how these systems should be evaluated, both from a technical point of view and HCI and Social Sciences perspectives. In this scope and despite current publications like Chorianopoulos (2006) it is believed that the methodology used for the evaluation of 2BeOn can still be a relevant contribution to this area.

The goal of the 2BeOn evaluation phase went beyond simple usability tests. Actually, it was aimed to enable regular users with an adequate level of hands on experience with the system, so that it would be possible to extract a wide set of results. This purpose meant, therefore, a set of 3 goals that the evaluation was intended to attain, namely:

1. Validate the usability of the prototype, to check if the various functional solutions were suitable and if they ensured a straightforward use of the system. This goal was mainly centred on the verification of the proper operation of the DGPs in order to analyse potential correlations between demonstrations of dissatisfaction from evaluators and technical implementation errors.
2. Evaluate the conceptual model of the system and the suitability of its functional solutions. This second goal implied the validation of the various hypotheses that sustained the

Table 1. Evaluation process

Phase 1	
Purpose:	to characterize a broad range of potential users (TV viewing habits and dynamics of communication around television) enabling the selection of three groups of non-expert evaluators with different levels of digital literacy. This allowed the application of a causal-comparative methodology with a non-probabilistic sample.
Method:	quantitative and qualitative analysis of the gathered data.
Tools:	questionnaire, enabling the selection of 3 groups of 5 evaluators with the following levels of digital literacy: low (first year students attending the course of Pre-School Teachers) - EdI; average (2nd year students attending the course of New Communication Technologies) - NTC; high (professionals in the area of information and communication technologies) - EXP.
Phase 2 (this phase occurred in two stages)	
Stage1 (3 non-consecutive sessions for each group of evaluators)	
Purpose:	to supply a substantial and uniform knowledge of the prototype (through the interaction with a remote moderator performing a series of tasks presented by a local assistant) and to test usability analysing the evolution of evaluators' performance (learning curve), throughout hands on sessions (Preece, 2002).
Method:	Journaled Sessions (Hom, 1998).
Tools:	internal mechanism for the registration of the evaluators' interactions; synchronized recording (video - what was happening on the screen; audio - evaluator's verbalizations during the evaluation of the prototype); opinion screen (that appears at the end of the session, inviting the evaluator to write down his opinions); and direct observation (using observation tables filled in by a local assistant).
Stage2 (at the end of the 3rd session)	
Purpose:	to allow free use of the prototype, in a close simulation of a real environment; verify the sort of actions and reactions that evaluators (who already knew the prototype) assumed in these conditions.
Method:	this stage started with a small set of interactions to integrate all evaluators in a similar situation. From then on the moderator remotely promoted the use of the application and evaluators did not have to follow a script of tasks. The assistant examined the type of problems evaluators faced.
Tools:	the tools for data collection were the same used in the first stage of this evaluation phase.
Phase 3	
Purpose:	to check the opinion of evaluators on qualitative factors of usability, suggestions and their level of interest regarding a future use of the application.
Tools:	final questionnaire and interview.

Figure 3. 2BeOn evaluation phases (left) / hands-on sessions during phase 2 (right)

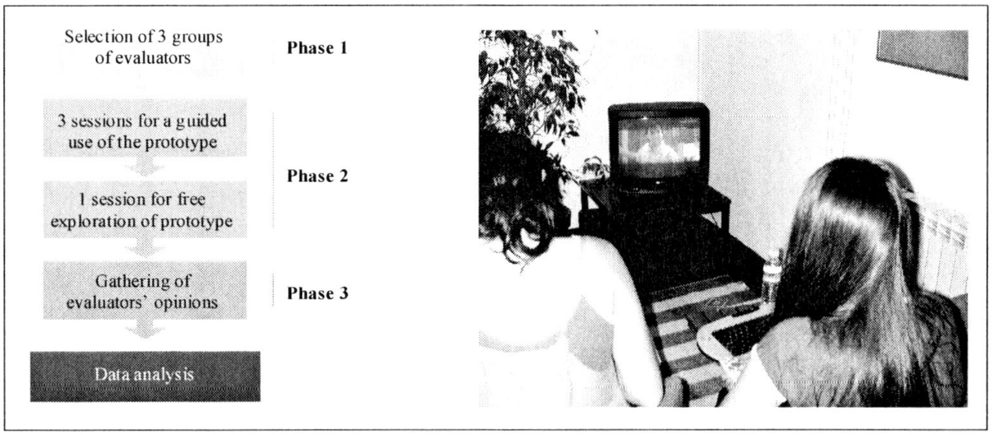

main functional requirements. It also aimed to identify the user's level of interest regarding future use of such an application.

3. Synthesise a number of opinions and suggestions, raised by the use of the prototype, to enable future improvements.

Considering these goals, the evaluation process was structured in three distinctive, but complementary, phases (each with its purpose, method and tools)—see Table 1 and Figure 3.

Evaluation Results

1st Goal: Usability of the Prototype

The validation of proper use of the DGPs was made from the analysis of results gathered by different combinations of the aforementioned instruments (tools) for data collection. Exemplifying, the solutions adopted to face *different levels of users' attention to TV programs* were verified by two questions on the final questionnaire, one in the interview and through the observation tables.

Only two minor problems related with two DGPs were detected. These are here described to show their importance on the usability of the prototype. They regard the solutions adopted to:

- Face different levels of users' attention to TV program—the volume of the sound alert linked with the incoming messages was to low, so some of the received messages were not immediately detected.
- Guarantee the consistency of the interface—a mismatch in the colour of the exit icon in two services resulted in an inconsistency and consequently in some interaction errors.

Correlation between Users' Digital Literacy and the Usability/Learning Curve

Considering the mean time spent by each group at each of the 3 evaluation sessions and the mean number of uses of the "back" key, it was possible to observe:

- An evidence of a correlation between digital literacy and ease of use in a first contact with the prototype (users of the EXP group took less time).
- A stabilization of the learning curve as sessions progress - evaluators with different levels of digital literacy tend to have, after some practice, identical levels of ease of use of the prototype.

Figure 4. The progress of the average time spent per session

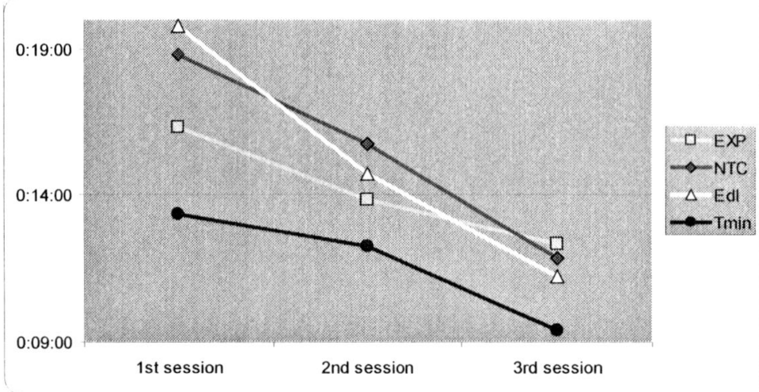

2nd Goal: (a) Validation of the System's Conceptual Model

Based mainly on the analysis of the final questionnaire and the interview it was possible to conclude that all the research hypotheses that sustained the main functional requirements proved to be valid.

2nd Goal: (b) Achievement of the Main Objective

This refers to the evaluation of the suitability of the application's features in supporting and promoting interaction between viewers, enabling an increase of communication and sociability practices. It was possible to conclude that:

- Information of presence (status and TV channel) acts as a catalyst of communication - most users agreed that, having this type of information from their contacts (in the FriendsOn section), they would have a higher propensity to start a conversation. This can be either encouraged by the fact that users detect they are watching the same channel of the person to be contacted, or by the need to alert about something they find interesting.

Some evaluators said this type of awareness also gave them a feeling of comfort and togetherness, stating, for example, that:

"it creates proximity..."—Evaluator NTC-C; "Maybe comfort in knowing that a person is not alone (...) know that there is someone there who is in the same situation"—Evaluator EdI-E.

- IM and TVChat were the services users found to be more suitable to support interpersonal communication (classified as "high" support). ClipEmails and the feature for sending TVPRs were classified as "middle" support.
- IM and TVChat use was also considered to be very compatible with television reception with minor intrusion on the attention to the TV program.

"I can monitor both (communication and television) perfectly"—Evaluator EdI-E.

- Users were also asked to predict the level of distraction to TV reception if voice communication was used instead of text messages. It is important to notice that on average they all said that text communication was less distractive than voice would be and only one of the evaluators agreed with a future

Figure 5. Influence of the information in section FriendsOn as a catalyst of conversations

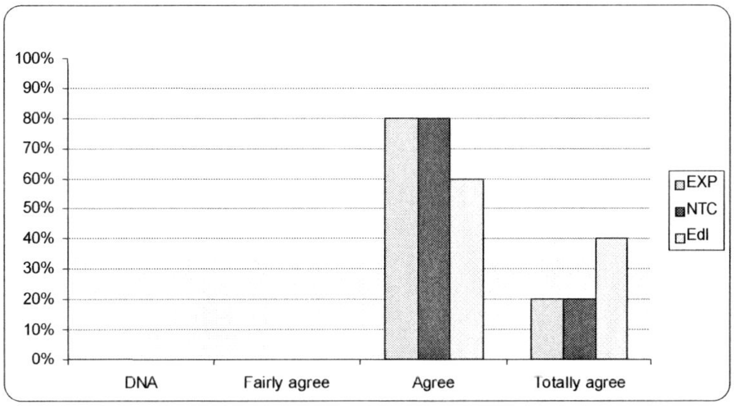

Figure 6. Classification of interpersonal communication support by various services

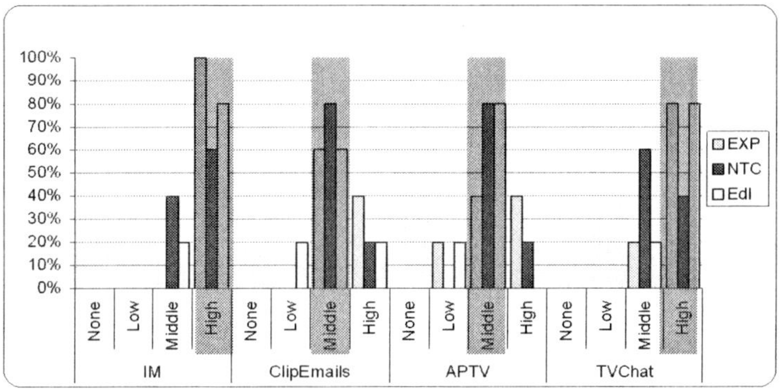

integration of voice communication in the system.

- Most evaluators, when asked if they would speak more about TV content with a Social iTV application like 2BeOn, answered that they would do it up to four times more than without such an application.
- Regarding the user's level of interest concerning a future use of such an application, very positive answers were given. 80% of the evaluators of NTC and 60% of the evaluators of the groups EXP and EdI expressed having "great interest". Comparing the answers given before and after the use of the prototype, a relevant change in opinion was observed, with an increase of 40% in the option "high interest".

Correlation between Users' Digital Literacy and the Suitability of the Application to Support and Promote Interactions and to Increase Communication Frequency

Regarding the suitability of the various communication services there were no significant differences between different groups of evaluators. However, considering the predicted increase of conversations expected by having an application like 2BeOn, the evaluators with the lowest digital literacy where the ones that predicted higher increases.

Correlation between Users' Digital Literacy and the Users' Wish to Use Such an Application

Figure 7. Prediction of distraction on the attention to the TV program if voice was used instead of text communication

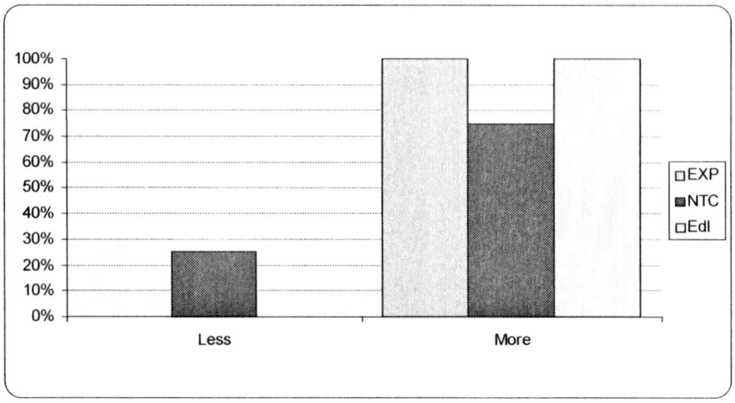

Figure 8. Prediction of the increase in conversations about TV content using 2BeOn

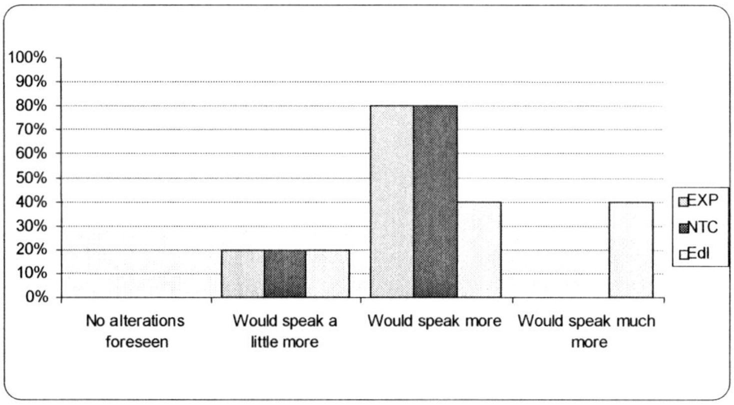

The results gathered allow predicting that, regarding the evaluators involved in the evaluation process (that are, within certain limits, a reflection of a generation between 19 to 31 years), there is a considerable interest in using this kind of Social iTV applications. However, no differences were detected between evaluators of different groups (with different levels of digital literacy).

NEW BEHAVIOURS AND PLATFORMS FOR SOCIAL (I)TV

The results from the 2BeOn evaluation along with results from other similar research projects show that users are willing to talk about the TV content they consume, and are ready to do it while they watch it. But before (a possible) widespread penetration of these kind of social features in iTV platforms occurs, users are already socializing, in large scale, using PC based AV distribution platforms or using other communication tools in parallel (like IM in PCs) while watching Webcasts or regular TV. Leurdijk (2007) and Lafrance (2005) refer not only to the ability of young people to multitask but also to consume different types of media simultaneously. Oumard et al (2008), in their ethnographic study, also concluded that "*Internet video consumption is already strongly integrated in social communication*".

Figure 9. Level of interest, shown after the use of the prototype, in an application like 2BeOn

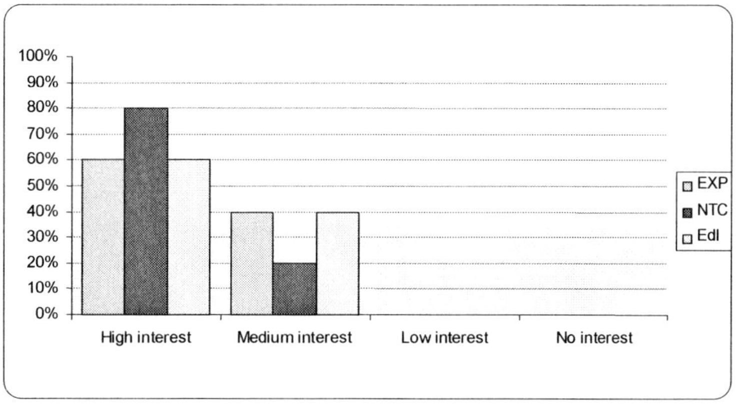

In this part of the chapter a description of these emergent scenarios, its tools and levels of sociability are described and a proposal for increasing social interaction around multiplatform distributed AV content is presented.

Where Users are Watching TV Content?

The online developments in Audiovisual (AV) distribution have been fast and diverse, mainly after the YouTube boom in 2006 and 2007.

Today, we can find multiple formats of platforms that broadcast AV content and promote different levels of participation for users. Some of the available platforms can be classified under the following categories:

- **You Tube like and User Generated Content (UGC) distribution platforms:** The YouTube like platforms are very popular and have been widely used for individual upload (UGC). But these platforms are also used for creating institutional, well planned and full budget broadcast channels, as it can be verified by the U.S. presidential campaign and the candidates' channels created in YouTube[3]. The media consumption in these platforms is very significant, as 75% of Internet users in the U.S. watched, in May 2007, more than 2,5h a month of online video (Lipsman, 2007). A report from Pew Internet shows that 27% watch videos from You Tube, 11% from Yahoo Videos and 6% from MySpace (Madden, 2007);
- **Stream+peer-to-peer platforms:** Joost is probably the most well known service of this category of TV distribution platforms. Using a combination of streaming video server and a peer-to-peer architecture, these platforms allow good video quality providing some popular (traditional) TV channels to be viewed in the computer along with some social and communication features. Babelgum and Zatoo are other important examples;
- **Online TV with UGC platforms:** Broadcasting powered by Internet platforms is no longer an exclusive of traditional media channels. Current Tv, Mogulus and Veoh TV, just to name a few, propose different online TV channels with some own production but mainly allowing users to contribute with their shows. These platforms are distinct from the You Tube like platforms by selecting the content to go online and presenting it in a pre-ordered playlist;

- **Online AV content archives and stores:** One example of a popular online music and video store is iTunes, being BBC iPlayer another important platform;
- **Other:** Other platforms with specific characteristics allow users to have easy access to AV content. Examples are SopCast and TV Ants (peer-to-peer TV channels streaming applications) or Messenger TV (like other IM applications with A/V support it allows users to share and watch a video while commenting on it. Examples of these tools are later described in this chapter).

What about Social AV/TV features?

Some of the platforms described in the previous topic were not originally designed with support for collaborative social media consumption. However, with the emergence of Web 2.0, many of these platforms have been upgraded with awareness, collaborative and communication features that support a set of social activities. These activities include:

- **Rating:** Classifying the video content quality and helping other users decide to watch it or not. The classification can be completed with comments associated with a video or channel. 13% of video consumers do this (Madden, 2007);
- **Forwarding:** Recommending a video content or TV channel to friends by sending the URL via email. This activity is very popular as 57% of video consumers on the Web share recommendations with others (Madden, 2007). As stated by Oumard (2008) some users even bookmark videos for future recommendation, engaging in chat/IM sessions around the video content;
- **Embedding:** An indirect recommendation feature by promoting the re-broadcasting of a video content in a blog or other platform;
- **IM or Chatting:** Some platforms allow users to chat with others watching the same channel or video content. Joost also allows users to talk to friends even if they are not tuned in the same channel;
- **Collaborative production:** The users' participation related to a TV channel or video content is not limited to the discussion or recommendation about content produced by others. The user can now be an active part in the production process. Current TV, Mogulus and other platforms allow users to create their own collaborative channel or send their programmes to be broadcast (if accepted) in the official channels.

These examples from Mogulus show that users can easily create their own channels using regular webcams and feeding them with the collaboration of others around the world. Al Gore, co-founder of Current TV, stated that they are trying to open the television business in order for viewers to be able to do television (Ribeiro, 2007).

As a Pew Internet (Lenhart, 2007) report states, "*The posting of content does not happen in a vacuum. Content is posted so that it might be seen by an audience (...). And often that audience responds to the content posted online, making the*

Figure 10. The set of Joost widgets available for interacting with content and other users

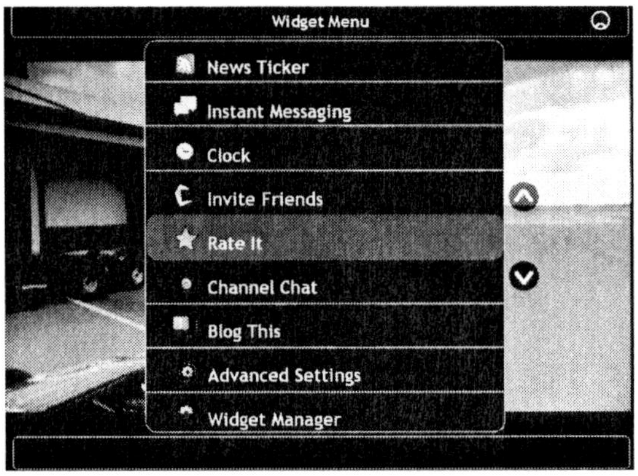

content as much about interaction with others as it is about sharing with them" (p. 15). The same report confirms that Internet users that share their own content online (39% of online teens) are more active communicators.

Challenges

These data concerning the use of new AV distribution platforms and the social activities of a "new" Internet generation of users bring out a new set of challenges when referring to Social (i)TV or, in a broader analysis, to social practices surrounding AV content:

- It is harder to have a specific content at a specific time as a common referential to viewers;
- Each platform offers its own social features, usually not compatible with each other;
- Users can have similar AV content consumption but watch it in different platforms and terminals (PC, TV, mobile, …);
- Users are now content (co)producers, searching for new ways to interact or influence content. As Leurdijk (2007) refers "*broadcasters' endeavours could also lead to highly valued new services for a general audience, (...) to new and fruitful exchanges between professional and amateur programme makers (content producers)*" (p. 94).

These challenges have restricted the Social AV/TV support in web based platforms to asynchronous tools. Users comment, forward, rate or discuss a video content in an asynchronous way precisely because they can watch the same content in any platform at any time. But what about synchronous Social AV/TV features? Platforms like Joost and a new set of applications are now offering some interesting features.

Tools for Supporting Synchronous Social AV/TV Activities

Recent developments have revealed some tools that promote Social AV/TV features for PC based users, such as the ability of different users to watch the same video collaboratively.

Some of the most relevant applications include:

- YouTube Streams (http://www.youtube.com/streams_main)—this feature from You Tube allows users to create rooms with their selection of videos (from YouTube of course), share these rooms with others and chat about them. It does not allow synchro-

Figure 11. The home-page of Mogulus showing some of the collaborative channels available (left) and the online editing tool (right).

Figure 12. The ClipSync interface[4].

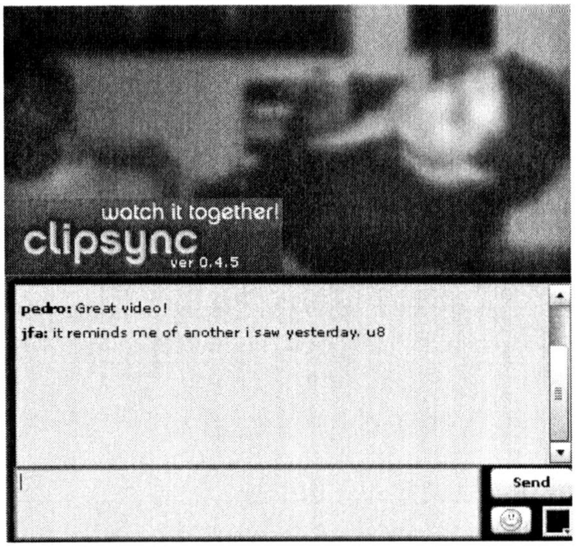

nization between users but allows different levels of participation such as the ability to add videos to the room;

- You Tube—Active Sharing—this feature allows users to know who is also seeing the same You Tube video. It does not provide features to manage a list of buddies or synchronous communication features to interact with other users.
- ClipSync (http://www.clipsync.com/)—this Flash based tool also allows creating and hosting rooms/sessions with video playlists (from different sources) and chat about them, but it adds some important features. It allows the synchronization of playback with the participants in the session and allows the integration of Skype friends lists;
- Messenger Zinc—a plug-in for the Yahoo IM service, introduced in 2007. This plug-in provides the ability to integrate a video player in the IM application allowing to share a video from popular video sites (like YouTube) for collaborative and synchronized viewing with others (Shamma, 2008);
- Microsoft Messenger TV—this new feature from Microsoft, introduced in May 2008, is part of the company's efforts to promote social video, using, in this case, the Windows Live messenger. It allows users to watch videos together from the MSN video service and chat about them.

THE USER MEDIA CONSUMPTION AWARENESS SYSTEM (UMCA)

If all AV viewers had Joost, a Social iTV platform like 2BeOn, host the same room/session in a social video tool or all had a compatible iTV system, it

Figure 13. Microsoft Messenger TV—a social video tool[5]

would be easy to conceive that one could check what one's friends were watching and interact with them based on that referential.

But in online video platforms, considering their huge numbers and diversity[6], it is hard to track what your friends are watching and promote instant interaction based on this. Messenger TV or Zinc allow users to share an AV consumption session but this feature depends on the initiative of the user to share each new video he decides to watch collaboratively.

Taking into account this constraint, in this final section of the chapter, a proposal for a general conceptualization for a new system, the User Media Consumption Awareness system, is presented. The proposed system could allow users to know (at any time and independently of the platforms used) what their friends are watching, allowing them to chat or to make recommendations based on that information.

Technical and Functional Model

Peeko Chat[7] for Firefox browsers is an example of a plug-in that tracks users in the same page and allows them to chat instantly. There is no content evaluation or correlation in this system, just URL tracking. Two or more users visiting the same domain can interact by chat.

The proposed *User Media Consumption Awareness* (UMCA) system extends the features of this type of browser chat, like Peeko, and of tools like ClipSync or Messenger Zync by allowing to:

- Track which video content or channel each user is watching and categorize it according to associated metadata (for example by examining an RSS feed and the tags associated with the video/channel)
- Keep a buddy list, like in a common IM service, allowing users to build their own network of friends and to get information about their instant AV content consumption[8]
- Give users a set of collaborative features that include an IM platform and a recommendation system.

System Configuration Proposal

For it to be successful, this kind of system should be simple and fully integrated in the operating system, like a regular IM tool is. It should act like an add-on to any browser. ClipSync, for example, allows some collaborative features but demands the use of a proprietary application to watch the videos.

The proposal for the UMCA system, as shown in Figure 15, should include the following components:

1. One component, acting as a plug-in for the browsers, detects in real time the URLs visited by the user and compares it with a list of popular video server sites. This list is permanently updated in the UMCA server. The structure of each new site is analysed to maximize compatibility of the sniffer component.

Figure 14. An example of the Peeko Chat in action

If the user is visiting a site, classified as a video site, the component can read the RSS feed related to the page to read the meta information associated with the video. The information provided by the feed allows the component to define the video being watched, the URL and related meta information.

2. Another component, close to an IM service, shares the information from the video with the buddies in the contact list. The user can define the level of privacy wanted for each video site or for each category of videos.
3. Users can know, in real time, which videos others are watching, their categories and URLs. Tools for video recommendation and synchronous text chat are also provided.

Another component of the system aggregates all the information to allow ranking buddies according to their video consumption, identifying the categories of preferred content (ex. Sports, news ...) and creating individual profiles for each user[9]. Finally, another component, derived from the previous one, defines a set of automatic recommendations based on buddies' video consumption. After watching any new video this tool can present the user with a contextual recommendation (received from UMCA) of other related videos, as in the following example:

- *User A just finished watching a video of a Cristiano Ronaldo goal in MSN videos;*
- *UMCA alerts the user with a list of related videos his buddies have watched (similar to Amazon's recommendation system). Ex. "User B watched "Manchester United—Euro champion" in You Tube; User C watched "Top ten goals" in Meta Café".*

The widespread use of a tool like UMCA could contribute to support social activities around online AV/TV consumption. Apart from having participatory viewers contributing with their own content, they could be more active in sharing videos, influencing others' choices and

Figure 15. Conceptual model of UMCA system

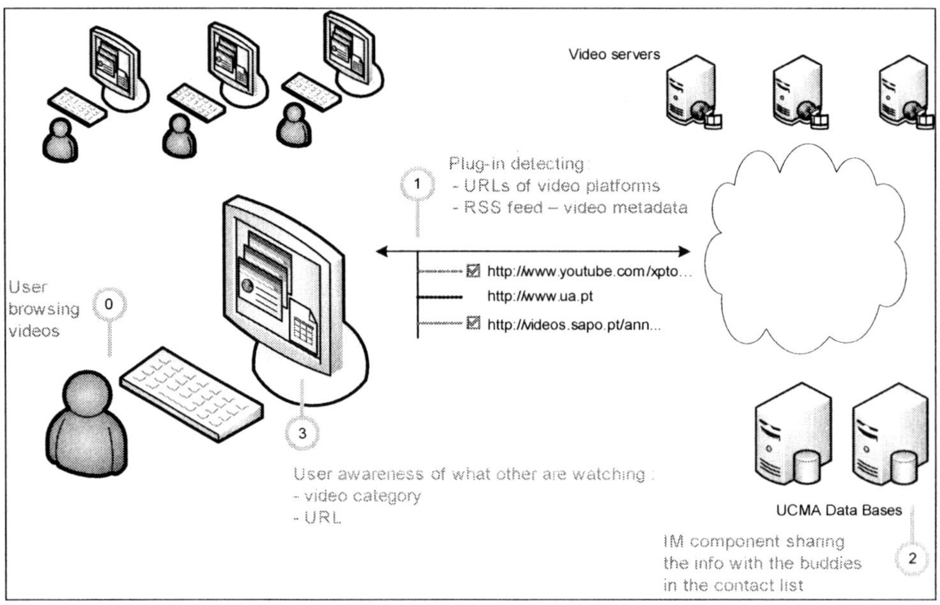

specially talking about what they are watching. This tool could have, potentially, the ability to recover to AV/TV online consumption some of the habits and behaviours we could experience with television decades ago.

CONCLUSION AND FUTURE WORK

Social (i)TV is a very relevant and wide-ranging field. Some of the platforms mentioned in this chapter try to promote social activities around TV consumption, but the transversal adoption of these activities is still in its early phases. The results from the evaluation of Social iTV systems, like 2BeOn, have shown the willingness of viewers to interact with their friends/colleagues while watching TV. Users want to talk about TV content and iTV providers are now alert to this reality and are trying to promote new ways of doing it.

In parallel, the proliferation of AV/TV online distribution platforms (PC, TV or mobile based) with consequent changes in the video consumption habits has added important new challenges to the Social AV/TV field. In the existing scenario, content distribution ranges from regular broadcasters to a set of websites and applications making it harder for synchronous interaction activities. The users no longer share a unique platform and a unique terminal, the TV set.

Although in its early stages of specification, the proposal for the User Media Consumption Awareness system presents a range of technical, design and interaction research concerns and challenges. It is expected that the UMCA system may trigger this research field further in searching for a new Social AV/TV add-on for online video distribution. Yet further software development is needed in order to have a prototype available for the validation of this proposal with real users.

Therefore the authors believe that in parallel with the development of social features for traditional iTV platforms that are currently taking advantage of the new IPTV infrastructures, important steps must also be done in online AV/TV platforms, since it is expected that these will enhance the potential for social interaction practices around AV online content.

REFERENCES

Abreu, J., Almeida, P., & Branco, V. (2002). 2BeOn: interactive television supporting interpersonal communication. In J. Jorge, N. Correia, & H. Jones (Eds.), *Multimedia 2001 - Proceedings of the Eurographics workshop in Manchester, UK* (pp. 199-208). New York; Springer-Verlag Wien.

Abreu, J. (2007). *Design de Serviços e Interfaces num Contexto de Televisão Interactiva: proposta de uma aplicação de suporte à comunicação interpessoal entre telespectadores (iTV Services and Interface Design: proposal of an iTV application to support interpersonal* communication), Doctoral dissertation, University of Aveiro, Portugal.

Andrea, G., Arianna, I., & Sonia, M. (2007). Living@room: a Support for Direct Sociability through Interactive TV. In A. Lugmayr, & P. Gołębiowski (Eds.), *Interactive TV: A Shared Experience - TICSP Adjunct Proceedings of EuroITV 2007. Tampere International Center for Signal Processing. TICSP series #35* (pp 207-212). Amsterdam, the Netherlands.

Boertjes, E. (2007). ConnecTV: Share the Experience. In A. Lugmayr & P. Gołębiowski (Eds.), *Interactive TV: A Shared Experience - TICSP Adjunct Proceedings of EuroITV 2007. Tampere International Center for Signal Processing. TICSP series # 35* (pp. 139-140). Amsterdam, The Netherlands.

Cesar, P., Bulterman, D., & Jansen, A. (2008). Usages of the Secondary Screen in an Interactive Television Environment: Control, Enrich, Share, and Transfer Television Content. In M. Tscheligi & M. Obrist (Eds.), *Changing Televion Environ-*

ments - Proceedings of 6th European Conference, EuroITV 2008, Salzburg, Austria, July 2008. LNCS, 5066, 168-177. Heidelberg, Germany: Springer-Verlag.

Chorianopoulos, K. (2004). *Virtual Television Channels: Conceptual Model, User Interface Design and Affective Usability Evaluation*. Doctoral dissertation. University of Economics and Business, Athens. Retrieved April 9, 2005 from http://uitv.info/about/editors/chorianopoulos/thesis/phd.pdf

Chorianopoulos, K., & Spinellis, D. (2006). User Interface Evaluation of Interactive TV: A Media Studies Perspective. *Universal Access in the Information Society*, 5(2), 209-218. Springer-Verlag (2006).

Chuah, M. (2002). Reality Instant Messenger: The Promise of iTV Delivered Today. *Proceedings of TV'02: the 2nd Workshop on Personalization in Future TV - Second International Conference on Adaptive Hypermedia and Adaptive Web-based Systems*, Spain (pp. 65-74), Malaga

Coates, T. (2005). Social Software for Set-Top boxes.... *Plasticbag.org*. Retrieved March 10, 2007, from: http://www.plasticbag.org/archives/2005/03/social_software_for_settop_boxes.shtml

Coppens, T., Trappeniers, L., & Godon, M. (2004). AmigoTV: towards a social TV experience. In J. Masthoff, R. Griffiths, & L. Pemberton (Eds.), *Proceedings from the Second European Conference on Interactive Television "Enhancing the experience"*, UK, University of Brighton

Dahlgren, P. (Ed.). (1995). *Television and the Public Sphere: Citizenship, Democracy and the Media*. London: SAGE.

Fiske, J. (ed.). (1990). *Introduction to communication studies*. Routledge.

Gauntlett, D., & Hill, A. (1999). *TV Living: Television, Culture and Everyday Life*. London: Routledge.

Gawlinski, M. (Ed.) (2003). *Interactive Television Production*. Oxford: Focal Press.

Geerts, D. (2006). Comparing voice chat and text chat in a communication tool for interactive television. In O. W. Bertelsen, S. Bødker, & K. Kuuti (Eds.), *Proceedings of the NordiCHI 2006*. (pp. 461-464). New York: ACM Press.

Harboe, G., Massey, N., & Metcalf, C., (2007). Perceptions of Value: The Uses of Social Television. In Cesar, P., Chorianopoulos, K. & Jensen, J. (Eds.), *Interactive TV: A Shared Experience - Proceedings of EuroITV 2007. Lecture Notes in Computer Science, 4471*, 116-125. Amsterdam, The Netherlands: Springer.

Harrison, C., & Amento, B. (2007). Collaboratv: Using asynchronous communication to make TV social again. In A. Lugmayr & P.Gołębiowski (Eds.), *Interactive TV: A Shared Experience - TICSP Adjunct Proceedings of EuroITV 2007. Tampere International Center for Signal Processing. TICSP series # 35* (pp 137-138). Amsterdam, the Netherlands.

Heß, J. (2007). Supporting community building in iTV environments. In A. Lugmayr & P. Gołębiowski (Eds.), *Interactive TV: A Shared Experience - TICSP Adjunct Proceedings of EuroITV 2007. Tampere International Center for Signal Processing. TICSP series # 35* (pp. 129-130). Amsterdam, The Netherlands.

Hom, J. (1998). The Usability Methods Toolbox. *Jameshom.com*. Retrieved June 6, 2004, from http://jthom.best.vwh.net/usability/

Kawamoto, D., Hu, J., & Miles, S. (2000). AOLTV service to hit market Monday *CNET*. Retrieved May 6, 2004, from http://news.com.com/2100-1023_3-241984.html

Kharif, O. (2006). PARC to Make TV Watching More Social. *BuzinessWeek Online*. Retrieved March 22, 2008, from http://www.businessweek.com/the_thread/techbeat/archives/2005/03/parc_to_make_tv.html

Lafrance, J., (2005). Le phénomène télénaute ou la convergence télévision/ordinateur chez les jeunes, *Réseaux, Paris, n° 129-130, 23,* 311-321.

Lenhart, A., Madden, M., Macgill, A., & Smith, A. (2007). Teens and Social Media. *Reports: Family, Friends & Community—Pew Internet.* Retrieved February 19, 2008, from http://www.pewinternet.org/PPF/r/230/report_display.asp

Leurdijk, A. (2007). Will Broadcasters Survive in the Online and Digital Domain? In P. Cesar, K. Chorianopoulos, & J. Jensen, (Ed.), *Interactive TV: a Shared Experience 5th European Conference, EuroITV 2007, Amsterdam, The Netherlands, May 24-25, 2007. Proceedings* (pp. 86-95). *Lecture Notes in Computer Science, 4471.* Springer.

Light, A. (2004). A need to Commune. *ACM - Interactions, XI.2,* New York, March-April, 2004, (pp. 74-75).

Lipsman (2007). 3 Out of 4 U.S. Internet Users Streamed Video Online in May. *ComScore Press Releases.* Retrieved February 13, 2008, from http://www.comscore.com/press/release.asp?press=1529

Madden, M. (2007). Online Video. *Reports: Technology & Media Use—Pew Internet.* Retrieved February 19, 2008, from http://www.pewinternet.org/PPF/r/219/report_display.asp

Mcquail, D., & Windahl, S. (Eds.). (1993). *Communication Models: For the Study of Mass Communications.* Prentice Hall

Nardi, B. A., Whittaker, S., & Bradner, E. (2000). Interaction and Outeraction: Instant Messaging in Action. In *Proceedings of the ACM conference on Computer Supported Cooperative Work* (pp. 79-99). Philadelphia, PA: ACM Press.

Oumard, M., Mirza, D., Kroy, J., & Chorianopoulos, K. (2008). A Cultural Probes Study on Video Sharing and Social Communication on the Internet. *In Proc. of DIMEA 2008.* ACM press.

Peng, C. (2002). *Digital Television Applications.* Doctoral dissertation, University of Technology, Helsinki. Retrieved July 7, 2005 from http://lib.tkk.fi/Diss/2002/isbn9512261723/isbn9512261723.pdf

Preece, J., Rogers, Y., & Sharp, H. (Ed.) (2002). *Interaction Design: beyond human-computer interaction.* New York: John Wiley & Sons, Inc.

Regan, T., & Todd, I. (2004). Media center buddies: instant messaging around a media center. In A. Hyrskykari (Ed.), *Proceedings of the Third Nordic Conference on Human-Computer interaction* (Tampere, Finland, October 23 - 27, 2004). NordiCHI '04, vol. 82 (pp. 141-144). New York: ACM Press.

Ribeiro, S. (2007). Current TV: a televisão feita por todos está na berra. *Jornal Público.* Retrieved February 29, 2008, from http://ultimahora.publico.clix.pt/noticia.aspx?id=1307974

Shamma, D. A., Bastea-Forte, M., Joubert, N., & Liu, Y. (2008). Enhancing online personal connections through the synchronized sharing of online video. In *Proceedings of ACM CHI 2008 Conference on Human Factors in Computing Systems,* April 5-10, 2008 (pp. 2931-2936). New York: ACM Press.

Strauss, R. (Ed.) (1997). *Managing Multimedia Projects.* Boston: Focal Press.

Weisz, J., Kiesler, S., Zhang, H, Ren, Y., Kraut, R., & Konstan, J. (2007). Watching together: integrating text chat with video. In M. Rosson, & D. Gilmore, (Eds.), *Proceedings of the SIGCHI conference on Human factors in computing systems* (pp. 877-886). New York: ACM Press.

Wolton, D. (Ed.). (1997). *Penser la communication.* Paris: Flammarion.

ENDNOTES

1 The impact of television in the social interaction among televiewers has been researched for many years. J. Klapper (1960) quoted by McQuail (1993); Katz, Gurevitch and Hassem (1973) quoted by Fiske (1990); and Wolton (1997) are only a few of the authors that reflected upon this subject.

2 In recent years, authors like Geerts (2006) and Weisz (2007) have studied the pros and cons of using textual versus voice communication. At the time the 2BeOn prototype was developed, IM sounded more compatible and less disturbing for TV reception than voice communication, despite the existence of a buddy list that helped an initiator of a communication to judge when the recipient was apparently available, enabling a minor number of interruptions (Nardi, 2000). As discussed in the "evaluation process" section this hint was corroborated by the participants during the evaluation phase.

3 E.g.: The Barack Obama channel: http://www.youtube.com/user/BarackObamadotcom Retrieved May 23, 2008.

4 Source: http://www.masternewmedia.org/pt/video_televisao_pela_internet/producao-video/videos-on-line-sincronizados-remotamente-assista-a-qualquer-video-em-sincronia-com-qualquer-pessoa-atraves-da-internet-conheca-melhores-ferramentas-20070527.htm . Retrieved May 23, 2008.

5 Source: http://arstechnica.com/news.ars/post/20080512-first-look-microsoft-enters-social-video-with-messenger-tv.html Retrieved May 23, 2008.

6 For an extensive list of online video platforms please visit http://www.everybodygoto.com/2007/05/21/the-ultimate-online-video-list/ Retrieved May 23, 2008.

7 For more information visit: http://peekko.com/HomePage.html Retrieved May 23, 2008.

8 Windows Live Messenger from Microsoft allows users to share information about the music they are listening on their computers. However, this feature relies only on the media player information. The new Messenger TV allows sharing video consumption but only if consumed from MSN video archives.

9 Oumard (2008), regarding the promotion of interpersonal communication around video content, also refers to the importance of keeping a public personal video profile of the watched videos attached to an IM application.

Section IV

Thinking Out-of-the-Box:

Social TV on Large Screens, Mobile Devices and the Web

Chapter XIV
Examining the Roles of Mobility in Social TV

Konstantinos Chorianopoulos
Ionian University, Greece

ABSTRACT

Mobile TVs have been available for many years, without ever becoming very popular. Moreover, the first wave of research has been mostly concerned with technology and standards, which are necessary to ensure interoperability and market acceptance. Although, there has been a significant body of computer supported co-operative work (CSCW) and mobile human-computer interaction (HCI) research findings, there is limited investigation in the context of leisure activities, such as TV. In this chapter, the author proposes three concepts that drive the main paths for research and practice in mobile and social TV: (1) Mobile TV as a content format, (2) Mobile TV as user behavior and (3) Mobile TV as interaction terminal. Further research should elaborate on these three concepts and highlight the cultural impact of mobile TV.

INTRODUCTION

One explanation for the slow diffusion of interactive TV (ITV) in the information society is that the difference between the broadcast and the telecommunications mentality has imposed an artificial distinction between content distribution and interpersonal communication. As a result, content has to be distributed and consumed through broadband, unidirectional and inflexible TV channels and interpersonal communication takes place over low-bandwidth bidirectional channels. However, the convergence of the telecommunication and content distribution platforms could be beneficial for viewers, as well as the commercial TV stakeholders. In addition, new devices and new types of content facilitate the emergence of novel consumer behaviors. In this article, we explore the interplay of these three concepts (device, content, behavior) in the context of mobile and social TV

TV content gradually finds its way through Internet and mobile platforms. Besides triple-play services, which offer integrated access to voice, content and data services, there are opportunities for new services enabled by the mobile infrastructure. Indeed, the convergence of broadcast, mobile and data platforms has offered many opportunities for integrated content and communication services, which we refer to as 'Social TV'. We define social TV as a socio-technical system that involves more than one user and networked audiovisual devices.

Previous definitions were focused only on the technological aspects and ignored the fact that even traditional TV is inherently social. For example, viewers compete mentally with quiz show participants, or between co-located groups. Moreover, viewers react emotionally to TV content, they record and share TV content with friends and discuss about shows either in real-time, or afterwards. In this context, it is necessary to not only pay attention to usability issues, but also to sociability. Indeed, ethnographic and survey studies have documented the social uses of TV (Duchenaut et al 2008, Lee and Lee 1995), but they have not described the user requirements of applications that facilitate the social uses of TV. For this purpose, we explore the related academic literature, we identify the user interface requirements of those computer mediated communication applications that enhance the social dimension of TV.

The rest of the article is structured as follows. We begin with an analysis of the social uses of mobile TV. In Section 3, we outline the multiple roles of mobility in Social TV. Finally, we describe the implication of such systems for practice and future research.

SOCIAL AND TECHNOLOGICAL ASPECTS OF MOBILE TV

This section explores the social and technological dimensions of TV and other related audiovisual media. In addition, we present a brief overview of technological support for TV sociability.

Social Value of TV

Although TV has been blamed for the reduction of social interaction within the family and the local community, there is a significant body of previous research that considers TV as a social medium, because it provides opportunities for shared experiences and group viewing. In particular, mobile phone applications that support sociability within families or distant groups might enhance the attractiveness of ITV as a leisure activity. This section draws on interdisciplinary literature and empirical research in order to raise the main research issues of the multiple roles of mobility within Social TV.

Despite the many criticisms on the quality of TV content and on the passive nature of the watching activity, the social uses of TV have been documented in acclaimed research (Gauntlett and Hill 1999, Kubey and Csikszentmihalyi 1990). It has also been established that viewers have adapted TV in many ways to meet their everyday life needs (Lee and Lee 1995, Rubin 1983). The findings of these works frame a set of opportunities for the design of social communication services in mobile TV.

The majority of previous research on ITV has over-emphasized the benefit of increased choice of content and of the interactivity with content. Instead, a worthwhile effort would be the fulfillment of seamless communication over, or about televised content. Such services could support human connectedness (Agamanolis 2006) over a distance (e.g. synchronous communication over a TV program between diasporic households), or enhancement of the shared experience that comes with TV co-viewing. For this reason, we explore an integrated view of the interpersonal communication together with the shared experience of mass communication.

There is a growing academic interest on Social TV systems, which consist of technological solutions for integrated interpersonal communication and content distribution. Although, there has been a significant body of computer supported co-operative work (CSCW) research on supporting interaction among geographically distributed co-workers, there is limited investigation in the context of leisure activities, such as TV. Similarly, research on interpersonal communication in the human-computer interaction (HCI) field has regarded video-mediated communication at work (Veinott et al. 1999). As a matter of fact, there is not much knowledge on designing applications for leisure or informal content enriched communication.

Cross Media Infrastructure

In the past, TV content in the living room has been provided either by broadcast, or optical discs. A basic ITV system includes a set-top box (STB) that decodes the signal and provides processing and storage capabilities that enable interactive applications. Nevertheless, the disagreement on a common open middleware platform has been an obstacle for the development of sophisticated interactive applications that are independent from the STB hardware. On the other hand, there is agreement over the specifications for the digital video broadcasting (DVB-S/C/T/H specifications satellite, cable, terrestrial, mobile). Furthermore, TV content can be efficiently distributed over peer-to-peer (P2P) networks. In this way, the variety of video content has been increasing with the support of new internet technologies, which allow new ways of distributing video (e.g. broadband connected TV set-top-boxes). Thus, ITV applications are neither limited to the traditional TV device and broadcast delivery, nor to the typical channels of satellite, cable, and digital terrestrial networks. Alternative and complementary devices and distribution methods should be considered, such as mobile phones (mobile DTV).

Social TV builds upon the convergence between different technological infrastructure, such as broadcasting, telecommunication, and internet. The convergence has been realized in different forms. On the one hand, Internet content may be accessed through television web browsers, or linked to ITV programs (e.g. interactive advertisements). Communication applications such as messaging, chatting, or voting during certain programs (quizzes, contests etc.) strengthen viewer's loyalty to the specific program. However, Internet access via television may disrupt current viewing patterns. Besides user interaction, at the network-level, internet connection facilitates video transfer over P2P networks. Moreover, the distribution of TV content over IP-based platforms, known as IPTV (Internet protocol TV), provides additional opportunities for the delivery of a wide variety of TV programming. In addition, 3G mobile networks could be used to distribute and control TV content.

Related Research in the HCI and CSCW Fields

One of the first approaches for a closer integration between TV content and social communication was the "Inhabited TV" research effort (Craven et al. 2000), which developed a collaborative virtual environment, where viewers could interact with other viewers or virtual objects. In this case, viewers were watching TV within the virtual environment and not within physical space. Thus, the TV experience was extended by enabling social interaction among participants and increased interaction with content. In an Inhabited

TV application, the television becomes an actor and a part of a group interaction within a virtual online world.

There are various approaches to integrate social communication features into TV, such as chat, IM and email. There has been particular commercial interest on integrating the SMS into TV. Indeed, SMS TV is very popular, which is based on the familiarity with SMS and the availability of the technical infrastructure. Besides SMS services, there is a growing body of research and development, which is presented next.

Coppens et al. (2005) have reported the development of a 'Social TV' system, but their description focuses on the technical details, the features and the potential of the system for end-users. The 'Amigo TV' system provides a technological platform for integrating content delivery, communities, and interpersonal communication (Coppens et al. 2005). In addition, the content of the broadcasts can be personalized by sharing personal photos and home videos. Amigo TV supports online user meetings and buddy lists. Interpersonal communication is based on voice, text, and video formats, as well as animated avatars.

Regan & Todd (2004) describe a system for messaging over TV content. The Media Centre Buddies system integrated TV technology into an instant messaging application. The main aim was to allow multiple users to log into an instant messaging client that was running next to a TV channel.

User Generated and Distributed Content

TV content production has been regarded as a one-way activity that begins with the professional TV producers and editors and ends with post-production at the broadcast station.

As a matter of fact, television viewers have long been considered passive receivers of content, but a new generation of computer literate TV viewers has been accustomed to make and share edits of video content online. Furthermore, the wide-availability of video capture (e.g. in mobile phones, photo cameras) and easy-to-use video editing software (standard in many desktop computer operating systems), opens up additional opportunities for wider distribution of home made content (e.g. through peer-to-peer, portable video players, etc). User generated content and social communication about media content has been also proposed by Resnick (2001), who suggested that interactions could create productive resources, which he refers to as socio-technical capital. This capital may consist of artifacts created from the interactions or relationships and practices developed through repeated social interactions. Such capital can enable future social interactions.

Mobile media players transform the traditionally solitary media consumption into a social experience. Mobile and wireless technology open up opportunities for new interesting social practices, where media consumption and sharing can take place in a variety of social, physical and temporal contexts. Advanced mobile phones are equipped with digital cameras, multimedia processing and multiple mobile communication technologies (such as short or long-range, low or high-bandwidth). Since a mobile phone remains constantly with the user, it could potential store a large amount of details about social interactions. Then, search and sharing of media content could benefit from this social dimension of smart phones: the user could share audiovisual content with those related (in terms of place or terms of social proximity) without investing effort on selecting these people. TunA (Agamanolis 2006) is one example of a mobile application where users can tune in to eavesdrop on the playlists of nearby users and listen to the same music in a synchronized way.

Content Enriched Interpersonal Communication

Social TV systems offer one or more computer mediated communication features, which are closely integrated with the TV watching experience. Computer mediated interpersonal communication over distance, or over time could employ various communication modalities such as: (1) audio, (2) text, (3) video, (4), photos and (5) non-verbal cues (e.g. emoticons, avatars). We refer to integrated content and communication services as 'content enriched communication.' Content enriched communication over a distance refers to two types of sociability: (1) synchronous, when viewers get together and watch the same show at the same time and (2) asynchronous, when viewers interact after the show has already been seen by each one, independently and at different times. Communication between spectators is realized at two levels: (1) direct communication, such as chat or instant messaging and (2) indirect communication, such as cooperating in a team to win a quiz. In brief, there are four basic scenarios of Social TV (Chorianopoulos and Lekakos 2008):

- Synchronous viewing over a distance: This is probably the most interesting scenario, because the requirement it poses is to recreate the experience of co-located group viewing, when the viewers are located in two or more distant places. For example, distant viewers should be able to watch together popular social TV content, such as sports, quiz shows, series, reality shows. A good starting point is to consider ways to disclose presence and status of viewers, to continue with support for multiple interpersonal communication modalities (non-verbal most notably), and to summarize the social experience with automated highlight production, which could motivate further discussion and social bonding between the distant viewers.
- Asynchronous viewing over a distance: This a feasible scenario if we consider that distance viewers might have very different time-schedules, patterns of daily life activities, or even live in distant time zones. Then, the probability of synchronous co-viewing is rather limited. In this case, a Social TV system could record and share shows and viewing habits with the members of the social circle. In addition, a Social TV system should allow annotation of content and recording of interactions, such as pausing, skipping, replaying and content browsing. In this way, each time a particular TV program is accessed keeps a trace, which is exploited at the next access, in order to personalize the content and most notably to provide a motivation for asynchronous communication. This could be rather subtle, such as visual annotation of the content highlights, or could be more explicit such as audio and text comments.
- Asynchronous viewing at the same place: The main motivation for the development of Social TV systems is based on the need to bridge the distance between social circles of people, but there is also the case that co-located groups of people do not manage to meet as often as they wish for a social TV night. A subset of the functionality that was described in the previous case might be the most appropriate here.
- In addition to the above, Social TV designers should consider the traditional TV watching scenario, where a group of viewers gathers in the same place to enjoy a favorite TV program. Although this is a case that content enriched communication is least needed, there might worthwhile benefits in employing a Social TV system. In all cases, designers should consider extended functionality for user generated content. For example, the ability to upload personal music, photos and videos might be used to achieve communication through content. In particular, the automated production of personal TV channels that keep track of individual life streams captured with a

mobile device (e.g. music, photos, personal videos) could be multiplexed with broadcast TV watching behavior. Indeed, Kubey and Csikszentmihalyi (1990) have found that everyday life experience is correlated with TV watching behavior. Thus, interpersonal communication could start with a screen displaying media use of each party during the past few days or hours. In practice, this scenario is rather feasible to implement, because the respective services have been very popular (e.g. YouTube, MySpace, Flickr)

Although mobile and social TV is thought to be suitable for the distant and synchronous communication scenario, there are several other opportunities. For example, multiple mobile terminals could be employed at the same place to control content on a shared big-screen. Moreover, user generated mobile TV content could be posted online and latter-on be annotated by other mobile TV users, when triggered by a particular location or other condition.

MULTIPLE ROLES OF MOBILITY

In this section, we propose three main directions of mobile TV research, and we offer suggestions for future research and market developments. Besides TV watching on-the-move, mobile TV has significant potential, both as a personal TV set and as a tool to establish a closer interaction with the television programs (e.g., TV voting).

Mobile TV as a Content Format

Digital mobile TV systems have been designed to complement mobile networks with broadcast and multicast capabilities for spectrum-efficient delivery of multimedia services on mobile devices in both outdoor and indoor environments. In particular, the DVB-H standard is based on the widely deployed series of DVB standards (DVB-S/C/T) and includes enhancements for mobile terminals, such as reduced power consumption and reception while on the move. Although the technical standards are suitable for mobile TV reception, it is clear that mobile TV prospects should be examined not as an alternative but as a complementary service to traditional living-room TV. This is because the perceived quality of TV on a mobile phone and the solitary experience are not the favored mode of watching TV, at least with regard to popular living-room content formats (e.g. TV series, sports).

For some time television has been the only major media format that has been missing from mobile phones. Technological advancements in wireless broadband (e.g. WiFi, 3G, 4G, DVB-H) and multimedia mobile terminals (e.g. multimedia mobile phones) have made a reality the reception of digital TV on the move. The distribution of TV content to mobile devices over broadband wireless raises the issue of video quality. Video quality depends on many aspects of the video encoding systems, such as bit rate and algorithms that model human perception of video on small screens. Most of the research on the effect of screen sizes in the field of consumer electronics has examined the impact of increasing the image size in the viewer's visual field by means of large physical displays or projection areas. The results show that larger image sizes are more arousing, better remembered, and generally preferred to smaller ones (Reeves et al. 1999).

There are many services that aim to provide users with audiovisual content while on the move. Although many of these services sound appealing, the end-users' subjectively perceived quality is an important factor for their success. The properties of video quality have many similarities between the different application domains (e.g. internet, broadcast), but the characteristics of mobile devices define a special set of constrains. The biggest differences to other application domains are the limited bandwidth, which leads to high-level requirements of compression and the limitations of the mobile devices such as display size, power resources, processing capabilities and memory. In addition, the wireless transmission of the content is error prone. Accordingly, the production of video under these special requirements should regard the possible distortions in the subjectively perceived quality (Knoche et al 2008). For this reason, subjective quality evaluation tests during product development are necessary, in order to ensure acceptable quality of service. In particular, the subjective quality of service for mobile TV depends on the perceived audio-visual quality of the consumed content and the interaction through which the user has to go to access it (e.g. the delay between selecting content and start of play).

Further research in mobile TV should investigate authoring tools that enable automatic post-production of video that is targeted for viewing on the move. Currently, mobile service providers encode and deliver existing broadcast material and interactive applications without additional editing, because it is more cost-effective than re-editing. Future research should improve on intelligent cropping mechanisms that present only a part of the original shot. On the application side, cross-media multimedia authoring tools should consider the diversity of screen formats and sizes in mobile devices. Besides content adaptation, further research should investigate the uses of user generated content and provide templates that facilitate creation and distribution (sharing) of end-user content.

Mobile TV as User Behavior

Early studies on user behavior and mobile TV systems have indicated short watching sessions (Södergård 2003), which are suitable for particular TV genres, such as news and sports highlights, and music videos. More recent research by the same group (VTT, Finland), has tracked the evolution of mobile TV usage (Oksman et al. 2007). They have identified that in mobile TV there is no prime time, only 'prime place', such as while commuting.

In contrast to living-room TV sets, which are shared displays for audiovisual content, mobile phones are natively social devices. Since their introduction, users have learned to use them as social connectivity tools, with voice-calls and text messages as basic functions. Therefore, the mobile TV user behavior might be shaped by established practices of interpersonal communications over a distance. Indeed, researchers have identified that for some users mobile TV might be a rather personal activity (Cui et al. 2007). In particular, they have identified that mobile TV is employed to privately watch content, at places and situations that are not socially appropriate for TV watching (e.g., business meeting, school lectures). They have also reported that mobile TV is also employed at home, when other TV sets are employed for different programs than the ones preferred by the mobile TV users.

Mobility means that content consumption takes place in various dynamic mobile contexts e.g. on the go, in the bus or at work, a direct contrast to the static embeddedness of living-room TV. Mobility thus means limited attention spans, but with increased user readiness for interruptions and interaction. Furthermore, mobile displays are considerably smaller than living-room screens. The consequence of these contextual factors is that static TV is superior in creating immersive, passive media experiences. Mobile TV might never be able to lull people in the same way as high definition television, but on the other hand it allows for more interactive and intimate experiences. Although there is a significant body of research on sharing content such as photos and music through desktop and mobile media, there is not much research on video sharing and content enriched communication through mobile devices (Schatz et al. 2007). Therefore, further research should consider the practices of sharing user generated video content.

Mobile TV as Interaction Terminal

Mobile phones include some kind of standard and familiar input and output facility. The most common input device on a mobile phone is a simple numeric keypad, a few function keys and navigation keys. In short, in terms of input capabilities, a mobile phone is very similar to a common TV remote control. Some contemporary phones have removed the numeric keypad in favor of a larger touch screen, which might dynamically render a numeric keypad or many any other input arrangements depending on the application. Moreover, mobile phones feature media rich output capabilities, such as full-color high-resolution (in comparison to size) screens and audio support. As a matter of fact, the output capabilities of contemporary mobile phones are equal or better to early TV sets. In addition to user input and output devices, mobile phones have several data networking capacities. Text messages are a common standard in mobile phones and they have been successfully exploited by TV channel operators as voting and chatting input devices.

The use of the input and output facilities of mobile phones as alternative communication channels for TV programs (e.g. voting, chatting, TV on the move) has been a straightforward and expected development. Cesar et al. (2007) has explored the use of mobile multimedia touch screens to augment the living room TV experience. They have demonstrated that besides remote control, personal mobile terminals could provide additional content, as well as annotation of content. Mobile TV broadcasts transmit content to all mobile terminals within the footprint of a base-station, which is relatively narrow when compared with terrestrial broadcasts. The presence of multiple base-stations is the main advantage of mobile broadcasting, because the content could be personalized to fit both the terminal and the context of use (e.g., time of the day, geographic location).

CONCLUSION

Mobile and Social TV applications could be feasibly offered through triple play infrastructures, which combine content delivery, voice, and data services. In this way, the network operator can provide interaction between the TV viewers on TV channels using an interactive broadband link. Triple-play services have been introduced on the assumption that telecommunication, content and data services could be delivered over the same technological infrastructure thanks to the convergence of the respective technological platforms. Although the convergence of previously distinct technological platforms is a significant benefit both for consumers and service providers, there are also additional benefits from a closer integration of platforms at the user-level. Content providers could be benefited by metered communication services, while telecom providers could be benefited by content distribution. In both cases, the users could gain access to intuitive content enriched communication.

In addition, mobile DTV infrastructure offers many opportunities for converged personal communication and content services. In particular, the availability of broadband wireless technology is rather suitable for the delivery of content enriched communication services (e.g. active content sharing, synchronous co-viewing or asynchronous notifications over a distance, discussion and annotations about shared content). Wireless network operators have invested in broadband licenses and infrastructures, but most of the services offered are only video communication, or only video on demand. The introduction of content enriched communication services is a worthwhile direction, because it offers an excellent balance between the basic need of users to communicate with a mobile device and the need of network providers for increased revenue by added value broadband services, such as mass media content distribution.

In summary, multimedia phones are essential elements of the next generation of Social TV services. They are established social connectivity providers, personal media interfaces, content capture and sharing tools, and thus complement stationary interactive TV setups very well. The proposition of mobile TV has a major difference with the analog predecessor. Most notably, it has the potential to offer localized and interactive programs and not just the same broadcasts as seen in living-room TV. In conclusion, while counter-intuitive to many, the activities that happen during television watching can be a very sociable. Therefore, the ultimate objective is to develop technological support and content for the social practices that surround mobile TV viewing, while retaining the centrality of TV as a leisure pursuit.

REFERENCES

Agamanolis, S. (2006). At the intersection of broadband and broadcasting: How ITV technologies can support Human Connectedness. *Proceedings of the 4th European Interactive TV Conference.*

Cesar, P., Bulterman, D. C. A., Obrenovic, Z., Ducret, J., Cruz-Lara, S. (2007). An Architecture for Non-intrusive User Interfaces for Interactive Digital Television. *EuroITV 2007,* (pp. 11-20), Springer LNCS.

Chorianopoulos, K., & Lekakos, G. (2008). Social TV: Enhancing the shared experience with interactive TV. *International Journal of Human-Computer Interaction, 24*(2), 113-120, Taylor Francis,

Craven, M., Benford, S., Greenhalgh, C., Wyver, J., Brazier, C., Oldroyd, A., and Regan, T. 2000. Ages of avatar: community building for inhabited television. In Proceedings of the Third international Conference on Collaborative Virtual Environments (CVE '00). ACM Press, 189-194. 2000.

Coppens T., Vanparijs F., Handekyn K. AmigoTV: A Social TV Experience Through Triple-Play Convergence. Alcatel, White paper, 2005.

Cui, Y., Chipchase, J., Jung, Y. Personal TV: A Qualitative Study of Mobile TV Users. EuroITV 2007: pages 195-204, Springer LNCS 2007.

Ducheneaut, N., Moore, R. J., Oehlberg, L. Thornton, J. D., & Nickell, E. (2008). Social TV: Designing for Distributed, Sociable Television Viewing. *International Journal of Human-Computer Interaction, 24*(2), 136–154. Taylor Francis.

Gauntlett, D., & Hill, A. (1999). *TV Living: Television, Culture and Everyday Life.* Routledge.

Girgensohn, A., & Lee, A. (2002). Making web sites be places for social interaction. *In Proceedings of the ACM Conference on Computer Supported Cooperative Work.* ACM Press.

Knoche, H, McCarthy, J. D., & Sasse, M. A. (2008. How low can you go? The effect of low resolutions on shot types in mobile TV. *Multimedia Tools Appl., 36*(1-2), 145-166.

Kubey, R., & Csikszentmihalyi, M. (1990). Television and the Quality of Life: How Viewing Shapes Everyday Experiences. Lawrence Erlbaum.

Lee, B., & Lee, R. S. (1995). How and why people watch TV: Implications for the future of interactive television. *Journal of Advertising Research, 35*(6), 9–18.

Oksman, V., Noppari, E., Tammela, A., Mäkinen, M., & Ollikainen, V. (2007). Mobile TV in Everyday Life Contexts - Individual Entertainment or Shared Experiences? *EuroITV 2007,* (pp. 215-225), Springer LNCS.

Preece, J. (2000). Online Communities: Designing Usability, Supporting Sociability. England: John Wiley & Sons, Ltd.

Reeves, B., Lang, A., Kim, E., & Tartar, D. (1999). The effects of screen size and message content on attention and arousal. *Media Psychology, 1*, 49-68.

Regan, T., & Todd, I. (2004). Media Center Buddies: Instant Messaging around a Media Center. *In Proceedings of the Fourth Nordic Conference on Human-Computer Interaction* (pp. 141-144). ACM press.

Resnick, P. (2001). Beyond Bowling Together: SocioTechnical Capital. In J. M. Carroll (Ed.), *Human-Computer Interaction in the New Millennium* (pp. 647-672). New York: Addison-Wesley.

Rubin, A. (1983). Television uses and gratifications: The interaction of viewing patterns and motivations. *Journal of Broadcasting & Electronic Media, 27*(1), 37–51.

Schatz, R., Wagner, S., Egger, S., Jordan, N. (2007). Mobile TV Becomes Social - Integrating Content with Communications. Information Technology Interfaces, 2007. *ITI 2007. 29th International Conference* (pp.263-270), 25-28 June.

Södergård, C. (2003*). Mobile television - technology and user experiences Report on the Mobile-TV project (Rep. No. P506) VTT Information Technology.*

Veinott, E. S., Olson, J., Olson, G. M., & Fu, X. (1999). Video helps remote work: speakers who need to negotiate common ground benefit from seeing each other. *In Proceedings of the SIGCHI Conference on Human Factors in Computing Systems (CHI '99)* (pp. 302-309). ACM Press.

Chapter XV
Watching in Public:
Understanding Audience Interaction with Big Screen TV in Urban Spaces

Kenton O'Hara
CSIRO ICT Centre, Australia

Maxine Glancy
BBC, UK

ABSTRACT

In this chapter the authors present a field study of BBC Big Screens Public Space Broadcasting initiative. Under this initiative, large screens have been installed in several urban locations across the United Kingdom and used to screen a range of television content and interactive applications. The chapter discusses a number of different findings and themes. These include different types of screen use such as viewing of standard television content, event-based use of the screens and interactive use of the screen. The chapter then goes on to discuss the content in relationship to its placement in a particular local context. Following this, the chapter looks at how architectural features shape the way the screens are used and impact on audience behaviour. Finally the authors explore issues of health and safety that impact on content choice and pragmatics of scheduling.

INTRODUCTION

Over the last decade we have seen a proliferation of digital displays within our urban environments. Such urban displays, using a variety of form factors, are being used for a wide range of commercial and cultural purposes, from advertising and branding through to entertainment, art and community support (Fatah gen. Schieck, 2005). The potential of these screens to transform the urban environment and people's experiences of the public spaces within which they are located is enormous. Understanding this potential, its impact on public spaces and on the people who inhabit those spaces is an important consideration for us if we are to successfully design valuable urban screen deployments and content.

In recent years, we have begun to see a concerted effort to develop such an understanding through both public experimentation in the arts (e.g. Lozano- Hemmer, 2002; 59th Minute -www.creativetime.org) and through the commentaries of cultural, architectural and new media theorists (e.g. Virilio, 1997; Manovich, 2006; Taylor, 2006). Much of this work though has focused on grand themes and issues, and while the insights provided are important, we often lose sight of the detailed material practice of people's engagement and behaviour with these technologies and content.

With this in mind, we conducted a study of the BBC Big Screens Public Space Broadcasting project. Within this project, large screens of approximately 25 metres squared have been placed in major city centres in the UK such as Manchester Liverpool, Leeds, Birmingham, Hull, Bradford, and Derby. The screens broadcast continuously around the clock and show a wide variety of content including nationally sourced BBC television programs and more locally sourced content. The content includes news bulletins, soap operas, sports events, live music, short films and interactive experiences. Having been running now for several years in several different locations, these public interactive television installations provide us with a unique opportunity for understanding how such technology fits within people's lives and the new behaviours afforded by the technology relative to traditional interactive television delivery into people's homes.

Our interests are with different use modes of the screens. These include *everyday TV broadcast* (e.g. soap operas, news), *event-based* usage (e.g. the Live Earth Concert,) and more *interactive* applications (Red-Nose game, Penalty Shootout, or sending pictures from mobile phones to the screen). We are interested in how people use and engage with the display in these different modes and what social, psychological, technical and spatial factors shape this use. Our exploration is also concerned with the practical aspects of content curation and how this is managed across key stakeholders such as the screen manager and the city councils. An important example here is the impact of particular content decisions on health and safety in public spaces and, vice versa, how health and safety management issues shape content choices.

STUDY

The study comprised a series of observations and interviews conducted across different cities where a BBC Big Screen was located. The interviews and observations took place over a period of several weeks during the summer of 2007. The sites studied were Leeds, Manchester, Liverpool, Bradford and Birmingham (see Figure 1).

Observations at the different sites were conducted at various different times of the day, on different days of the week. The observations primarily focused around people's behaviour within the immediate vicinity of the screens but some more general observations were also conducted to get a sense of the surrounding area within which the screens were immersed. In conducting our observations, the aim was to get a sense of what people were doing in and around the space where the screens were located and any ways the screen and content played a role in shaping these behaviours.

Interviews were conducted on the same days as the observations with people in the vicinity of the screen deemed by the interviewers to be or have been engaged with the screen in some way. The interviews focused initially on the particular visit to the screen that day but broadened out to more general experiences with the screen where applicable. A broad range of participants was interviewed and covered a wide age range from

Figure 1. The Big Screens at (a) Birmingham, (b) Bradford, (c) Manchester, (d) Leeds

(a) Birmingham (b) Bradford (c) Manchester (d) Leeds

young teenagers through to senior citizens.

We also interviewed the screen managers from the Manchester, Liverpool, Birmingham and Bradford sites. The interviews were concerned with the role of the screen manager and the things they have to do on a day-to-day basis in order to realise a particular schedule of content for these situated screens. Through this it was possible to elicit factors that influenced their decision-making and identify key stakeholders involved in the way these screens come to be adopted when immersed in these urban environments. These interviews also provided a longer-term perspective about audience behaviour with the managers having observed it and spoken to people about it over the course of several years.

FINDINGS

We organise the findings around a number of different themes. These include different types of screen use: viewing of standard television content; event-based use of the urban screens; and then behaviour around interactive applications on the screen. We go on to discuss the screen content in relation to its placement in a particular local context. Following this, the chapter turns to look at how architectural features shape the way these screens come to be used and the impact on audience behaviour. Finally we explore issues of health and safety that impact on content choice and the measures that need to be in place when presenting content in such a public setting.

Standard TV Content Viewing

With regards to standard TV viewing, three different types of viewing are explored:

- Walk by viewing
- Opportunistic viewing while resting waiting
- Viewing by appointment and structuring activity around schedule

Walk By

People come into city spaces for a wide variety of reasons—to work, shop, eat, drink, meet and socialise etc. In entering the particular spaces where the Big Screens are located, many people were there for purposes other than watching the screens. This had consequences for how people in the vicinity of the screens were able to engage with the content. In our observations many people in these spaces simply passed through on the way to somewhere else towards their primary purpose. While some of these paid no noticeable attention to the display, others, would glance up at the screen and watch as they were walking. A small percentage would stop on the basis of what they saw or heard and stand around for a few minutes in a more active viewing of the screen. These people remained standing as they watched indicating a temporary commitment of attention to the screen rather than committing to a more prolonged and settled watching. This can be contrasted with traditional television viewing. Potential engagement with content on the screen was fleeting and not a primary focus of attention—in the order of tens of seconds to a few minutes. People's ability to visually engage with the screen as they walked was compromised as their visual attention was required elsewhere. As we observed, casual glances were possible while walking but this was very much dependent upon the positioning of the screen with respect to the direction of the main walk ways through the area and the direction people were travelling. If we consider the screens in Birmingham and Bradford, for example, some of the key walkways through the area are perpendicular to the screen. As such, those walking towards the screen were able to visually engage with the screen from a distance and potentially watch as they walked. For those walking away from the screen, visual engagement was not possible except by deliberately turning round to watch. In Manchester, where the walk way ran parallel to the screen, people from either direction could only visually engage once they were close enough for the viewing angle to become acceptable.

Within these contexts, where visual attention on the screen is constrained there are particular implications for the presentation of content. Audio plays a particularly important role in supporting engagement for this kind of fleeting audience. People can simply listen as they walk, or the audio is the thing that attracts their attention as they walk (resulting in them glancing up at the screen or stopping for a more prolonged visual engagement). Providing visual information that can be extracted from quick glances is a way to allow further engagement from such people, e.g., having elements of static content or succinct headlines can be important here and content that is based around very short narrative.

Such a viewing mode of course does not characterise all Big Screen audiences, but are an important aspect to consider at particular times when such behaviour is predominant. These times are dependent upon the basic rhythms of the city. For example, morning and evening commutes are dominated by this kind of viewing as people aim to get to work or go home. For this reason, at these times, many of the Big Screen Managers tend to schedule "magazine" type content and news programs where content is segmented into small narrative units that that people can quickly dip into and out of as they move through the space. Key here, then, is developing an understanding of these city rhythms with respect to the different locations where the screens are located. As with the scheduling of television programs for home-based television viewing, particular rhythms of family life come to bear on content schedules. But these rhythms of home are very different from those that govern behaviour within public spaces such that one cannot simply transfer schedules and content directly from one domain to another. Having a good local knowledge of these rhythms with respect to the spaces is a significant issue in determining appropriate content scheduling for these places. It is clear from talking with the Screen Managers that their work "in situ" is an important part of understanding these rhythms and determining choice of content. This suggests that a more centralised content management structure, such as with national broadcasts, are not necessarily appropriate.

Viewing while Resting and Waiting

Other people in our observations were involved in longer viewing engagements than the "walk by" behaviour characterised above. Such engagements were typically of the order of magnitude of 5-20 minutes though there were examples of longer periods. In understanding the nature of these more prolonged engagements it is important to consider the underlying motivations behind these sojourns and the role the screen plays in shaping their decisions to be there.

Generally speaking, people again had not come along specifically to watch television on the screens. Rather they were there to have a break and rest from their primary activities (e.g. shopping) or to wait for other people and things. The nature and social underpinnings of these resting and waiting activities in public spaces again distinguish this everyday viewing behaviour from that of traditional television. For example, for people watching the screen alone, the screens played a role in how these individuals related to the others within this public space (cf. Goffman, 1966). For individuals waiting, it was important to present themselves as being visibly occupied to other people around. To just sit or stand around waiting would look odd in certain places and consequently was socially uncomfortable for the person concerned. Watching the screens allows lone individuals to present themselves as occupied and with purpose, making waiting more psychologically comfortable. One of the individuals we spoke to was waiting round for a job interview. He had arrived early but wanted to prepare himself mentally for the interview. So while waiting he was staring at the screen but was actually focused on particular issues pertaining to his job interview rather than engaged with the content itself.

Families and small groups of friends would also use the screens as a place to wait or have a break often using this waiting as an opportunity to have something to eat or drink. In particular when it was fine weather, it was a place where people would sit and enjoy the sunshine. The screen was something that accompanied these primary behaviours—a resource which contributed to the general atmosphere of the place. Within these social contexts, the nature of the engagement with the screen cannot be characterised in any homogenous way but rather was dependent upon the particular social dynamics of the groups concerned. For some groups, the screen was backgrounded with glances occasionally interleaved with conversation. For other groups, the watching was more predominant, potentially a way to avoid talk. A key issue here was the use of the screen in this way was opportunistic. These people were typically not coming to these places because of the

Big Screen but when they arrived, opportunistically decided that it would be a nice to place to wait and rest. The screens were a component that contributed to their decisions to rest there—but only one of a number of factors.

A second issue here was the screens came to be used as a resource by people within the social context of these waits contributing to people's decisions to stay longer in these places than they otherwise might have done. For example, one mother was in Birmingham with her children. The mother decided to rest at the screen in part because there were children's television programs being shown on the screen. The mother thought it would be a good way of occupying the children enabling her to rest without having to concern herself with them. We saw this too in our observations where another mother was getting her family ready to leave. Getting "*ready to leave*" was an activity that involved putting coats on her younger children. While she was engaged with one child, her attention could not be devoted to another child. However, the screen was used to occupy the other children as she concentrated getting one child's coat on.

Being outdoors was an important part of shaping this behaviour around the BBC Big Screens. Several people were there primarily as a way of being outside and enjoying the sunshine as opposed to waiting inside. The content being viewed was not something these people would necessarily be watching at home in similar circumstances. In this respect, the content was often secondary relative to the importance of being outside. One couple we spoke to said if they were at home they wouldn't be watching this content on television. Rather they would be outside in the garden. For others too, the outdoors and fine weather added to the experience and encouraged them to stay longer. What was interesting was that by considering the primacy of the outdoor and sunshine experience, it created opportunities for thinking of new types of content and use of the displays. A good example here was when the Birmingham screen with a city beach attached was used to display photos of beach scenes accompanied by the sounds of waves. In this sense the content was used as a backdrop to the primary activity of being outdoors as opposed to trying to be the primary activity demanding people's attention. Again this is distinctive from what might be regarded as traditional television viewing.

In contrast, poor weather also affected engagement opportunities with the screen by encouraging people to wait indoors or to discourage them from waiting at all. In Birmingham, however, the rain on one day of observations was so heavy that it forced people to stop walking and take shelter. While the rain moved people from the area immediately in front of the screen where they would usually congregate, many people decided to avoid the heavy downfall by waiting under the sheltered fronts and doorways of the surrounding buildings. While these people were waiting in the shelter for the rain to die down many of them began to watch the screen. Ironically then in this instance, the rain actually encouraged people to wait and watch the screen.

Viewing by Appointment: Structuring Activity Around Schedule

Over time, some people were able to learn regularities in the content schedule of the Big Screens, such as those who regularly passed through as they moved to and from work, or those who worked in nearby office locations and visited the screen during their break times. Of interest here was how this allowed people to structure their daily activities and routine so they could watch particular content. For example, one person who worked in an office nearby the Manchester screen had shifted the time of her lunch break so she could eat her lunch while watching the lunchtime episode of *Neighbours*. Another example concerned watching the evening episode of *Eastenders*. In this instance, the person in question would make a judgement as to whether, after finishing work, they would be able to get home in time to watch it at home. On occasions when they deemed it unlikely they would get home in time, they would stay and watch the episode on the Big Screen on the way home from work. Other appointment based viewing centred on more event type content such as sports events or dedicated screenings of local content. We go on to discuss these in the next section.

Event-Based Content and Behaviour

Big-screens host and support many types of events. In this section we start by looking at the different types of events. We discuss the role screens play within an event and consider the impact of other factors on the event. A number of different events have taken place on and around the big-screens. Sporting events such as football matches and live tennis would always draw crowds. Music festivals also worked well around the big-screen. Events ranged from a linear broadcast of large concerts, such as the Princess Diana concert and the Live Earth concert to national music festivals such as the annual opera season which consisted of a series of operas/concerts streamed live from the Royal Opera House in London. This would regularly draw a crowd of several hundred people at each screen. Local arts and music events also utilised the screens, with stages set up in front of the screen where local musicians and artists perform. In theses instances, the screens were used as a backdrop to the performance. This use of the screens in conjunction with live performance is another important point of contrast with traditional television and provides new opportunities for thinking about content design that is complimentary as opposed to standalone.

Cultural festivals also worked well on the screens. The White Rose Festival was celebrated with stalls and activities around the vicinity of the Bradford and Leeds screens, while on the screens themselves were films with 'local' themes. Bradford and Leeds also played host to the international Asian (Bollywood) Film festival in June 2007. The Big Screen audience in Bradford viewed a selection of Bollywood films, while stage-shows and an Asian market were in the square in front of the screen. Eight thousand people visited the square on one evening during the festival. Important to note here is how these events went beyond the screen experience itself. The screen and its contents were a component in the overall event package—the screens were again complimentary to the overall experience.

Events are set apart because they are 'special' self-contained occasions, which have a limited occurrence. While many were planned, we also saw more emergent of event-type gatherings in which large numbers of people spontaneously gathered in relation to particular content and activities. For example, in Manchester, the screen was showing BBC's BoogieBeebies, a children's TV program that teaches children how to dance. As children in the crowd began to gather and dance along with the presenters, more and more people gathered around to watch. In this sense, the children's interaction with the content became a spectacle that created an audience. This kind of behaviour arose only by virtue of broadcasting this content within the context of a public setting.

Another interesting aspect of event-based use of the screens was the use of compères to support events. The compère, positioned in front of the screen would use a microphone to rouse public interest in the event and/or screen content. Compères provided information to explain the event and content. They also structured engagement of the audience throughout the content - a powerful factor regarding the audience behaviour towards events and the screen, stimulating and maintaining interest over pro-longed period.

Event-based viewing of the screens provided new values when compared with watching similar content at home. Watching in a larger crowd provided an important sense of *belonging* and *community* in events such as the broadcasting of the D-Day celebrations that invoked a sense of national pride among the crowds watching the screens. Likewise, in Leeds, the funeral of terminally ill fundraiser Jane Tomlinson was broadcast live to the Big Screen. An audience of 400, who were unable to attend the service in the Cathedral, gathered in the square to pay their respects and collectively remember a person of particular local significance. This collective viewing provided a completely different experience to watching at home, with people crying and hugging each other. Watching as a crowd was important too, in sporting events, where a sense of competitiveness and being part of a larger group, transformed the atmosphere of viewing. As sociological and social psychological research has documented, crowds encourage people to engage in different kinds of behaviors with respect to a spectacle. Within the context of a crowd people can lose themselves and becomes anonymous, liberating them to engage with broadcast content in new ways.

Interactive Content Behaviours

Interactive applications were an important part of the Big Screens. These ranged from simple SMS/MMS based applications in which people used their mobile phones to send text or images to the display such that they appear on the screen or affect a voting mechanism (e.g. in Manchester there was an application in which people could vote for their favourite book). Other applications were more game-like with both individual and team-based play, e.g., the *'Red-Noses'* game (see figure 2). In this game, the screen was splattered with small 'blobs' (red noses), overlaying a live feed from camera above the screen pointing to the area directly in front of the screen. Image processing techniques would detect when people on the camera feed "touched" a red blob, causing it to move in the direction they pushed. When a blob touched another blob they would merge to make a larger blob. The aim of the game was to manoeuvre all blobs together until only one large blob was left—this then scored a point and the next round begins. Many people could play the game at the same time. The game could be played both locally but could also be set up so that different cities with big screens could compete against each other– this competition and city rivalry could be an important motivating factor for continued game play.

Figure 2. Red Nose Game

Other examples included the *'Seaside Postcard'* game (see figure 3) and a "Formula One driving game. In the Seaside Postcard application, the screen was covered with a cartoon postcard image of a seaside scene. Where the faces of the cartoon characters were, a cut-out allowed the image from the live camera to be seen. The aim of the game was for people to position themselves so that their faces on the live feed appeared in the cut-out region of the cartoon heads; In the "Formula One" driving game, people controlled the cars through verbalisations of engine noises into a microphone.

Placing such interactive applications in public settings raised interesting issues in terms of social values and factors affecting participation. In understanding these experiences, we cannot simply look to the interactive qualities and mechanics of the game as the underpinning of people's participation. People's participation was something highly visible to other people in the setting. It became a public spectacle in which the audience played a key role in the experience. For some an audience was an inhibitor to participation, with people not wanting to be subject to public judgement. For others, it was a social facilitator; an opportunity for showing off and public expression. Having themselves on screen was akin to being locally famous for a brief period of time. With these factors in mind, initiating participation in an interactive public spectacle sometimes presented difficulties. Unlike interactive applications on traditional television within the home, there was less opportunity to explore and practice game play at length due to the *evaluation apprehension* arising from an audience. Therefore game design within this context needs to be immediately understandable and playable relative to more complex and challenging video games that make sense within the comfort of the home. People would often wait around for others to start playing and only join in when there were a sufficient number of people playing—a phenomenon labeled the "Honey-Pot" effect (Brignull and Rogers, 2003). The Honey-pot effect is a social affordance that sees a progressive increase in the number of people in the vicinity of the screen activity, creating a social 'buzz'. Just by being in the space, and showing an interest, people give a signal to others watching that they were doing something interesting. The Honey-Pot

Figure 3. Seaside Postcard application

effect allowed potential participants to understand the social standing of the activity within the immediate community and the required etiquette surrounding participation. The time allowed the potential participant to comprehend what others thought of it, the type of people using it, and how they behaved towards it, and in turn make a judgement about the value and benefits of their own potential input._

Another issue for interactive applications in these public settings concerned potential player coordination. Some games were essentially individualistic in nature and involved basic turn taking. Others, such as the Red Nose game allowed multiple people to play at the same time. Coordinated game play among players who are strangers can be a difficult task that requires careful consideration for the game mechanics. With the Red Nose game, there was a flexibility in terms of the coordination required that simplifies any coordination required. For example, it was possible for individuals or groups of individuals unknown to each other to work on different areas of the screen in parallel and still work towards the goal of the game. The game was not dependent upon close coordination for success. Once players become more comfortable with each other, though, more sophisticated coordination of game play could evolve.

Again, compères played an important role with the interactive applications. They encouraged participation through explanation of game play and inviting participation to overcome social inhibitions. They also played a role in coordinating the game play, helping players work together in ways that might not have happened spontaneously among public participants who are strangers to each other. In essence, they provide a leadership role to group play. For example, in Birmingham while playing the Red Nose game, the compère got all the people playing to join hands and then "sweep" across the screen to gather all the blobs together—a very efficient strategy that would be difficult to evolve without leadership.

The importance of compères also highlighted another issue in such game play, namely the audience experience. We have already discussed how presence of an audience created an interesting social context for the player experience. But the audience experience itself was actually an important feature of these games. What this pointed to is that game play and interactions should be designed to be visible to an audience. With the Red Nose game for example, one can imagine other input devices that could be used to control the red blobs. However, such manipulations might not be visible to the audience. By using movement of the players to push the red noses around, the interactive qualities of the game were very visible to the people in the vicinity. This made it much more engaging for the audience. Indeed, several interviewees suggested it was far more interesting than watching ambient content (such as the news, etc). Some audience members would also actively encourage game play, cheering on friends and family. Compères also played a role here through their amplified commentary of the game play. By overlaying their narrative over the game play, they helped project the game out to the audience, drawing their attention to particular features and encouraging their engagement.

Big Screen and Situated Content

Particular screens acquired additional significance by virtue of their specific locations. Political, social and practical concerns were at play here that impacted on the kind of content displayed, and the meaning it came to acquire. Being situated in particular city locations provided each of the screens with a particular catchment area that defined its potential audience—a "locally" defined audience made up of the people who lived and worked in the city or who visited the city.

An important issue here was that placement of these screens in the city centre was subject to regulation by the relevant local councils. The local councils were responsible for a large part of the funding for these screens but they also had ultimate responsibility for the impact of the screens and content on city-centre life. The installation of the screens comes with certain conditions in terms of the content shown. One such remit set in place by these governing bodies was that a certain amount of content shown on these screens should be of relevance for the local community.

As the screens become a more established feature within the community, more people are becoming aware of it as a resource for screening content they have made or want to make. As a consequence, more people are willing to approach the screen managers with content and ideas for content to be screened. Important here is how the screen, as situated within a particular local area, provides a forum for community produced content. Much of this content would be unlikely to be screened on national TV because the competition with other content would be higher. But within the catchment area of a city centre, the opportunity for being on these public "TV" screens was a real one that could have enormous value for those that choose to get involved. A good example here concerned a local group of Muslim women in Birmingham. The screen manager in this city encouraged the women to make a documentary about themselves and their local community to be aired on the screen. The screen manager's engagement with these women was important in structuring the supporting resources needed in terms of skills and tools. But perhaps as important was the supporting role of the screen manager in giving these people the confidence to do this. Putting something on these public screens is much more of a public performance that simply posting something to YouTube—the apprehension is potentially much higher as the content appears on a screen that also shows more professionally produced content such as the national news, nationally aired soap operas etc.

The significance of screening such content on these screens was further highlighted by the relationship with the audience. Again taking the above example of the documentary created by the Muslim women, the screening of the film became an event in itself. The film makers organised for friends and neighbours to come to the screen to watch it together as a community. Several hundred people showed up to watch the film. People from the local community in the audience would cheer as they recognised particular family friends and neighbours in the film. If such a film were shown away from the local context enabled by the screen, such a level of audience engagement would be unlikely. Such behaviour was not typical of traditional television viewing though there may be some parallels with watching local news stories and watching home videos.

Audience engagement with local content such as this also occurred more spontaneously. In one example form our observations, a couple of teenage girls were walking past the screen as a music video was playing. The music video had been locally produced by a school group and consisted of a collage of rappers arranged in a grid on the screen. The girls stopped to watch the video. When asked why they had stopped, they said that they recognised some of the rappers from their school, two being their classmates and one being a teacher at the school. They didn't think it was a very good song but were interested and amused nonetheless because it was people they knew. These kinds of encounters with content arose because the content was meaningful to the people who inhabited the local area.

Architectural and Spatial Characteristics and Influence on Behaviour

Important considerations in understanding behaviour around these Big Screens were the urban and architectural contexts in which they are immersed. The position of the screens within each of the cities had an important impact on they way they came to be experienced. The screens were components of larger urban spaces that influenced the rhythms and flows of people through the spaces and the activity contexts in which their encounters with the screen were experienced. This was about both the immediate architectural and spatial context but also the broader surrounding context of amenities and work places that shaped city flow.

The broader urban context of the screens was important, especially at times when the screens were not destinations in themselves (such as when it was part of an event) - in particular everyday use, when interaction with the screen was often opportunistic and done within the context of other activities. This was apparent in the experiences of people in our observations and interviews. Obviously these people were not making references to such abstract terms as flows or rhythms but the description of their experiences generally made reference to coming to or from amenities and locations in the surrounding areas. For example, an elderly couple we spoke described how they had just been to a jazz concert at a venue a few streets away from the big screen and were passing time while waiting for their bus, the bus stop being a short walk from the square in which the screen was located.

At a higher level of abstraction, people spoke more directly about the urban environments in which the screens were located making reference to the general characteristics of the area as it related to surrounding activities or lack of. One of the screens in particular was regarded as being in a poor location because there was no real reason to go to the space or to drive footfall through the space. It was viewed as a "dead end" in terms of people flow.

These spaces also had particular identity by virtue of their architectural surroundings that influenced the experience with the screens. This was illustrated in our discussions with one of the screen managers who described use of large screens in Birmingham during one of the major international football championships. As well as the Big Screen in the centre of Birmingham, another large screen was temporarily erected in another area surrounded by bars and pubs. The screen manager described two very different atmospheres at the screening of the same football matches, and very different kinds of audiences. The one by the pubs and bars attracted a drinking crowd with a volatile atmosphere, the other in the main square a more family-oriented audience and atmosphere.

But the immediate architectural features also played an important role in defining how the screen was appropriated. In Birmingham, for example, the screen was in a civic square with surrounding buildings. Within the square there was a large curved arrangement of steps. These steps acted as a natural seating area that faced the screen, providing somewhere where people could sit and rest and the opportunity to watch the screen either as individuals, couples or small groups of family and friends. Another key feature during the period of our study was the installation of a city beach within the square. This was a bounded area in front of the screen in which sand had been brought in to create an artificial beach with deck chairs, tables and benches. There was a low fence surrounding the beach area that followed the line of the curved steps. This beach was an attraction in itself to the area and thereby the screen too, but its presence had other notable impacts on the screen and audience behaviour. For example, it provided a "safe area" in which children could play and be monitored by their parents. We spoke to a number of parents sitting on the steps outside the beach area who were happy for their children to play in the beach area while they sat and watched television.

Health and Safety Issues with the Big Screens

Architectural features were also an important consideration with regards to health and safety, which in turn had an impact on scheduling practices and people's experience with the screens. In Manchester, for example, there is a natural seating area in front of the screen provided by a tiered set of low walls on which people can sit. But between this seating area and the actual screen is a pedestrian through way where a large flow of people walk through while going in and out of shops. For everyday use this did not really impact use since the tiered walls were sufficient to accommodate the kind of numbers that typically wait and watch. However, for larger events or interactive games, these architectural arrangements meant that barriers were used to separate the areas so as not interfere with the pedestrian flow. While these barriers proved useful in one respect, there were certain consequences in terms of audience development and involvement since a barrier portrayed particular meaning to passing crowds in the area. A notable example here was for the Live Earth concert in Manchester where numerous people were reluctant to enter the space because barrier made them think they would have to pay and entry fee. Some of this also occurred in Birmingham when a particular interactive game was being shown on the screen (Red Nose Game). Given the physical nature of this game, again it became a health and safety issue with the potential for the physical movements of the players to interfere with other members of the public. As a consequence, a special area in front of the screen within the beach area had to be cordoned off by barriers to separate the interactive area from the beach area.

Perhaps more significant though was the impact on content curation. Such health and safety arrangements created an additional effort burden extending beyond simply putting the content on the screen. This had particular budgetary considerations, as someone needs to pay for these barriers to be taken up and put down. The delivery of these experiences in public settings then is not as simple as uploading content and pressing a button to start the game as it might be with more traditional interactive television content but incurs all sorts of additional arrangements and costs. These are a direct consequence of the interactive qualities of a game and so are important factors to consider right at the outset of the design of such experiences. These concerns, too, impacted the broader scheduling process. In scheduling content, the screen manager had to make judgements about impact on the public. For example, would particular content attract a large crowd or will it encourage certain types of behaviour. Close negotiations about content schedule were therefore necessary with the local council to ensure public safety and have appropriate measures in place to mitigate any potential crowd issues. This was perhaps most obvious in relation to the screening of large sporting events such as football matches which attracted large numbers of people and were potentially volatile. The screening of these events required an appropriate police presence and strategy to be organised. In certain cities, the architectural characteristics of the areas around the screens presented particular difficulties to policing around the large sporting events. This has led to these events not being shown on the Big Screens, with a preference to erect temporary large screens in areas outside the city where the policing would be easier. Other non-sporting content too was subject to crowd concerns. For example, A Noam Chomsky lecture at the Manchester screen was considered to be a potential problem regarding crowd response. The "Stop the War" campaign had attached itself to the event for political purposes creating a concern from the Screen Manager that this could lead to some disruptive activity from elements of the crowd. Negotiations were necessary between the council, screen and organisers of the 'Stop the War' campaign to ensure the demonstrations that were to be held near the screens, passed peacefully. In the event of any trouble the screen manager had told the campaign that the broadcast of the talk on the Big Screen would be switched off. The architectural space (seating, thoroughfares) of particular urban locations where the Big Screens are located, then, has a bearing on public safety, and affects the types of methods chosen to deal with each situation. The solution is defined by the fit between content and the architectural space. If there is no 'fit' content is removed, if there is a 'fit', precautions around it are considered and implemented.

CONCLUSION

Through our findings, we have begun to characterise some of the everyday behaviours and practices that have developed around the Big Screens across a range of urban settings. The aim has been to take a closer look at the details of people's behaviour and motivations underlying these behaviours. In particular we have attempted to relate these behaviours and motivations to the social context created by situating these screens within the city environment.

Like traditional television, viewing behaviours are subject to particular daily rhythms. In contrast to traditional TV where such rhythms are bound up with home, the patterns of behaviour around the large screens are bound up with the rhythms of the city. An understanding of these rhythms is essential to determining appropriate content for these screens - some of this understanding is location specific depending on the architectural configuration of the location and surrounding environment.

As well as providing ambient content where viewers experience content for only tens of seconds as they pass by, there are also many examples of more prolonged viewing. The screens work as an important "waiting" technology, one that supports the act of waiting, or resting. Understanding the nature and purpose of these waiting activities can help inform decisions about content.

There was also viewing by appointment where some people restructured their activities to to view particular content. In particular though, these screens play an important role in larger event based scenarios. The screens allow a diverse set of people to come together and create a crowd atmosphere. It is this crowd atmosphere that is a strong component of people's enjoyment of the content and a pull for people to come to the screens. But there are issues of how this coordination of the crowd is achieved. Publicity of the schedule plays an important role here as do event compères.

In terms of interactivity, there are many important lessons to be considered. Most important are the effects of placing such interactivity in a public context. In doing this, interactivity becomes a public spectacle—people have an audience, which introduces a different set of social dynamics compared to using interactive content in the comfort of ones own home. For some, this is a source of evaluation apprehension and for others the opportunity to show off. People typically want the support of playing with someone else from a close-knit social group rather than a loosely coupled collection of strangers. In thinking about these games, it is also important to consider how these applications are not simply engaging for those playing but also how they create and engaging spectacle for those who are watching. It is also important to consider how these games can be understood and mastered by an audience that doesn't have instructions or the time to devote to playing the game.

Also noteworthy about these screens is how they acquire significance by virtue of their specific location. Social, political, and practical concerns impact on the kind of content displayed. The location of these screens essentially defines its audience in terms of a local community. This has implications for the kind of content that makes it onto these screens and also for the level of participation from local content providers. These screens create opportunities for local content makers to "broadcast" things that would not be appropriate for traditional broadcast television. The meaning that content comes to acquire and the relationship between the audience and the content are very different as a consequence of this local relationship. Consider, for example, the high levels of audience engagement when they recognise friends or family on the screen. Consider too, the role the "situated" screens play in celebrating local events/achievements.

A final issue associated with these screens concerned health and safety, which has important consequences for the scheduling of content. Particular types of content can have very real practical effects on shaping the interactions of large numbers of people within constrained urban environments. Sporting events or large national concerts can attract large numbers of people into spaces designed for smaller numbers during everyday use. Further, with interactive elements to the content, behaviours potentially interfere with the more everyday behaviours that take place in the spaces. These behavioural consequences, over and above their interest as a social phenomenon, actually affect the process and cost of scheduling particular content. For example requiring coordination with appropriate authorities and financing appropriate temporary architectural safety features as required. So rather than simply being the case of pushing a particular button to change schedules there are additional work processes that need to be followed.

REFERENCES

Brignull, H. & Rogers, Y. (2003). Enticing People to Interact with Large Public Displays in Public Spaces. In *Proceedings of INTERACT' 03*, Zurich.

Fatah gen. Schieck, A. (2006). Towards an integrated architectural media space. In *First Monday, Special Issue #4: Urban Screens: Discovering the potential of outdoor screens for urban society.* http://www.firstmonday.org/issues/special11_2/

Goffman, E. (1966). *Behaviour in Public Places.* Free Press.

Lozano-Hemmer, R. (2002). Alien Relationships with Public Space. In J. Brouwer & A. Mulder (Eds.), *TransUrbanism.* NAI Publishers.

Manovich, L. (2006). The Poetics of Urban Media Surfaces. In *First Monday, Special Issue #4: Urban Screens: Discovering the potential of outdoor screens for urban society.* http://www.firstmonday.org/issues/special11_2/

Taylor, K. (2006). Programming video art for urban screens in public space. In *First Monday, Special Issue #4: Urban Screens: Discovering the potential of outdoor screens for urban society.* http://www.firstmonday.org/issues/special11_2/

Virilio, P. (1997). *Open Sky.* NY: Verso.

Chapter XVI
Zync with Me:
Synchronized Sharing of Video through Instant Messaging

David A. Shamma
Yahoo! Research, USA

Yiming Liu
Yahoo! Research, USA

ABSTRACT

This chapter details a real world case study of a synchronized video-sharing tool, Zync. Zync was integrated into a popular instant messenger program that enables a virtual co-present "watch together" and "on-the-couch" viewing experience. Users can watch videos together in sync with each others, while they chat via text messages. Using logged data from 2,814 users, three types of people who share videos via IM are identified, as well as the different kinds of videos commonly shared and how users shared these videos with their IM friends. The authors follow up on these trends with user studies and interviews, which further explains how and why people are motivated to share videos. Based on their findings, the authors identify and define future design areas for synchronized sharing of video; to help keep people connected through media.

INTRODUCTION

We go to the movies with friends and we invite others to our homes to watch special events on TV. When we watch videos together, we often glance at a friend to see if they laughed at a joke. Afterward, we converse about the movie over coffee, ice cream, or a drink. Watching television and movies with friends is a regular social activity.

Web video—particularly its principle representative, YouTube—does not support synchronous social conversation. Without co-presence, videos are watched alone. Our social interactions with videos are limited to asynchronous comment threads, emails, or instant messages (IM). When we do invite friends to watch a video with us, sharing and conversation often takes place by crowding around a monitor to watch a funny clip or huddling around a small screen on a video iPod.

In the presence of friends, having a conversation while watching a video can be arduous. Our attention is split between watching a video and talking to our friends. While some work has begun to investigate inter-personal connectivity and distraction created by watching together (Weisz, J. D, Kiesler, S, Zhang, Ren, Kraut, & Konstan, 2007), we know of no study which describes design principles of a synchronous video sharing system or people's motivation to use it.

This chapter investigates the design, deployment, and study of Zync, a tool to provide synchronized video sharing for online videos through Instant Messaging. Unlike other sites (such as clipsync.com's short pilot run in 2007) that offer their own infrastructure on the web, Zync is built inside of Yahoo! Messenger, an instant messaging client, using a publicly available API. Using an existing IM application provides access to an implicit social network (a friends/contact list) as well as lowering the barrier of entry to usage. Additionally being inside Yahoo! Messenger also allows Zync to be a tool for enhancing relationships and social capital amongst friends (Powazek, 2002). Merging a synchronized video-sharing tool with an IM client creates a single sharing experience and merges the media object with the conversation around it.

In addition to enabling a new sharing experience, Zync is built as a research platform to identify several unanswered questions for integrated chat and video sharing applications. Of particular interest is people's interaction with Zync while sharing (when people chat, how they interact with the video player while sharing, etc.) and their motivations for sharing via Zync as opposed to conventional means. While research studies have shown people may enjoy media more and feel more connected when watching online together (Weisz, et. Al, 2007, Harboe, Mecalf, Bentley, Tullio, Massey, Romano, 2008), further questions still remain about the overall experience of synchronized sharing. How do people express a moment of interest in a video? Does conversation ensue over the course of a running video or do people chat when a video is paused. These questions must be answered through the study of real world tools, such as Zync, as the pragmatics and design of such systems have a direct outcome in people, their behaviors, and their motivation. With a greater understanding of the design interaction and people's sharing motivations, we can inform future iterations of video sharing systems, understand how people remain socially connected and how personal relationships are maintained through media sharing.

BACKGROUND

The growth of online video has lead people to share videos. In some cases, the video is sent via a link (like an email or an instant message). Online chatting along with a live stream or broadcast has also recently become popular. In a study, Weisz, Kiesler, Zhang, Ren, Kraut, and Konstan (2007) showed people feel closer to each other and will enjoy the media more when they can chat with others at the same time, even though they found the chat distracting.

Another way of sharing and chatting about videos is waiting until a friend drops by their home and showing them the video in person. Many people leave video clips on their cell phone to share with people they might run into (Kirk, Sellen, Harper, Wood, 2007). Despite the distractions inherent to listening to two concurrent audio streams at once (Welford, 1967), face-to-face sharing provides a rich social engagement, one where people discuss while the video is playing or pause and rewind to review or talk about the funny or critical moments.

Usually, the video content itself is analyzed to find summarizations and to select poster frames. While significant advancements have been made in the field of content analysis for this use, Zync provides new data to assist in this task: sharing patterns and behaviors. This bears similarity to the work of Syeda-Mahmood and Ponceleon (2001) who analyzed video browsing patterns for key-frame generation. Using videos as conversational objects brings new behaviors in an IM context. Synchronous communication of shared video changes people's motivations, and hence, the video's sequence behaviors. Additionally, Zync leads towards an understanding of overall sharing and communication behaviors in this new medium of web videos by examining both browsing behavior and chat communication as one experience.

Web applications, such as http://live.yahoo.com and http://www.justin.tv, allow for group chat during a broadcast or live stream. A moving stream encourages rather telegraphic conversation of short quips. Other sites provide chat with unsynchronized video playback (YouTube Streams and http://meebo.com) where conversation is not correlated with a given time or piece of content. CollaboraTV (Harrison & Amento 2007) introduced chat from a virtual audience while watching a video. The October 2007 release of Mac OS X 10.5 included "iChat Theater" which allows users to stream video from one Macintosh to another while chatting. In May of 2008, Microsoft released "Messenger TV" which bears a strong similarity to Zync. As unified sharing experiences continue to grow, videos will become media objects centered on sharing and conversation.

DESIGN

Traditionally, when a video link is sent over IM, people become disconnected. Viewing the video requires people to leave the chat window or, in the case of inline players, begin watching but not typing too much as the inline player will scroll off the screen. The sender is sent little to no feedback on what the remote is doing and may resort to asking "did you watch it yet?" or "have you seen that cool part?"

Near Sync

Zync is a peer-to-peer video player that keeps two people in sync across the Internet. When a video URL is dropped into a conversation (see Figure 1), a player loads next to the conversation in both the sender's and receiver's chat window. Both players then automatically start the playback together. Watching together allows people to identify and share meaningful moments in the video. By integrating into Yahoo's instant messaging system using their free plugin API, web videos are transformed from a collection of the media objects into a collection of conversational interactions shared within an implicit social network, the buddy list.

Both viewers have access to the "remote control" and can play, pause, and rewind the video while chatting. The video player is 'docked' next to the text conversation chat in an IM window. Either participant can enter a video URL from a popular video sharing site (Yahoo! Video, YouTube, Google Video, etc.) to begin sharing. The video loads for both participants and automatically starts playback once both client machines are ready.

Having a truly synchronized video sharing experience is not possible with most web videos, since they are generally served via the web as a progressive download. True sync would require streaming (not progressive) servers, which most web video sites do not offer. To stay in sync, video URLs are loaded on the sender's machine and sent to the remote friend's machine where it is also loaded. Once either machine (say the sender's) has buffered enough content to begin

Figure 1. A synchronized video chat using Zync. Here, the remote user (Ralph) pauses the video when he asks a question. The local user (Ayman) answers the question and resumes playback himself.

playback, it sends a silent notification to the remote and waits. Once the remote has satisfied its buffer, it sends a silent note back to the sender to begin playback then starts its own playback. The only artifact here is the latency between the sender and the remote machine which may cause the two players to be 'off-sync'. In practice, this offset is about 300ms and tends not to interfere with the 'in-sync' feeling of sharing video.

The control of the synchronization is open for either user. Zync can only have two people at a time in a session so there is no need for token passing of the playback control. In Figure 1, the remote user (Ralph) pauses the video to he asks a question. Ayman then answers the question and resumes playback. This model facilitates interaction that is similar to the face-to-face video sharing experience.

For the initial pilot, Zync was seeded with three suggested videos: a short 90-second clip on coffee art, a 3-minute advertisement for a popular soda, and a 25-minute documentary program on the history of WuShu. These videos served as a baseline where we could collect usage data across a few single URLs of different content styles. These URLs are provided as a means to examine sequence behaviors. The activity around these URLs was removed in our studies.

Finding Metrics

To serve its purpose as a research instrument for investigating video sharing behaviors, we integrated a set of data collection and event logging mechanisms within Zync. Given the sensitive context of personal IM communication, this data collection is performed only on an opt-in basis. Any person who had opted in or out can change that preference at any time. Personally identifiable data, such as Yahoo IDs, are kept anonymous. At first load, Zync will present the user with a privacy-setting panel explaining our goals in us-

ing this data. The user can at that point choose to allow or disallow data collection. She can also return to the privacy settings panel at any time to change these preferences.
With data collection active, Zync records information about the video, user interactions with the synchronized player, and other meta-data about the video sharing session. More specifically, we log the following types of interaction events:

- **Chat messages:** For privacy reasons, Zync does not record the actual conversation. It records chat volume as the number of words and the number of characters typed in each message.
- **Watch commands:** Instructions for the player to load a video.
- **Control commands**: Instructions for the player to play or pause the video.
- **Scrub commands:** Instructions to fast-forward or rewind the video.
- **Plug-in state:** General plugin loaded, unloaded, and errors.
- **UI Clicks:** Mouse clicks to share the video with a friend, obtain the original video, post to del.icio.us. (Liu, Shamma, Shafton, Yang, 2007)

Each event is identified with some additional contextual information, including:

- A unique session hash, identifying a particular exchange of IMs between two users.
- Unique hashes representing the sending and receiving participants.
- The URL of the video being shared.
- A time stamp of when this event occurred.
- A player time, or the time on the video at which this event occurred.
- For chat messages, the number of characters and the number of words in the IM.
- For chat messages, any emoticons contained in the message.
- Whether the event was triggered by typing out a command in the chat window, or a mouse-click on a UI button. (Liu et. al, 2007)

With this data, the general effectiveness of our design can be verified. Further, by applying social network analysis to each video-sharing conversation, one can distill models of usage and offer some insights into user behavior during synchronized video experiences over IM. Zync recreates the 'with you, on the couch' media sharing experience. We predicted synchronization and social co-presence would promote online conversation, engagement with the shared media, and improve the social aspects of media sharing. Zync further informs an understanding of how people share and interact with media as a conversation, sharing patterns within social networks, types of media shared among different networks, and people's (conversation and interaction) behaviors.

BEHAVIORS AND MOTIVATIONS

The first public prototype of Zync was released on March 13, 2007. For a period of 6 months, from March to September of 2007, we logged 2,814 unique users who obtained Zync and consented to data collection. Of this collection, 895 (32 percent of) users returned to use the plug-in in more than one IM session.

In all, Zync was used in 3,010 IM sessions, accounting for 8,909 video-sharing conversations (A conversation is defined from the start of a video until another video is started, or the plug-in is stopped) and 5,825 unique videos (excluding our three suggested videos). The majority of these videos came from YouTube. For each conversation, participants scrubbed (fast-forwarded or rewound) an average of 1.42 times, though the median number of scrubs was 0. Chat volume averaged 263 characters per video. The median is 108 characters. Overall popular videos averaged 197.7

seconds in length. Videos that were heavily chatted averaged 304.9 seconds and heavily scrubbed videos averaged longer at 345.9 seconds.

To help interpret this usage data at a qualitative level, we interviewed Zync users to find out more about their practices and motivations for using Zync. Users were recruited via a call for participation link posted to the main interface of the plugin during the last month of our quantitative sample. There were three male and two female participants with ages ranging from 16 to 31. All interviewees resided in the United States modulo one Serbian resident. Five IM interviews, between 1.5 and 2 hours in length, were conducted, covering users' online and offline video practices, content preferences, and Zync usage details.

Interaction Behaviors

Zync allows two input modalities with its interface: keyboard and mouse. Videos must be started via text entry by sending the single URL as a message to a buddy. Interaction commands involving the video player, such as play, pause, stop, or rewind, can be invoked by either typing a literal command into the chat window, or by clicking the playback controls on the player UI.

IM chat is typically a text-oriented environment requiring heavy keyboard use. The use of mixed modality is unique to Zync and is not found in Messenger TV. Of the commands with both keyboard and mouse modality, users only demonstrated a clear input modality preference for the Play command. In this case, a significant ($p<.05$) number chose to type out the "Play" command on the keyboard rather than clicking the Play button in the player panel. From watching people's interaction with Zync, if the video is not being played, users are typing chat messages. With the hands on the keyboard, typing 'play' might offer some convenience. The remaining commands showed no significant difference between input modalities, which poses an argument for designing for both mouse and text interactions.

Scrub Now, Chat Later

How users interact with the shared video content—and with each other during the sharing experience—is also of interest. Each recorded Zync interaction contains an absolute time stamp for when the event was generated. The player time, relative to the time in the video, is also recorded when an event is triggered.

Using this data, and aggregating events across conversations between any pair of users with respect to a particular video, timelines of activity for that video were created. The timelines were annotated with the types of user activity occurring at any given point in the video, as well as relative frequencies of their occurrence.

Similar to previous work on video browsing patterns (Powazek, 2002), these visualizations were used to examine patterns of video usage. Unlike the work of Syeda-Mahmood and Ponceleon, we focused on user behaviors during synchronized IM video sharing (both scrubbing and chatting), and how they interacted with both the video and each other during the experience. In particular we found (see Figure 2) that scrubbing activity generally occurred close to the start of the video, while chat activity tended to dominate nearing the end of a video.

Some videos had event distributions that contained spikes of scrub or chat activity at various points on the time line. This might be correlated with video content, and the particular nature of that content may tend to induce user interaction and activity modeled as sequence behavior. However, no conclusive sequence or event pattern could discerned. This is largely due to the relatively limited number of users who share any individual video.

This tendency for chat to occur at the end of the video is particularly interesting. Weisz et. al (2007) saw that simultaneous chatting and video watching distracted users from following the video. Our results suggest one possible explanation in light of this finding: users may be

Figure 2. Top: This chart shows the activity (as a percentage) of 391 people sharing a single video (in pairs) using Zync, broken down by activity type. Bottom: the same chart shown by volume. The spike in chat at the end of the video reflects how conversation ensues after the playback has finished.

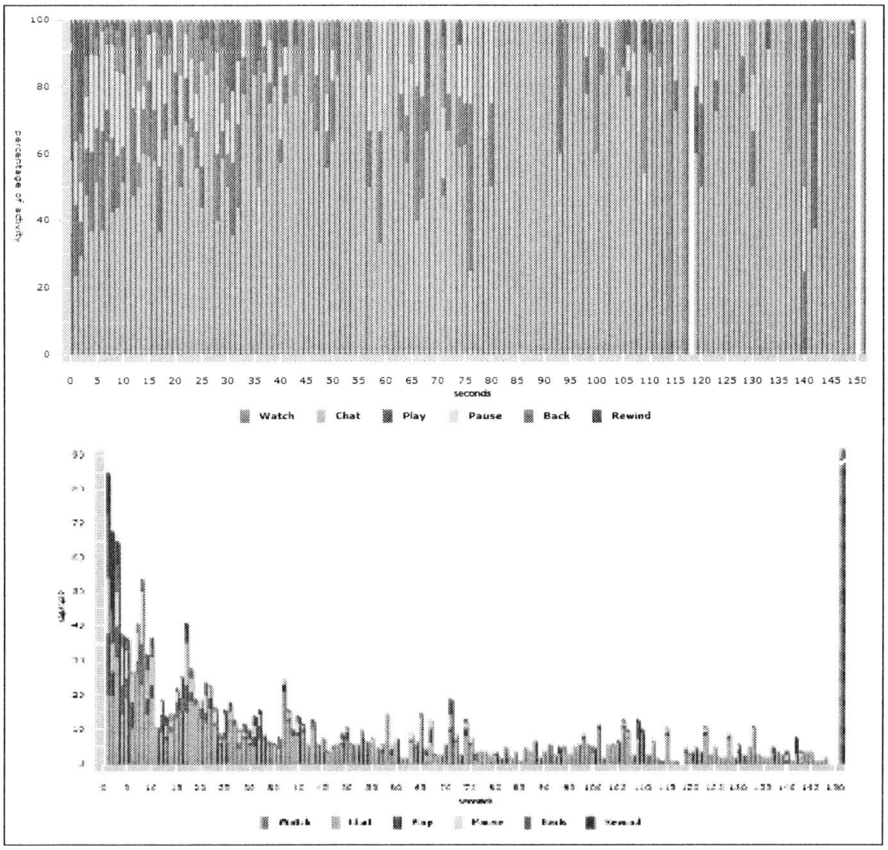

naturally holding back on verbose conversation until the video concludes, except for spots of specific interest that drew immediate feedback or remarks.

The qualitative interviews appear to support these results. Users discussed two types of interactions that can be described as *light* and *heavy*. Light interactions call attention to a particular moment in a video or act as small presence indicators. These varied from emoticons and small expressions like "LOL" (laugh out loud) to typing a direct quote from the video. When used to convey emotion, light indicators become virtual expressions and social interactions like, as one interviewee described "[looking] at each other when something happens and know what each thinks at that moment."

Less frequent were heavy interactions, where interviewees discussed wanting to discuss something in more detail. This typically involved pausing and rewinding the video to locate the moment of interest. Additionally, discussions involved general conversation as well as predictions where the discussion would "talk about what just happened and try to figure out what will happen next."

One pair of interviewees, in particular, reported that they typically and deliberately saved the bulk of their discussion until the end of a video—sometimes letting it play a second time while they discussed. Another pair, who were particularly interested in watching and sharing anime, did not like to pause shows, as one of them felt that "anime is best watched the way it is." Instead they generally limited their chat while watching the show. In all cases, the "lighter" type of interaction such as moment indicators (e.g. LOL) was accepted as standard, real-time interactions and considered generally non-interrupting and non-disruptive.

Conversation and Control

A pairwise, synchronized video sharing experience creates a social interaction conundrum. By design, both participants in Zync have equal control over the video's playback. Either participant may play, pause, stop, scrub, or send a new video to watch at any time. This would be analogous to someone watching TV with a friend and each friend holds her own remote. Sharing of player control in these instances is unlike the natural experience of co-watching, in which one person guides the joint experience.

The question becomes whether there is sufficient evidence to believe one of the two IM participants would emerge to control the synchronized experience on behalf of the pair, or if both participants would cooperate and/or jockey for that control. We examined scrubbing commands of the sender and receiver as a measure of this "holding of the remote control." In 65 percent of cases, neither participant issued any scrubbing commands. Examining scrub activity by person and conversation in the remaining cases, the sending participant appeared to issue the majority of scrubbing commands over the receiving participant ($p<.05$).

The Influence of Time

Beyond the effects of social networks in propagating content, there are temporal influences on the kinds of videos being shared among Zync users. The vast majority of videos shared via Zync are from the popular video hosting site YouTube, user-generated meta-data—such as category and tags—were available from YouTube's API. This allows us to associate some semantic attributes with the video URLs. While the tags applied to YouTube videos were generally noisy, the video categories suggest some patterns in both the content being shared and temporal influences on sharing behavior.

Table 1 lists the videos shared via Zync by their categorical bins from the YouTube set of categories. The music and comedy categories are by far the most popular. Entertainment and film/animation are the second most popular although their volume is considerably less than music and comedy. The remaining shared video categories show no significant variation.

Referencing the timestamps on each sharing event, it is possible to construct a temporal view of sharing events, for each day of the week. Early analysis of video sharing via IM is consistent with those results. Fewer people shared videos during weekends than weekdays (Figure 3). These pre-

Table 1. The count of videos shared via Zync identified by their YouTube of category.

Category	Count
Music	1968
Comedy	1269
Entertainment	963
Film & Animation	685
People & Blogs	254
Gadgets & Games	254
Sports	132
Pets & Animals	114
News & Politics	65
Howto & DIY	55
Autos & Vehicles	40
Travel & Places	26

Figure 3. The average number of Zync users shown by day of the week shows a drop in weekend usage.

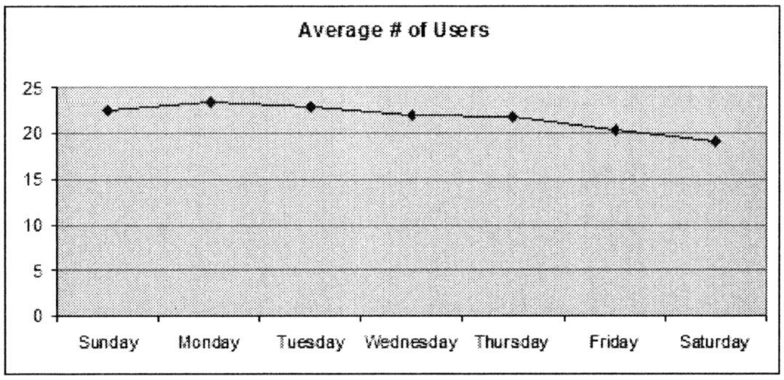

Figure 4. The average number of URLs shared in Zync per user per session by the day of the week show.

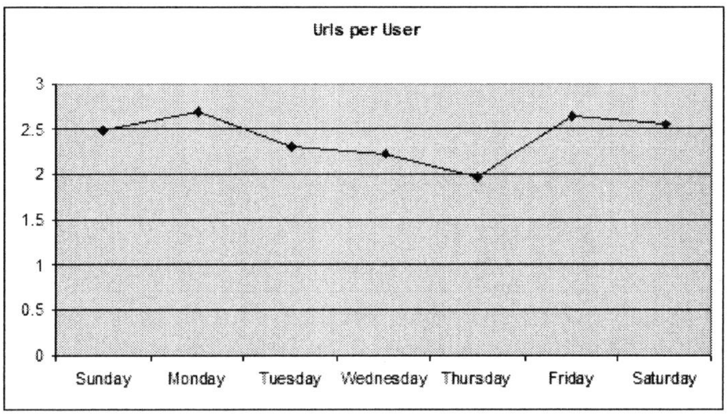

liminary results are similar to that of other social sites' weekend behavior; a study of Facebook users suggested that online socialization was at its lowest over the weekend (Golder, Wilkinson, Huberman, 2006).

The number of videos being shared on the weekends was generally equal in volume compared to the weekdays; weekend Zync users often shared more videos per session. Figure 4 shows the number of shared URLs per user by the day of the week. This suggests that some Zync users increase their sharing during the weekend, which compensates for the overall decrease of active users on those days.

Associating this data with the categorical labels, more comedy videos were apparently shared at the beginning of the workweek (on Monday and Tuesday) than the other days of the week. By contrast, while music-related videos are generally dominant, sharing activity over the weekend is especially music-focused, accounting for nearly 36 percent of weekend video shares. By comparison, comedy, the next highest category, accounts for only around 21 percent of shares. When graphed (see Figure 5), the percentages of categories watched across the week shows music to be the most dominate category shared on the weekend. If the week were broken into Friday–Sunday,

Figure 5. The percentages of categories watched across the days of the week.

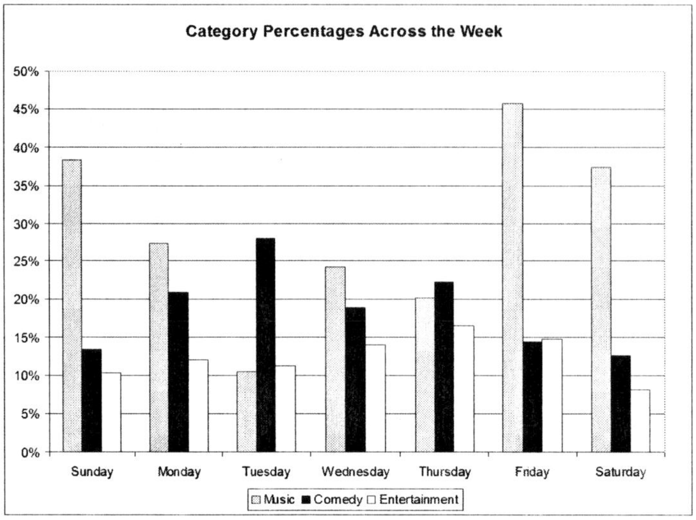

Monday–Tuesday, Wednesday–Thursday pieces, one sees the change from music to comedy move significantly ($p<.14$) at the start of the week.

Social Structures of IM Video Sharing

As a plug-in to an instant messenger system, Zync provides a natural platform for social network-based analysis of video sharing. An information-oriented network analysis examines the content and patterns of relationships, to provide a model of resource flow among a set of actors (Haythornthwaite, 1996). In this case, we characterize each user as an actor and the video as the information resource. The sharing of video constitutes information relationships that connect pairs of actors and establish a flow of information from the sender to the receiver.

This creates a social network graph, with users as nodes, video sharing conversations as edge. This graph is used to derive a set of empirical characterizations of sharing behaviors and users' roles in sharing networks. Nodes are then chronologically labeled, to add a temporal dimension to the graph. This reveals diffusion effects as users begin using Zync and propagate both the plug-in and videos to their friends.

For the initial prototype, the network graph that emerged is sparse. The main network structures identified within the graph falls into one of three general classes: hubs, chains, and dyads. Figure 6 shows a typical sub-network with the overall Zync user set.

The most common type of sharing activity can be characterized as dyadic relationships pairs of people exchanging video shares with each other but are otherwise unconnected with the rest of the graph (see Figure 7). This sharing can be unidirectional, with one partner consistently sharing with one other partner, or bidirectional/reciprocal, in which both partners send shares to each other. The magnitude of bidirectional information relationships may be largely equal, in which case the number of shares from participant A to B is roughly comparable, or unequal, in which case one participant shares significantly more content than the other. Most dyads tend to have little to no sustained activity. However, a few dyads account for an unexpectedly large amount of activity, with over 100 videos shared between the two partners over the course of a month.

In line with this analysis, the users interviewed reported that they shared predominantly with one other person. Though we did not initially set out to recruit sharing pairs and the call for participation was only visible when users loaded the Zync plugin (i.e. users were getting ready to share), once a user decided to participate, she would tell the intended sharee about the study. Since much of the network consists of dyads, Zync tend to be dyadic sharing scenarios.

A second class of sharing structures can be characterized as "chains", or "line graphs" in terms of centrality (Wasserman, Fause, 1994) (see Figure 7). This is a version of the "small world" chain (Milgram, 1967), in which the resource is passed down from actor to actor, or the colloquial "grapevine" structure of information sharing. In several instances, we observe a video being sent

Figure 6. A group in the overall Zync network graph. Nodes have been re-labeled chronologically, from smaller numbered nodes being earlier Zync users. Edges are labeled as a, b, c, d, e, where a is the number of shares between these nodes, b is the number of scrub commands issued by the sending node, c is the number of scrub commands issued by the receiving node, d is the number of characters in chat typed by the sending node, and e is the number of characters in chat typed by the receiving node. Emoticon usage is listed adjacent the edge label.

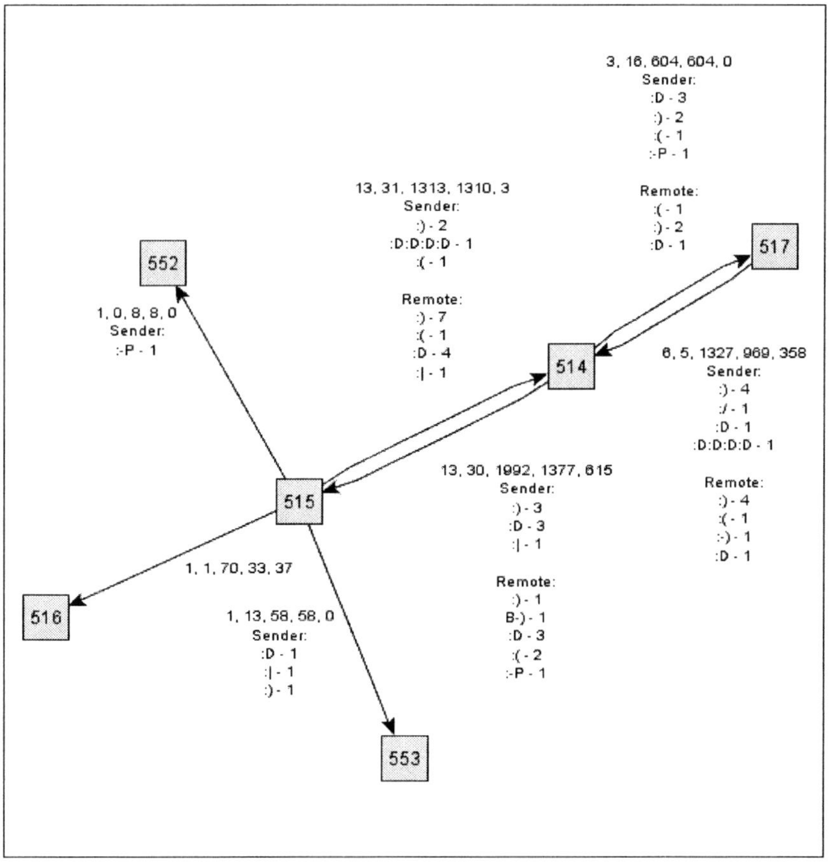

by one Zync user to a second user, who in turn shares this same video with a third user (who is otherwise unconnected to the first user), and so on. The occurrence of this pattern of sharing is surprisingly rare in our data set due to the sparsity of the graph. For this study, the Zync prototype was published to the Yahoo! gallery as a third party plug-ins (not an official Yahoo! product). This means it would have to first be discovered on the web gallery, downloaded, manually started and then sent to a friend. This additionally startup difficulty increases the "small world" effect.

The third, and traditionally one of the most interesting sharing structures, is the "hub and spoke" or "star graph" (Wasserman & Faust, 1994 and Watts, 2003) (Figure 7). The "stars", or the central nodes in these hubs, act as information brokers and broadcasters. The relationship may be bidirectional; in that case, the receivers also share videos with the sender on many occasions. There

Figure 7. Left: A unidirectional sharing of video between two users in a dyad—otherwise diverse groups are linked by such chains, are not observed. Right: Three users, passing a single video down a chain or the "grapevine." Bottom: An information broadcaster in the center of a hub sends a video.

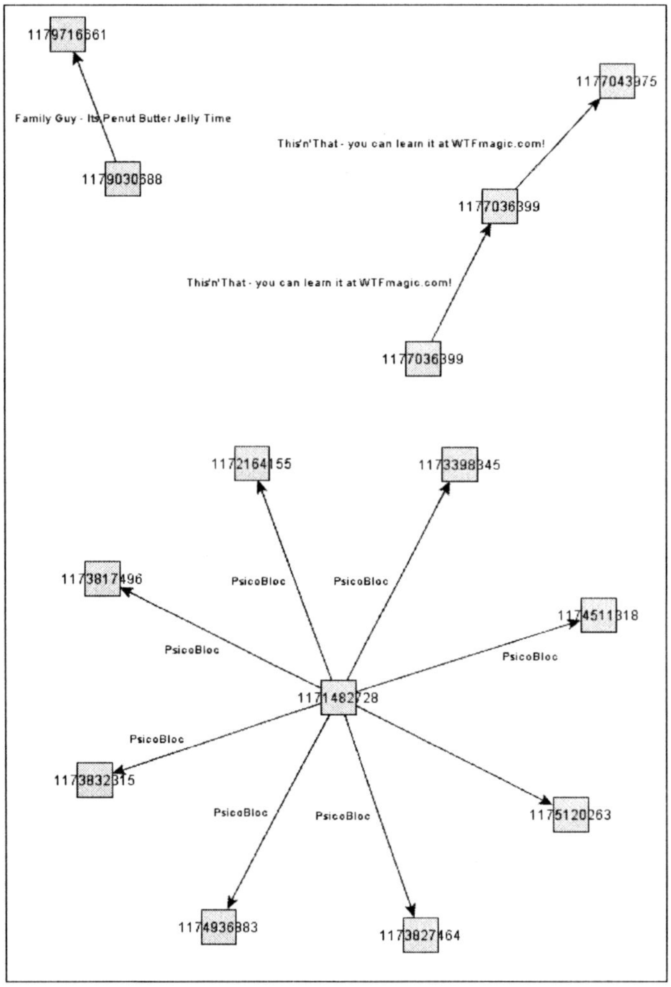

Figure 8. Two distinct dyads of Zync users, sharing a variety of URLs between themselves. Both sets of users have shared a common video, "amazed-lonestar". This video is a boundary object between the two user sets.

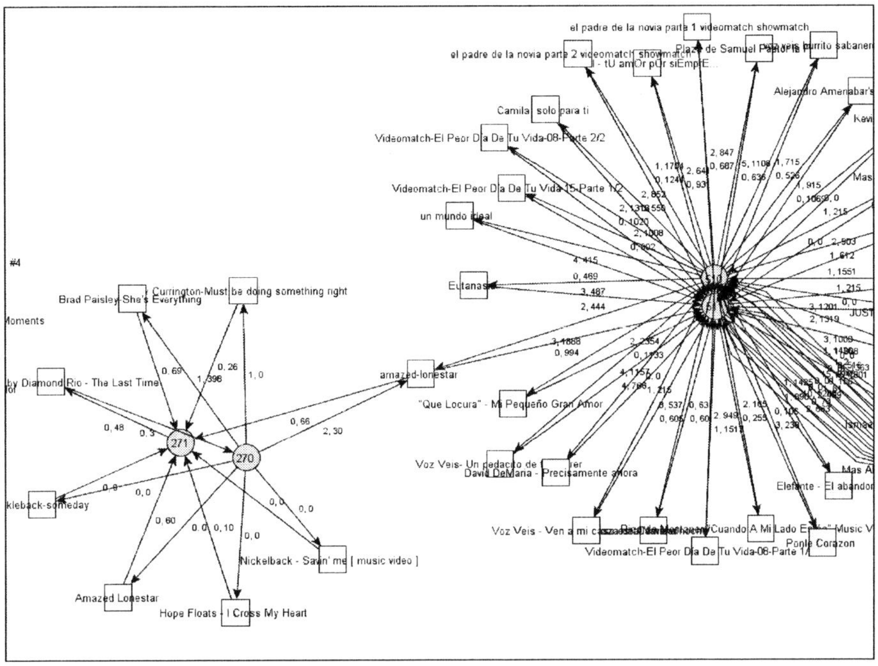

are usually no shares directly between the various receivers. In several cases, the hub received a video share from a friend, and within the space of days, this video was sent out to a number of other users in the hub's network.

By adding the shared videos (associated with their YouTube categories) as first class nodes into the network graph, patterns of propagation are illuminated for each category. Two patterns are of immediate interest.

Certain groups of people share almost exclusively the same category of content. This is most apparent in some dyads, where the plurality or majorities of videos being exchanged are of one particular category. These genre-specialists tend to hold a single-category sharing pattern over time and across many different IM sessions.

The videos shared in these scenarios were of variable lengths. When interviewed about their general video consumption behavior on and offline, all participants were surprisingly comfortable with watching longer video online. Three interviewees regularly watched feature length movies. Two people would watch multiple 22-minute TV episodes a day. While all the interviewees had TVs, they did not express a clear preference with regards to TV usage versus online video consumption aside from the simple *availability* of particular content on one platform or the other.

Also of interest is a pattern in which once-distinct networks of people would be connected through common interest via a particular video. In Figure 8, we see that otherwise separate groups of Zync users appear to share a common interest in "amazed-lonestar". Such connections imply some property of this video that makes it popular with unrelated groups of people.

Motivations

Zync usage appears to be driven by two types of motivations. In the first, Zync serves as an "enhancement" to conversations and supported ongoing dialog on a particular topic. For example, one user shared a series of videos chronicling the different periods of a celebrity's life with her friend over the course of a few days. Another user reported that when she chats with her brother, the conversation topics trigger memories of relevant videos, which they then share.

In the second, Zync was used as a deliberate means for sharing particular video content. Users entered into conversation wanting to watch particular videos. One user said that when he realized that his sister was available to talk for more than a few minutes he decided to share a video. Several pairs discussed making arrangements to watch videos together.

Watching with Close Friends

Users reported a predominant use of Zync with their closest friends. The pairs interviewed in our study reported very strong relationships: a brother-sister pair and two best-friend pairs. In each case, relationships began offline when the individuals lived near each other. They adopted Zync only after becoming geographically separated. While our small sample of users is not safe to generalize from, given the structure of the Zync usage network, it seems likely that Zync serves a relationship maintenance function.

One interviewee, who used to watch a particular show with her friend before he moved away, stated, "[it] just wasn't the same without someone to pause and laugh at [it] with." This is inline with other studies which show the effect of in-group laughter during co-present viewing (Platow, Haslam, Both, Chew, Cuddon, Goharpey, Maurer, Rosini, Tsekouras, Grace, 2005), particularly laughing with someone will make content appear more humorous.

In another instance, one user watched an extraordinary amount of full-length anime episodes with his best friend, despite having typically seen the episodes before hand. He simply stated, "the feeling of watching [video] together is great." The other sharer in the pair stated he was only interested in watching videos with his best friend and generally ignored videos sent to him by his other contacts. These responses suggest that for these users, the Zync experience tends to be an intimate social interaction.

Ambient Virtual Co-Presence

The aforementioned scenarios evoke the concept of "ambient virtual co-presence", a technosocial situation described by Ito and Okabe (2004) to characterize Japanese teens' mobile email world. Such co-presence is "a space of shared awareness," which consists largely of low substance, tap-on-the shoulder type messaging, within a small intimate social network. In a similar fashion, Zync provides users—who have a shared awareness of each other by virtue of Yahoo! Messenger—the opportunity to engage in what our usage and interview data suggests to be a highly intimate virtual experience. Once a Zync session starts, the etiquette and social practices for instant messaging seem to change in favor of a much "lighter," emotive, interaction that users describe as an experience akin to physically going to the movies with a friend. However, frequent moment indication across pairs suggests that there is still an expectation of attention and focus. Zync provides an experience that both supports pre-existing social video consumption practices and provides a new "ambient virtual co-presence" experience, which has its own set of social practices.

FUTURE WORK

Synchronized sharing heightens engagement and lengthens conversations when built into an existing IM client. Congruent with previous research, people felt more connected and found the media more enjoyable while sharing in sync. Users also felt more comfortable sharing long form content and were happy to selectively share media from many sources (YouTube, Yahoo! Video, etc.). Our interviewees all had strong relationships with their sharing partner and all of them pointed to having a heightened sense of closeness while using Zync.

With more synchronized video and TV systems emerging, sharing structures can inform how research is conducted and how systems will operate in the near future. The transformation of media from content to conversation informs recommendation, summarizations, and thumbnail selection techniques. Each of these techniques is an area of research on its own; however, work has now begun to explore how people are connected through videos and how conversation ensues. This continued study through practice and new experiences will improve and change how we think of media viewing.

Moving forward, we note four design areas which will affect future iterations of Zync as well as the design of synchronized video sharing applications: low entry barriers, concurrent sharing, enhanced presence, and long-form content support. (Shamma, Bastea-Forte, Joubert, & Liu, 2008)

In Zync, 16% of all videos watched had zero conversation. This suggests that users were simply using Zync to share video and likely using some other channel to communicate—possibly another IM program, VoIP, or telephones. A more universal video synchronization application, accessible from any chat client on any operating system is desirable for lightweight, ad-hoc sharing of media.

With the amount of content available and the increasing role of instant messengers in maintaining contact with friends, a user may carry on concurrent IM conversations with multiple people. Further, a user might want to share the same video with a number of different people at the same time. Zync is designed primarily for sharing media in one-to-one chats, though it can also be used in a one-to-many fashion. It is potentially rewarding to extend this functionality by allowing any number of users to join a synchronized video session, in a many-to-many context. This allows a group of friends to get together and watch some video content together, with everyone having access to the "remote" control.

When using text chat, all emotions and physical actions (such as leaving your computer) must be conveyed explicitly through text messages. Users often sent indicators such as "LOL" as a means to convey reaction or appreciation to a shared video and to indicate co-presence. While enabling a text channel for discussion as well synchronized sharing is a step forward, it may be useful to go further. Enhancing this presence, with audio or video conferencing capability, will allow participants to see when someone is paying attention, laughing, or smiling, without the added weight of conveying this via text. This may create a more engaging and compelling interaction for most users.

In developing Zync, we are still investigating new user interactions. With these new interactions, new questions and new technical limitations arise. As these new systems grow and become commonplace, they shed light on existing research questions and expand our design framework to create new experiences through synchronized sharing.

REFERENCES

Golder, S., Wilkinson, D. M., & Huberman, B. A. (2006). Rhythms of social interaction: messaging within a massive online network. *cs/0611137*. Retrieved May 28, 2008, from http://arxiv.org/abs/cs/0611137.

Harboe, G., Metcalf, C. J., Bently, F., Tulio, J., Massey, N., & Romano, G. (2008). Ambient social TV: Drawing people into a shared experience. In *CHI'08: Proceedings of the twenty-sixth annual SIGGCHI conference on Human Factors in computing systems* (pp. 1–10), New York, NY, USA.

Harrison, C., & Amento, B. (2007). CollaboraTV - Making TV Social Again. In *EuroITV 2007* Adj. Proc.

Haythornthwaite, C. (1996). Social network analysis: An approach and technique for the study of information exchange. *Library & Information Science Research, 18*(4), 323-342. doi: 10.1016/S0740-8188(96)90003-1.

Kirk, D., Sellen, A., Harper, R., & Wood, K. (2007). Understanding videowork. In *Proceedings of the SIGCHI conference on Human factors in computing systems* (pp. 61-70). San Jose, California, USA: ACM. doi: 10.1145/1240624.1240634.

Liu, Y., Shamma, D. A., Shafton, P., & Yang, J. (2007). Zync: the design of synchronized video sharing. In *Proceedings of Designing the User Experience.* Chicago, USA.

Milgram, S. (1967). The small world problem. *Psychology Today, 2*(1), 60-67.

Okabe, D. (2004). Emergent Social Practices, Situations and Relations through Everyday Camera Phone Use. In *Proceedings of the Conference on Mobile Communication and Social Change.*

Platow, M. J., Haslam, S. A., Both, A., Chew, I., Cuddon, M., Goharpey, N., et al. (2005). "It's not funny if they're laughing": Self-categorization, social influence, and responses to canned laughter. *Journal of Experimental Social Psychology, 41*(5), 542-550. doi: 10.1016/j.jesp.2004.09.005.

Powazek, D. (2001). *Design for Community: The Art of Connecting Real People in Virtual Places* (p. 307). New Riders Publishing. Retrieved May 28, 2008, from http://portal.acm.org/citation.cfm?id=516336.

Shamma, D. A., Bastea-Forte, M., Joubert, N., & Liu, Y. (2008). Enhancing online personal connections through the synchronized sharing of online video. In *CHI '08 extended abstracts on Human factors in computing systems* (pp. 2931-2936). Florence, Italy: ACM. doi: 10.1145/1358628.1358786.

Syeda-Mahmood, T., & Ponceleon, D. (2001). Learning video browsing behavior and its application in the generation of video previews. In *Proceedings of the ninth ACM international conference on Multimedia* (pp. 119-128). Ottawa, Canada: ACM. doi: 10.1145/500141.500161.

Wasserman, S., & Faust, K. (1994). *Social Network Analysis: Methods and Applications* (p. 825). Cambridge UniversityPress.

Watts, D. J. (2003). *Six Degrees: The Science of a Connected Age* (p. 368). W. W. Norton & Company.

Weisz, J. D., Kiesler, S., Zhang, H., Ren, Y., Kraut, R. E., & Konstan, J. A. (2007). Watching together: integrating text chat with video. In *Proceedings of the SIGCHI conference on Human factors in computing systems* (pp. 877-886). San Jose, California, USA: ACM. doi: 10.1145/1240624.1240756.

Welford, A. T. (1967). Single-channel operation in the brain. *Acta psychologica, 27,* (pp. 5-22). doi: 4863177.

Chapter XVII
Online Video as a Social Activity

Justin D. Weisz
Carnegie Mellon University, USA

ABSTRACT

Online video is one of the Internet's most popular services. In addition to entertainment, it can provide a social experience. This chapter describes several design decisions related to incorporating a live chat feature alongside online video, and how these decisions could influence the entertainment and social dimensions of the viewing experience. Two laboratory studies explore issues of distraction, entertainment and sociability when integrating live chat with online video. Surprisingly, these studies show that despite being distracting, chat with video can be enjoyable and sociable. Researchers and practitioners need to explore the generality of this finding both for different genres of content and interaction media and for small and large viewing audiences. The chapter concludes by looking at new models for online video and chat and the implications of those models on sociability.

DESIGNING SOCIAL ONLINE VIDEO EXPERIENCES

The Internet is a highly engaging medium. People from all over the world use the Internet not only to have fun but also to get information about everything from dog training to football scores to health issues. But the Internet is not only an information medium – it also enables people to socialize, both in real time and asynchronously. These social interactions take place in thousands of online communities – networked spaces where people having a shared interest or relationship gather together. In this chapter, I demonstrate that online video is not just about being entertained or keeping informed; rather, by combining video with interactive communication, such as text chat, online video can be a social experience as well. I also discuss several technologies used to power online video and their implications for sociability.

Why Make Online Video Social?

Currently, most video content is watched on the television. For those who watch with others, television can be a companionable activity, as when a group of friends get together to watch the latest episode of Lost, a Netflix movie, or the Super Bowl. Television content often facilitates conversations, both in the real world and online (Baym, 1999; Brown & Barkhuus, 2006). However, watching television alone is an isolating experience (Putnam, 1995). Watching alone is usually done at home and precludes getting out and socializing with people. Watching television alone may also induce passivity and discourage people from thinking critically about important issues (Postman, 1986). Online video, which reaches far beyond the living room, could overcome this social isolation and stimulate conversation among those who are watching. Indeed, interaction drives the liking of others (cf. Cartwright & Zander, 1953; McKenna et al., 2002; Utz, 2003) and synchronous interaction is associated with increased levels of intimacy (Powazek, 2002). Potentially, communities of viewers could engage and interact while watching videos online. Oumard et al. (2008) conclude that heavy Internet video users are already integrating online video as part of their daily television watching, and are watching, discussing and sharing video content with their friends. Providing viewers with a chat feature may turn their experience into one that is not only entertaining but also socially engaging.

However, this optimistic vision of viewers socializing while watching videos online runs into a fundamental problem: watching and chatting are two separate activities. Watching a video and socializing both require attention and mental processing, which are limited cognitive resources. Thus, watching a video and chatting with others can be distracting and unpleasant; this is one reason why movie theatres remind people to turn off their cell phones and not talk during the show. When your friend does talk during the movie, it is difficult to follow both the on-screen dialogue and what your friend is saying. In this case, you may miss part of the video (by attending to your friend) or what your friend is saying (by ignoring him or her). Either way, this is likely to cause a decrement in the entertainment value of the movie (if you missed part of it) or in the social experience of watching with others (if you ignored your friend). Keeping this element of distractibility from spoiling the experience is a major design challenge for combining chat with online video.

This chapter asks whether online video with chat can provide a context for positive social interaction, or whether the attentional demands of watching a video and simultaneously chatting with others presents an insurmountable barrier to a satisfactory experience. I present results from two laboratory studies conducted at Carnegie Mellon University. The first experiment establishes that watching online video with chat is distracting, and measures how distracting it is. The second shows that watching and chatting are compatible activities (for the short, entertainment-focused videos we used) and this combination may be useful for building and promoting online video communities. This research is a start at informing the design of online video sites that wish to provide both entertainment and social value to their viewers.

DESIGN CONSIDERATIONS

Online video sites and systems currently offer several different features and types of experiences. Some sites offer videos of a particular genre or theme; others contain videos from many genres. Some allow only short video clips; others offer movie-length videos and documentaries. Some encourage community among viewers; others do not. These different features constitute design decisions that can affect the entertainment and social value of online video.

Content

Many different types of videos are offered online. Video genres include cartoons, music videos, political speeches, "reality TV" and many others. Videos also vary in running time, and some sites place explicit limits on the length of content. For example, YouTube (http://www.youtube.com/) limits videos to about 10 minutes in length. Google Video (http://video.google.com/) and Veoh (http://www.veoh.com/) allow videos over an hour long.

The purpose of the site and its video content also varies. Some sites host videos created purely for entertainment or humor, such as videos typically hosted on CollegeHumor.com or Break.com. Other sites create or encourage videos for informational, educational or political purposes, such as the 2008 CNN/YouTube debate partnership, talks on TED.com, and the video lectures of MIT's OpenCourseWare.

Video sites also vary in the production quality of the videos they present. Some sites focus on professionally produced videos where scripts are pre-written and rehearsed before recording, or where professional equipment and lighting is used during production. Examples of these sites include online TV shows like The Scene (http://www.welcometothescene.com/), PurePwnage (http://www.purepwnage.com/) and Red vs. Blue (http://www.redvsblue.com/), as well as the Rocketboom news podcast (http://www.rocketboom.com/). Other sites present videos taken on the spur of the moment, which are often recorded with low quality equipment such as a web cam or cell phone.

These differences in video content are important to understand because they serve to define the type of community an online video site fosters. Video content affects the types of viewers the site attracts as well as the interactions that take place among those viewers. For example, a site that presents political and newsworthy content tends to attract an audience interested in politics, and will promote political conversation. A site that specializes in short, amateurish and humorous video clips may promote a myriad of conversational topics, including nonsensical or humorous remarks and banter. In the research studies presented later in this chapter, I utilize brief entertainment content similar to that found on YouTube, because entertainment videos comprise the most popular segment of the online video space, and because social experiences often take place in this kind of entertainment context.

Delivery

Online video sites employ two models of video delivery. The most common is the *download model*, whereby viewers download videos and/or watch them as they download. This is the model currently used by sites like YouTube and Yahoo Video (http://video.yahoo.com/). This model gives viewers the flexibility of watching a video at some future time and of watching on a portable device such as a video iPod.

The *streaming model* is familiar from classical (DVR-less) television. In this model, a video is streamed live as the viewer watches it: tune in the middle of the stream and you've missed the beginning. Sites such as Justin.TV and UStream.TV currently use this model.

Each of these models has implications for the entertainment and social value of an online video site. The download model allows for extended, asynchronous conversations. Viewers can post comments and get responses on any video they have watched. They can also chat in real time about the videos. The download model also gives viewers greater flexibility in selecting content to watch. Since viewers are not constrained by having to watch the same video content as everyone else, they are free to explore a wider range of content, picking and choosing what to watch at their leisure. However, this model has implications for sociability. Since viewers do not watch the same content at the same time, synchronous chat may be less effective at providing a social experience for viewers, because they have less common ground than their streaming-video peers. For example, when watching a soccer match, a viewer who sees his team score a goal might want to cheer and celebrate, but those cheers may fall on confused ears if no one else is watching the same soccer match at the same time.

Compared to the download model, the streaming model is constrained by its synchrony. All viewers see that stream near-simultaneously in time (although network delays may cause small asynchronies in playback). The most natural choice for integrating social interaction into this model is synchronous chat. This allows, for example, viewers of a soccer match to share their reactions and excitement as a player scores a goal the moment it happens. Sharing these moments is one important reason why people enjoy watching sporting events live, because events are revealed to the viewers together and they can share their reactions (Vosgerau et al., 2006). An asynchronous chat medium would fail to capture such an experience.

Another aspect of video delivery relates to the technology used to deliver video data. The choice of technology has implications for the size of the viewing audience. Three of the common methods used to deliver video data are shown in Table 1. Each of these methods can be used to implement either a download or a streaming model.

The client-server model is the simplest and least expensive to implement. It works by having a server (or group of servers) deliver video data to each viewer. In this model, the size of the audience is limited by the amount of available bandwidth on each server. Because of this, the client-server model only supports a few hundred to a few thousand simultaneous users.

Content distribution networks (CDNs) can be used to reach the next order of magnitude of users. CDNs are essentially multiple instances of the client-server model, but with one key optimization. Videos (or video streams) are replicated across many different servers located in different geographic regions, and video data are delivered to viewers from a proximate server. This allows the CDN model to scale to hundreds of thousands or millions of viewers. However, the costs to create and maintain such an infrastructure are high, and thus so is the cost of delivering video to a large audience. These costs are typically prohibitive to an individual user seeking to reach a large audience.

An alternative to using a commercial service is to use a free video hosting service such as YouTube or Yahoo Video. These sites allow members to upload videos for free and make them available for watching by millions of viewers. Underneath the hood, these services are also CDNs, in that video files are replicated across many servers and delivery is optimized by geography. However,

Table 1. Methods for online video delivery

Distribution Method	Description	Tradeoffs	Implications
Client-server	One or more servers send video data	Inexpensive; easy to implement; does not scale well	Smaller audiences; few opportunities to meet new people
Content distribution network (CDN)	Multiple servers used to send data; delivery optimized by geography	Expensive; difficult to implement; scales well	Large audiences; many opportunities to meet new people
Peer-to-peer	Viewers distribute video data to each other	Inexpensive; difficult to implement; scales well	Large audiences; many opportunities to meet new people

this approach still requires a significant investment in bandwidth (on YouTube or Yahoo's part). Moreover, as of this writing, they do not currently support streaming videos to a large audience. Thus, while this solution does allow millions of users to download a video, it does not yet allow them to watch simultaneously. Further, these services are experiencing a pressing need to generate revenue in order to remain financially solvent. Their hope is to be supported by serving advertisements alongside their videos, but this may not be acceptable to some users.

The peer-to-peer (P2P) model overcomes the limitations of the client-server and CDN models by distributing the work of delivering video data across all viewers. Peer-to-peer models support both downloading and live streaming for video. Many different P2P protocols and systems exist, including BitTorrent (http://www.bittorrent.com/), Joost (http://www.joost.com/), End System Multicast (http://esm.cs.cmu.edu/), Sopcast (http://www.sopcast.com/) and TVU (http://www.tvunetworks.com/). These technologies are just beginning to enter into mainstream use (cf. Grigonis, 2008) and they have the potential to revolutionize the distribution of online video by allowing individual users to broadcast video to audiences of thousands or millions. As peer-to-peer technologies become more popular, the potential for online video events with millions of users becomes more realistic. It is important to begin exploring the implications of an audience of this size and the challenges it presents to interaction among viewers.

Viewing Device

The phrase "online video" has strong connotations that the video is, in fact, being viewed on a computer connected to the Internet. This connotation is somewhat misleading, as there are many different options for viewing "online video". Although this chapter focuses on watching video on a computer, it is important to understand that online video is blurring the distinction between the computer, the television, and the real world.

Research in interactive and social television has focused on creating systems that integrate social interaction during the television-watching experience. AmigoTV combines voice chat with an overlay display on the TV showing avatars of other viewers, and allows users to see what their friends are watching (Coppens et al., 2004). Media Center Buddies allows viewers to send instant messages and chat with one another while watching television (Regan & Todd., 2004). Telebuddies encourages interaction by allowing content producers to add quiz games and trivia contests to their broadcast (Luyten et al., 2006). Reality IM combines social interaction with additional contextual information about a TV program, such as advertisements and sports statistics (Chuah, 2003). Social TV 2 uses an ambient display to inform users of when their friends are watching television, and it allows them to chat with each other when they are watching at the same time (Harobe et al., 2008).

This idea of interacting around television content is spreading to online communities as well. Fan-based and officially sponsored communities exist for popular TV shows and movies. Some of the major broadcasting companies in the US, including NBC, ABC, and CBS, have dedicated discussion boards online for their television programs. Fan-run sites like Cinema Blend (http://www.cinemablend.com/) and First Showing (http://www.firstshowing.net/) have discussion boards and blog commentary for the latest movies. As these examples show, the line between the television and the computer has become increasingly blurred. Further, some communities have created a new viewing experience by offering live chat as a television show is aired, augmenting the television with a computer to transform a solitary viewing experience into a socially engaging one. One example of this is the Lostpedia community (http://www.lostpedia.com/), which hosts live chats for fans of the show Lost as it airs.

Mobile devices are also becoming a popular way for watching "online video". These devices, which include video-capable cell phones and iPods, allow viewers to watch video whenever they want (time-shifting) and wherever they are (space-shifting). Thus, they allow people to watch videos on the go, away from their computers. Cell phone providers also offer streaming video services to video-capable cell phones using high-speed cellular networks. The iPhone and iPod Touch allow users to watch videos directly from YouTube on their devices wherever they are, through either wireless or cellular networks. The proliferation of video-capable mobile devices will likely expand the number of design choices to be made when creating an entertaining and social video experience.

Model for Interaction

One way to understand the interaction space of online video is through the time-space taxonomy heretofore applied to groupware systems (Ellis et al., 1991). Table 2 describes four types of interactions among viewers that online video is capable of supporting, and gives an example of each.

In this time-space framework, viewers watching at the same time can have synchronous live chat or they can communicate asynchronously whether or not they are in the same place. Those watching at different times can communicate only asynchronously.

Sites that use a download model for video delivery typically have a place where members can post messages or comments in an asynchronous fashion. However, it is also possible to add synchronous communication to a download model. For example, in YouTube Streams (http://www.youtube.com/streams_main), each "stream" is a chat room with its own video library. Viewers can build a playlist of videos they want to watch from this library. They can also chat with each other in real time while watching videos from their own playlist. This model blends synchronous interaction with unsynchronized content.

Online video sites can also differ in the type of media they use for interaction. Although many sites implement text-based interactions, both for live chat and posting messages or comments, some sites have been experimenting with alternative media. For example, Gaia Online (http://www.gaiaonline.com/) allows its members to watch videos in a 2D graphical chat room. Viewers are represented as graphical avatars and chat is displayed using chat bubbles. Second Life (http://www.secondlife.com/), a 3D virtual environment, allows its members to embed videos in virtual televisions and movie screens. Chat in Second Life is also text-based, and uses both chat bubbles as well as a text box to display chat. Other chat media are also possible, such as audio or video chat while watching. These examples illustrate the wide variety of possibilities in designing interactive experiences for viewers.

In the next section, I describe two laboratory studies I have conducted that systematically evaluate the effects of a real-time text chat feature on the experience of watching online video. Asynchronous communication related to video

Table 2. Time-space taxonomy applied to online video interactions

	Same time	Different times
Same place	Face to face interaction; e.g. a group watching tennis on a mobile device or computer together	Asynchronous interaction; e.g. recording a video for someone else or leaving comments on a public video display
Different places	Synchronous interaction; e.g. dispersed friends watch a tennis match together and chat while they watch	Asynchronous interaction; e.g. dispersed friends watch a YouTube video and post comments to each other

has long been studied in the computer-mediated communication research community (cf. Baym, 1999). Synchronous chat while watching is much newer and has become popular due to the maturation and adoption of technologies that allow videos to be rapidly transmitted and viewed online. These studies focus on two short-term aspects of the experience: the social impact of simultaneously watching and chatting and the distraction present between these two activities.

EMPIRICAL STUDIES OF DISTRACTION AND SOCIABILITY

I have conducted two laboratory studies to explore the issues of distraction and sociability in online video with live chat. The first study deals with distraction, and it considers the question of how distracting it is to watch a video and simultaneously read the chat of other people. The second study deals with sociability, and it considers the question of whether online video with chat promotes social relationships between people. These studies demonstrate that watching together is an enjoyable activity and that while it is distracting to watch a video and chat with other people, this distraction does not seem to be detrimental to the overall experience.

Distraction Study

The distraction of chat while watching a video can be understood through multiple resource theory (Wickens, 2002). People have a very difficult time paying attention to two simultaneous visual or auditory sources of information. However, people are able to attend to one visual and one auditory source simultaneously – this is why we can watch a movie and follow what the actors are saying. Adding a text chat feature while watching a video would result in having to pay attention to two visual sources of information, and multiple resource theory would predict that this would be distracting. However, the extent of this distraction is less clear. Would people be able to manage their attention between chat and a video, or is the video (or chat) so immersive that they ignore the chat (or the video) altogether?

To answer this question, we considered the case of watching videos and reading chat from other people. Participants were brought into the laboratory and watched a series of eight short videos in random order. After each video, they were asked two questions about what they had watched to see how much they remembered. The videos were taken from YouTube and included cartoons, short segments from movies and television, stop-motion animations and amateur productions.

We compared two chat conditions in this study. Participants without chat watched only the videos. This provided a baseline measure for how much participants would remember from the videos. Participants with chat had a text box on their screen below the video, where chat was displayed from a pre-recorded chat transcript. To display synchronized video and chat, as well as ask questions about each video immediately after it was shown, we implemented a custom application in Adobe Flash. Screenshots of our application with and without chat are shown in Figure 1.

To control for the amount of chat seen by each participant, we created a pre-recorded chat transcript for these videos by having a group of friends chat with each other while watching them. This allowed us to make quantitative comparisons between participants with and without chat, and ensure that any statistical differences between the conditions were not caused by variations in how much chat each participant saw. This also allowed us to ask questions about the content of the chat to measure the distractive effects of the video on reading the chat. Since the videos were shown in a random order to each participant, chat in the transcript was keyed to the video that was playing when that chat was originally spoken. In other words, the chat that originally occurred during the Ali G video was always shown during

Figure 1. Screenshots from the distraction study. (a) No chat condition. (b) Chat condition

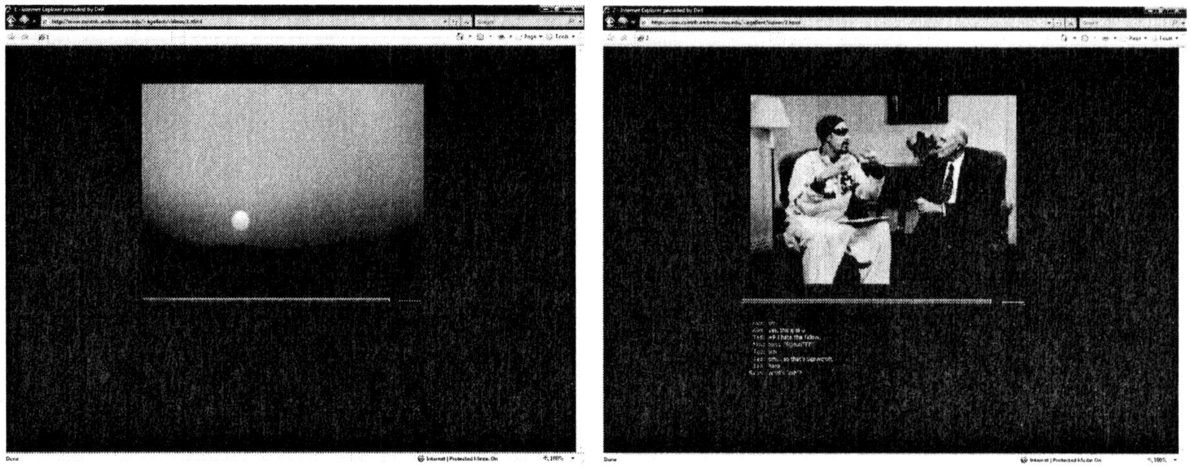

(a) (b)

the Ali G video. This was done to ensure that any references made in the chat to a video were correctly aligned to that video.

We recruited 28 people for our study, of whom 11 were female and 17 were male. The average age of our participants was about 25 years. Participants were split evenly between the two chat conditions.

Our primary measure in this study was memory, which objectively quantified how much attention participants paid to the video and the chat. Four questions about each video were asked: two about the video and two about the chat. As there were eight videos shown, the maximum attainable video memory and chat memory scores were 16, each. Participants were asked the video and chat memory questions immediately after watching the video to avoid difficulties in recall. Participants without chat did not have a chat memory score.

The average video memory scores for participants with and without chat are shown in Figure 2. Participants without chat got about 13 questions correct, and participants with chat got about 11 questions correct. No one in this study got a perfect video memory score. The difference between participants without chat and participants with chat is statistically significant, demonstrating that participants without chat remembered more about the videos than participants with chat.

Participants with chat were tested for their memory of the chat, and they got about 10 of 16 questions correct. As with the questions about the videos, no one with chat got a perfect chat memory score. Interestingly, there was a weak (but statistically insignificant) correlation between the video and chat memory scores for participants with chat ($r = .27$). Since the correlation is positive, it suggests that those participants who recalled more about the videos also recalled more about the chat. This runs counter to the expectation that paying attention to the video results in not paying attention to the chat. Rather, participants who paid closer attention to the videos may also have paid closer attention to the chat. Or, because the chat included several lines of history, participants may have used "down times" during the videos to catch up in reading the chat.

Discussion of the Distraction Study

This study shows that chatting while watching is distracting, and that this distraction does cause people to miss parts of the video when they attend

Figure 2. Video memory scores for participants without chat (N = 14) and participants with chat (N = 14).

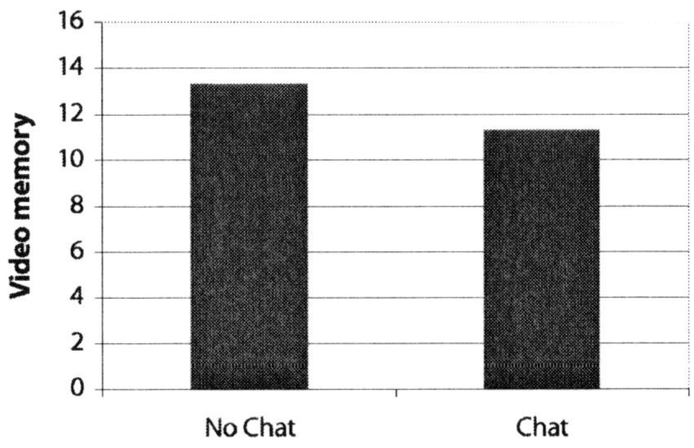

to the chat. This distraction caused participants with chat to score about two points lower on the video memory test than participants without chat. However, it remains unclear if participants felt overwhelmed from this distraction, or if they felt they were able to manage their attention between the video and the chat. Further, as this experiment only measured the distraction from reading chat, we can only speculate that distraction would increase when people are invested in the chat, and able to compose chat responses to each other.

Interestingly, there was a positive (but statistically insignificant) correlation between the video memory and chat memory scores. With a negative correlation, we might conclude that attention to the video shifted attention away from the chat, or that attention to the chat shifted attention away from the video. However, with a positive correlation, it seems that participants' attention is focused on the entire experience – both chat and video – and the more attention they put into the experience, the more they remembered from it. Or, perhaps some people are better at remembering what they see, hear and read than others.

Sociability Study

In the distraction study, we saw that reading chat while watching is distracting: it causes people to miss things that happen in the video. In the sociability study, we consider what happens when we allow people to chat with one another. Our first question is about whether or not people will chat with each other: is chat while watching so distracting that people simply ignore the chat? Or are they able to multitask between chatting and watching? If people do chat with one another, what do they chat about? Intuitively, the videos would provide a continuous source for conversational topics. However, if people simply remain on topic in their chats because the wellspring of topics suggested by the videos is too strong, then the likelihood of building social relationships through online video is diminished. In online communities, bonds between people typically form as they veer to off-topic subjects in their communications (Sassenberg, 2002). Therefore, the first questions we answer in this study are about whether or not people chat, what they chat about, if it is fun to chat while watching, and if the presence of the video is so strong that it discourages people from talking about alternative topics.

Watching together is a social experience, and one of the goals of this study is to quantifiably measure the effect of chatting on sociability. We examine how chat affects the sociability of the viewing experience by measuring two psychological constructs: people's liking of one another and their feelings of closeness to one another. We make this examination by comparing groups of people using a text chat feature while watching videos to groups of people who watch videos without chat. We also consider both groups of friends who already know one another (and presumably like each other), as well as groups of strangers. This is to mirror the social dynamics present in many online communities: some community members know each other already, and some are newcomers.

We recruited 30 small groups of participants for this study, evenly split between groups of friends and groups of strangers. Each group had between 2 to 4 people in it. We had 85 total participants. The average age was about 24 years; approximately half of our participants were female. The majority of our participants were students (75%), although other populations were represented as well, including faculty, staff, alumni and retirees.

Groups of participants were brought into the laboratory and watched a series of seven short videos on separate computers, shown in random order. These videos were cartoons on themes of education, humor, love, character struggles, and music. None of our participants had previously seen these cartoons before this study.

We examined the streaming model in this study since it is the most natural fit for synchronous interaction. To ensure that everyone watched the same content at the same time, we used the Windows Media encoder and player software to stream video from a server on the same LAN as the experimental machines. We also used the mIRC IRC client (http://www.mirc.com/) for text chat, and collected chat logs using an open-source IRC chat server, instrumented to log the timestamp, sender and contents of each chat message. A picture of the typical arrangement of a participant's screen is shown in Figure 3. One consequence of this setup is that participants were able to rearrange the window positions and sizes to their liking, and about one-third of participants did so. Common alternative arrangements included positioning the chat and the video side-by-side, or making the video full screen and placing the chat on top of the video.

Twenty groups were given the text chat feature, and 10 did not have chat available to them. We told the participants with chat that they could chat at any time during the study, and about any topic they liked. Participants were only told of the availability of the chat feature, not that it was a mandatory requirement of their participation.

To evaluate the experience of chatting while watching, we used several different measures in this study. The first was ratings of the cartoons. To avoid difficulties in recall, participants rated each cartoon immediately after they had finished watching it. Participants also completed a survey at the end that asked about their enjoyment of the chat, distraction, and feelings towards other people in their group.

Figure 3. Screenshot from the sociability study. Participants were allowed to move and resize the windows to their individual preference.

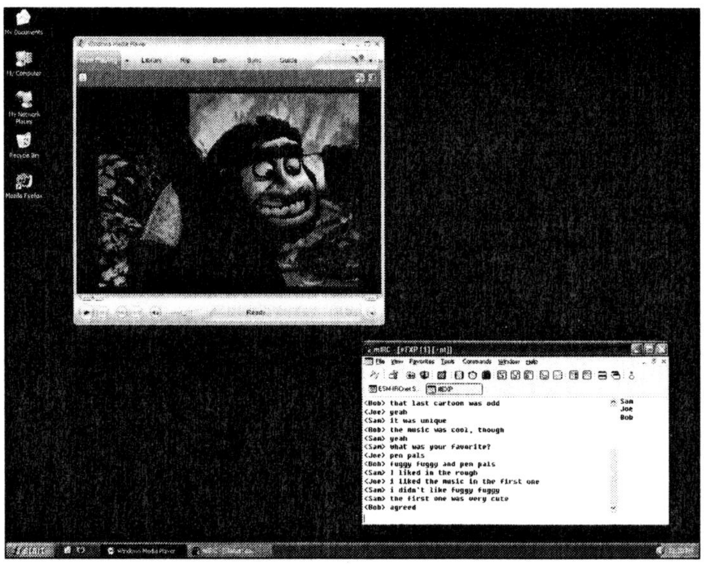

Results from the Sociability Study

In this chapter, I only present the results from this study pertinent to issues of distraction, enjoyment and sociability. Additional detail from the results of this study can be found in Weisz et al. (2007).

Chat Amount

On average, groups with chat produced an average of 233 lines of chat. Not surprisingly, groups of friends chatted more than groups of strangers (friends average: 337 lines; strangers average: 129 lines).

Chat Topics

For both friend and stranger groups, the cartoons were a major topic of conversation. Participants reported that their favorite chat topics included "the cartoons themselves" (participant P16), "the music and the quality of the drawings" (P22), "the rating" (P23), "how good each cartoon was" (P27) and "[the] artistry of videos" (P58).

Participants also made jokes and talked about their lives. For example,

"We discussed some stuff about our professors by comparing them to the characters. One was related to [two] professors who are a couple and that was hilarious." (P1)

"I liked chatting with my friends about our inside jokes. It may appear that we don't like each other, but there is so much love between the three of us that it is hard for a stranger to imagine." (P4)

Stranger groups were able to find common ground with each other beyond the videos, and some of their favorite topics included "information about graduate school" (P13), "our career plans" (P14), "smoothies at Lulu's" (P53, about a local restaurant) and "sports" (P23).

Enjoyment

Participants had fun in our study, and the average reported enjoyment was 4.0 on a 5-point scale. Enjoyment was significantly correlated with participants' average cartoon ratings ($r = .47$). There were no statistically significant differences in enjoyment between groups of friends and groups of strangers, or between groups without chat and groups with chat. There was also no correlation between the amount an individual chatted and their enjoyment.

Sociability

Our two measures of sociability in this study were how much participants reported liking one another, and how close they felt to each other. These were measured on 5-point and 7-point scales, respectively. Figure 4a shows the average amount of liking for groups of friends and strangers, and for groups with and without chat. All comparisons between groups are statistically significant: groups with chat liked each other more than groups without chat; friend groups liked each other more with chat; and stranger groups liked each other more with chat. Figure 4b shows the average feelings of closeness between groups of friends and groups of strangers, and between groups with chat and without chat. The same effects are seen here, as well: groups with chat felt closer than groups without chat; groups of friends felt closer with chat; and groups of strangers felt closer with chat.

Distraction

To measure distraction in this study, we asked participants how distracted they felt from using the chat feature. Participants reported an average distraction of 3.5 out of 7, suggesting that while chatting while watching was distracting, it was not overwhelmingly so. Distraction did not correlate with enjoyment of the experience, but it did significantly correlate with the amount an individual chatted (r = .38); thus, participants who spoke more in the chat felt more distracted by it.

Discussion of the Sociability Study

This study demonstrates that the tradeoff between chat and distraction may be manageable. Participants did report feeling distracted by the chat feature in this study, but there was no evidence that this distraction lead to a poor experience or that it dampened the breadth of social interactions that took place. Participants chatted with each other about a variety of topics, including the cartoons they watched and their own personal lives. This is encouraging, as one metric for when members are bonding in an online community is when they conduct off-topic conversations not related to the theme of the community (Sassenberg, 2002). Further, this study showed statistically significant increases in sociability – as measured by feelings of liking and closeness – when groups chatted with each other while watching. Therefore, these results support the argument for including a chat feature alongside online in order to promote social interactions among video viewers.

DESIGN DECISIONS FOR ONLINE VIDEO

Watching videos online has only recently become pervasive because of the advances made in broadband Internet, content delivery and video playback technologies. Because this is a comparatively new Internet application, not much is known about how to design interactive and social television experiences for the online space. Many online video sites are experimenting with different ways of presenting online video, and each of these sites offers a unique experience. While the current trend in online video is to provide an entertaining experience for the individual viewer, I believe that the future of online video lies in the shared creation of a social viewing experience. In this world, watching a video becomes a secondary or supplementary activity to a larger social interaction. Viewers are active participants and contributors, interacting with each other by creating, remixing, sharing, commenting, tagging, rating and discussing the videos they watch. Videos become a "social glue" that brings people together and provides them a context for social interaction.

The research presented in this chapter is the first step towards understanding how the design

Figure 4. (a) Effect of chat on liking. (b) Effect of chat on feelings of closeness.

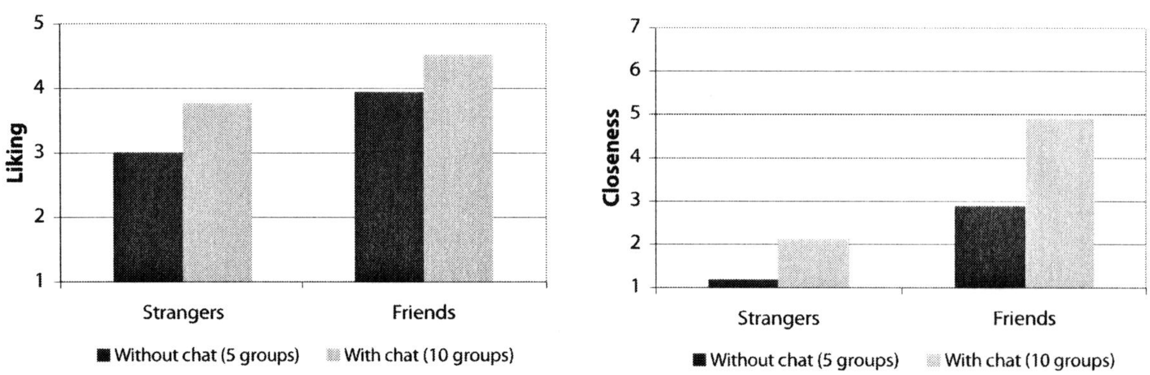

(a) (b)

decisions made in creating online video experiences enables such a future. The studies presented in this chapter focus on the small-group experience of watching online video and show that even though chat is distracting, it is an activity that people enjoy and can help people maintain and extend their social networks. However, our studies focused on only one type of online video experience – real-time text chat while watching short, entertaining videos in a streaming playback model. Researchers in interactive and social TV should consider alternative online video designs.

Adapting Chat to Different Video Content

There are many different genres of video content. Videos can be entertaining, educational, or informative. They can be about political or historical topics, news or current events, or the weather. They can also be highly personal, such as videos of a wedding or Bar Mitzvah, or highly public, such as videos of the presidential debates in 2008. In our studies, we have focused on the short, entertaining videos typically found on YouTube or Yahoo Video. However, the experience of watching videos from other genres may be significantly different. For example, in a political debate, much of the content comes from listening to the candidates speak. This abundance of verbal content may require more of one's attention to process than watching a cartoon of a caveman fighting a bear. Thus, chatting while watching a political debate may be more distracting than chatting while watching a cartoon. In this situation, an online video site may not want to provide a live chat feature during the video broadcast. Instead, they may want to hold a live chat for questions at the end, or they may want to engage their audience in a discussion on an online forum.

Another example is in education. Gibbons et al. (1977) examined the use of videotaped lectures on learning. They developed a method called Tutored Videotape Instruction (TVI) where small groups of students watch videotaped lectures together in a group. One key finding of their study is that students learned best when the lectures were paused so they could discuss the material. These pauses tended to be quite frequent and fairly short – pausing every 5-10 minutes for 3-5 minutes of discussion. Today, it is possible to apply the TVI method to distance learning online by designing a system that allows participants (or a moderator) to control video playback. This could be achieved by defining pre-arranged "intermission" periods during a video or by giving viewers a "shared remote" that lets them pause video at their own leisure. This feature would enable viewers to create opportunities for discussion without having to worry about missing video content. Indeed, such a system has already been built for distance education (Cadiz et al., 2000). However, the concepts of a "shared remote" or of inserting intermission periods during a video may be applicable to other video genres as well. Television content is already edited to have placeholders for commercials, sporting events tend to have natural breaks between plays, and musical concerts and stage productions typically have periods for intermissions.

Multimedia Chat

Media richness is the notion that communications media differ in their ability to convey rich information (cf. Daft & Lengel, 1986; Dennis & Kinney, 1998). Media that are rich, such as face-to-face conversations or video chats allow one to convey both linguistic and non-verbal cues, such as raising one's voice at the end of a sentence to indicate a question, or yawning when one is bored. Richer media also provide a greater sense of immediacy and intimacy. People expressing themselves through media low in richness, such as text chat, must resort to other means to convey these same cues, such as using emoticons. Table 3 gives a few examples of common chat media and their relative richness to each other.

Since richer media provide a heightened sense of awareness and presence and are more intimate and emotionally conductive, they may be beneficial in the design of online video experiences. One fun application of online video is in re-creating the experience of watching television on the couch with friends. With audio or video chat, people would be able to shout and scream and gesture their excitement (or disappointment) when their favorite football team scores a touchdown. This same excitement cannot be captured as efficiently or as accurately with text chat.

However, richer media may be additionally distracting. For example, one argument against using audio chat stems from research in listening and attention. Moray (1969) studied auditory attention, and found that in a dichotic listening task, where people listen to two audio sources simultaneously, people can only pay attention to only one audio channel and ignore the other. Because of this, they have little memory of what they heard in the channel they ignored. In the case of online video where viewers talk out loud to each other, this corresponds to listening to the video's audio or the speech of other viewers, and ignoring the other. This same argument applies to video as well; with two visual sources of information, one cannot process both simultaneously (Wickens, 2002).

In this chapter, I have only considered the case of either not being able to chat or having a text chat feature. In another study I examined the use of audio chat while watching videos and found that it is just as enjoyable to use as text chat, and it is not additionally distracting (Weisz & Kiesler, 2008). Thus, richer media may be quite successful when used with online video, and systems that provide richer interactions while watching are currently being developed. For example, Zync is a web-based tool that allows any number of people to watch synchronized video with each other while simultaneously having a video chat (Shamma et al., 2008). Researchers will continue to examine the impact of richer media – both audio and video – on the experience of watching online video, and specifically on how they affect distraction and sociability.

Mobile Video

Mobile devices capable of playing video are becoming extremely popular. One open research question with regard to mobile devices is about how to design a social mobile video experience. A social experience requires interaction, but the constraints of the device's size make this difficult. With only a small screen, integrating text chat together with video may be difficult. Further, audio chat may require a headset if using the speaker and microphone requires holding the device up to your ear (as with a cell phone). Researchers in the mobile video space have explored different interfaces and requirements for integrating chat and video on mobile devices (cf. Schatz & Egger, 2008; Knoche & McCarthy, 2005). Despite the constraints of the mobile platform, mobile devices open up new possibilities for online video. As most of these devices are equipped with cameras, the killer application for mobile video may simply involve recording and sharing videos. For example, dispersed friends or family members can send videos of themselves or their surroundings to keep connected. Or, they could send text messages to each other while watching a television show. Researchers will continue to design novel social experiences for mobile video,

Table 3. Media richness scale for online chat media.

Least rich			**Most rich**
No chat	Text chat	Audio chat	Video chat

as well as to examine whether these experiences result in an increase in feelings of connectedness and togetherness.

Supporting Large Audiences

One key benefit to technologies that allow for the widespread distribution of video content online is that they allow large audiences of online viewers to watch together. Unlike television, which generally restricts one's interactions to those in present company, online video allows viewers to communicate with any number of other viewers, all around the world. Further, high profile events that attract millions of television viewers, such as the CNN/YouTube debates and the State of the Union address in 2008, are now being broadcasted online. If online video is truly displacing television, then online video site designers will be faced with an interesting challenge as more viewers opt to watch these events online. How can we provide an engaging social experience to an audience of millions? Simply adding a live chat feature that allows people to communicate may result in an overwhelming, impersonal and unenjoyable experience. With millions of people chatting all at once, the immense amount of chat may be highly distracting while trying to watch a video. Further, people may be less inclined to chat with each other if they feel that no one will recognize them in the chat or see the things that they say.

The goal of this line of research is to explore methods and techniques for segmenting an audience of millions of viewers into smaller chat groups. This would allow online video site designers to provide the small-group experience that worked well in our studies to a much larger audience. One way to do this is to provide recommendations for chat groups for people to join. These recommendations can be based on a number of different criteria. For example, an individual may want to chat with people based on their geography (i.e. chat with people in the same country vs. a different country), their social network (i.e. chat with you friends, friends of your friends, or strangers), or about a particular topic. One open research question in this area is determining which criteria people care about when choosing a group to join, and whether recommendations based on that criteria result in an enjoyable and sociable experience.

Once an audience of millions has been segmented into many smaller chat groups, people will lose awareness of the entire audience. An audience of millions generates more chat than an individual could process. However, in order to feel part of something much larger than themselves, people should have a sense of how many other people are watching with them and what they are talking about. To do this, visualizations ought to be designed to summarize these chat groups and their activity. These visualizations would serve a dual purpose, providing awareness of the larger audience as a whole and allowing users to navigate among and select from the available chats. A visualization of chat groups could also be used in conjunction with a chat group recommender system. For example, an individual may request a recommendation for a chat group. The system can present him or her with several candidate groups, along with a visual summary of the popular topics of chat in those groups. This would allow users to make informed decisions as to which group to join. However, adding another visual component may increase distraction because viewers are also trying to watch a video and interact with others. An evaluation of these visualizations would consider their ability to keep users informed without significantly increasing distraction.

CONCLUSION

Online video has the potential to revolutionize how we experience video content. By incorporating social interaction, online video can provide viewers with opportunities for the collaborative construction of new entertainment, educational and social experiences. One argument against combining the social experience of chatting with the visual experience of watching video is that each activity requires one's attention and it would be distracting to chat while watching. This distraction may be overwhelming to the viewer, resulting in a poor experience.

The first study presented in this chapter explored this issue of distraction and found that when people watched a series of videos while reading chat, they recalled less about the videos than people who did not read the chat. In the second study, we explored the social benefits of chatting while watching. We found that the addition of a live chat feature increased feelings of mutual liking and closeness, both for groups of friends and for groups of strangers. Further, we saw that the videos did provide a source of conversational topics and strangers were able to use these to establish common ground. However, the videos were not the sole topic of conversation; people spoke about other topics as well.

These results are encouraging for online video sites that wish to incorporate a chat feature, as they show that even though chatting while watching is distracting, people are able to manage their attention between the two activities. Further, people do feel a connection with each other when they chat. Looking ahead to the future, I believe the vision of the "couch potato" – someone who sits passively on the couch watching videos – will fade away and be replaced with that of the "active viewer" – someone who engages with others in debate, dialogue, chatter and banter as they watch, transforming the act of watching into an active, engaging and social experience.

ACKNOWLEDGMENT

Many thanks to Sara Kiesler for her valuable feedback. This material is based upon work supported by the National Science Foundation under Grant No. 0435382. Any opinions, findings and conclusions or recommendations expressed in this material are those of the author and do not necessarily reflect the views of the National Science Foundation (NSF).

REFERENCES

Baym, N. (1999). *Tune in, log on: Soaps, fandom, and online community.* Thousand Oaks, CA: Sage.

Brown, B. & Barkhuus, L. (2006). The television will be revolutionized: Effects of PVRs and filesharing on television watching. In *Proceedings of the SIGCHI Conference on Human Factors in Computer Systems.* ACM, New York, NY, (pp. 663-666).

Cadiz, J. J., Balachandran, A., Sanocki, E., Grudin, J., & Jancke, G. (2000). Distance learning through distributed collaborative video viewing. In *Proceedings of the 2000 ACM Conference on Computer Supported Cooperative Work.* ACM, New York, NY, (pp. 135-144).

Cartwright, D., & Zander, A. F. (1953). *Group dynamics, research and theory.* Evanston, IL: Row, Peterson and Co.

Chuah, M. (2003). Reality instant messaging: injecting a dose of reality into online chat. In *Proceedings of the SIGCHI Conference on Human Factors in Computer Systems.* ACM, New York, NY, (pp. 926-927).

Coppens, T., Trappeniers, L., & Godon, M. (2004). AmigoTV: towards a social TV experience. In *Proceedings of EuroITV 2004,* University of Brighton, U.K.

Daft, R. L, & Lengel, R. H. (1986). Organizational information requirements, media richness and structural design. *Management Science, 32*(5), (pp. 554-571).

Dennis, A. R., & Kinney, S. T. (1998). Testing media richness theory in the new media: The effects of cues, feedback, and task equivocality. *Information Systems Research, 9*(3), 256-274.

Ellis, C. A., Gibbs, S. J., & Rein, G. L. (1991). Groupware: some issues and experiences. *Communications of the ACM, 34*(1), 38-58.

Gibbons, J. F., Kincheloe, W. R., & Down, K. S. (1977). Tutored videotape instruction: a new use of electronics media in education. *Science, 195*(4283), 1139-1146.

Grigonis, R. (2008). Introducing P2P-Next, a European Union Internet Data and TV Standard. *TMCnet.* Retrieved May 21, 2008 from http://iptv.tmcnet.com/topics/iptv-technology/articles/21510-introducing-p2p-next-european-union-internet-data-tv.htm

Harobe, G., Metcalf, C. J., Bentley, F., Tullio, J., Massey, N., & Romano, G. (2008). Ambient social TV: drawing people into a shared experience. In *Proceedings of the SIGCHI Conference on Human Factors in Computer Systems.* ACM, New York, NY, (pp. 1-10).

Knoche, H., & McCarthy, J. D. (2005). Design requirements for mobile TV. In *Proceedings of MobileHCI 2005.* ACM, New York, NY, (pp. 69-76).

Luyten, K., Thys, K., Huypens, S., & Coninx, K. (2006). Telebuddies: social stitching with interactive television. In *CHI'06 Extended Abstracts on Human Factors in Computing Systems.* ACM, New York, NY, (pp. 1049-1054).

McKenna, K. Y. A., Green, A. S., & Gleason, M. E. J. (2002). Relationship formation on the Internet: What's the big attraction? *Journal of Social Issues, 58*(1), 9-31.

Moray, N. (1969). *Listening and Attention.* Baltimore, MD: Penguin.

Oumard, M., Mirza, D., Kroy, J., & Chorianopoulos, K. A cultural probes study on video sharing and social communication on the Internet. To appear in *Proceedings of DIMEA 2008.* ACM, New York, NY.

Postman, N. (1986). *Amusing ourselves to death: Public discourse in the age of show business.* New York, NY: Penguin.

Powazek, D. M. (2002). Design for community: The art of connecting real people in virtual places. Indianapolis, IN: New Riders Publishing.

Putnam, R. D. (1995). Tuning in, tuning out: The strange disappearance of social capital in America. *PS: Political Science and Politics, 28*(4), 664-683.

Regan, T. & Todd, I. (2004). Media center buddies: Instant messaging around a media center. In *Proceedings of the Third Nordic Conference on Human-Computer Interaction.* ACM, New York, NY, (pp. 141-144).

Sassenberg, K. (2002). Common bond and common identity groups on the Internet: Attachment and normative behavior in on-topic and off-topic chats. *Group Dynamics: Theory, Research, and Practice, 6*(1), 27-37.

Schatz, R., & Egger, S. (2008). Social interaction features for mobile TV services. In *Proceedings of IEEE Broadband Multimedia Systems and Broadcast Symposium 2008*, (pp. 1-6).

Shamma, D. A., Bastéa-Forte, M., Joubert, N., & Liu, Y. (2008). Enhancing online personal connections through the synchronized sharing of online video. In *Proceedings of SIGCHI 2008.* ACM, New York, NY, 2931-2936.

Utz, S. (2003). Social identification and interpersonal attraction in MUDs. *Swiss Journal of Psychology, 62*(2), 91-101.

Vosgerau, J., Wertenbroch, K., & Carmon, Z. (2006). Indeterminacy and live television. *Journal of Consumer Research, 32*, 487-495.

Weisz, J. D., Kiesler, S., Zhang, H., Ren, Y., Kraut, R. E., & Konstan, J. A. (2007). Watching together: Integrating text chat with video. In *Proceedings of the SIGCHI conference on Human Factors in Computing Systems*. ACM, New York, NY, (pp. 877-886).

Weisz, J. D. & Kiesler, S. (2008). How text and audio chat change the online video experience. To appear in the *First International Conference on Designing Interactive User Experiences for TV and Video* (uxTV'08). ACM, New York, NY.

Wickens, C. D. (2002). Multiple resources and performance prediction. *Theoretical Issues in Ergonomics Science, 3*(2), 159-177.

Afterword
Television is Dead. Long Live Television!

Nicolas Ducheneaut
Palo Alto Center, USA

Mark Twain once said, "reports of my death are greatly exaggerated." Can the same be said about television? In the age of personal video recorders (PVRs, e.g. TiVo) and BitTorrent movie downloads, the stereotypical model of a television set broadcasting content at a relatively passive audience looks more dead than alive. And yet, television viewing remains strong: Nielsen reports that, in 2007, North American households spent a record number of hours (8 hours and 14 minutes per day, to be precise[1]) with a set tuned in to their favorite channels. What, then, can the future of television possibly be? In view of current trends, we can venture a few educated guesses.

The Web is widely mentioned as the main threat to television as we know it. While online content is often seen as competing directly with television for the audience's attention, current practices show that the two media are in fact quite symbiotic. Witness for instance the dynamic and resourceful communities built around popular shows like ABC's Lost. Consulting (and potentially contributing to) Lostpedia[2] online is as much a part of watching the show as viewing the actual episodes. By enabling a community of like-minded viewers to interact asynchronously prior to and after each viewing, online social networks sustain and expand the reach of televised content. The obvious next step would therefore be to make this marriage more explicit and, unsurprisingly, several startups have begun offering services in this area[3]. One could say that the future of television is therefore interactive, but not in the sense this term is usually understood: the point here is not to interact with the television but rather with its viewers and with additional content that enrich the experience of watching a show.

In fact, the involvement of the audience might not limit itself to discussions about a given show and can even extend to creating entirely new visual content, which brings us to our second trend. The ever-increasing popularity of video sharing sites like YouTube provides an outlet for viewers to re-enact, parody, or continue a show along lines that might not have been envisioned by its producers. With the widespread availability of "mass market" video editing (such as the iMovie software on Apple machines, for instance), it is quite possible that a wider share of content will be produced by and for the audience, instead of the current, more centralized model (which of course has potentially dramatic effects on the broadcasters' advertising revenue and explains their fear at this paradigm shift). This kind of "remix" has happened in other media before (the manga industry in Japan is a good example[4]), where the audience became part and parcel of the content creation cycle and, in the end, contributed to the survival and growth of the medium.

Finally, the advent of PVRs and peer-to-peer video sharing have both proven that audiences want more control over the way they experience televised content: control over time (deferred viewing), control over place (viewing content on different platforms, that is both traditional TV and on a mobile phone), and control over storage (keeping shows on recorded DVDs or large hard disks, either for future viewing or simply for collecting episodes of a series they are particularly fond of). PVRs now offer the first form of control, but the latter two are more experimental and require technical expertise (e.g. re-encoding a video stream) that most viewers do not have. We can expect this to change as technology is designed to address this latent demand.

So, in the end, the future of television looks brighter than it may seem at first, although we might wonder if the transformed medium that is emerging can still be called television. While television should become a more social and interactive medium, giving its viewers more flexibility about when and what to consume (the latter possibly including viewer-generated content), audiences will most probably still need "big ticket", high-quality shows to federate their interest and seed interactions and content creation. The popularity of hit series, some of which have become real cultural phenomena, shows no signs of abating: Lost, The Sopranos, Sex and the City, Battlestar Galactica… each season, large communities of viewers impatiently await the next development in their fictional universe of choice. This capacity to enable the emergence of a "fan culture" around televised content (which Henry Jenkins describes in his book, "Convergence culture[5]") seems to be a defining feature of television as we know it. This can be preserved, provided broadcasters leverage their strength in creating such compelling content and simultaneously find ways to recoup their investment that do not require viewers to seat in front of the screen at a pre-determined time and place.

There are signs that television networks understand this need to redefine television as a broader ecology of content and technologies that support and enhance their shows: many networks now generate revenue through the sale of "collector's edition" DVDs, advertising revenue from Web site tied to a particular series, downloads of specially-formatted videos for viewing on cell phones, and so forth. Our challenge as technology designers is to embrace this broader notion of what television really is, in order to let viewers easily consume, store, edit, share and interact around the televised content they enjoy as easily as possible.

ENDNOTES

1. Holmes, Gary (2007). Nielsen reports television tuning remains at record levels. http://tinyurl.com/34qwb3
2. http://www.lostpedia.com
3. See for instance Joost: http://www.joost.com
4. Pink, Daniel (2007). Japan, Ink: Inside the Manga industrial complex. Wired, 15.11, http://tinyurl.com/yo7twk
5. Jenkins, H. (2006). Convergence culture: where old and new media collide. NYU Press.

Compilation of References

ABI Research. (2006). *Worldwide DVR market analysis.* Technical report. Oyster Bay, NY: ABI Research.

Abreu, J. (2007). *Design de Serviços e Interfaces num Contexto de Televisão Interactiva: proposta de uma aplicação de suporte à comunicação interpessoal entre telespectadores (iTV Services and Interface Design: proposal of an iTV application to support interpersonal* communication), Doctoral dissertation, University of Aveiro, Portugal.

Abreu, J. F., & Almeida, P. (2008). From Scratch to User Evaluation: Validating a Social iTV Platform. In A. Lugmayr, S. Kemper, M. Obrist, T. Mirlacher, & M. Tscheligi (Eds.), *Adjunct Proceedings of EuroITV 2008* (pp. 31–32). Tampere, Finland: Tampere University of Technology.

Abreu, J., Almeida, P., & Branco, V. (2001). 2BeOn: Interactive Television Supporting Interpersonal Communication. In J. A. Jorge, N. Correia, H. Jones, & M. B. Kamegai, (Eds.), *Proceedings of the Sixth Eurographics Workshop on Multimedia 2001* (pp. 199–208). New York: Springer Verlag.

Adar, E., & Huberman, B. A. (2000). Free riding on Gnutella. *First Monday, 5*(10).

Agamanolis, S. (2003). Designing Displays for Human Connectedness. In K. O'Hara, M. Perry, E. Churchill, & D. Russell, (Eds.), *Public and Situated Displays: Social and interactional aspects of shared display technologies* (pp. 309-334), Kluwer.

Agamanolis, S. (2008). At the Intersection of Broadband and Broadcasting: How Interactive TV Technologies can Support Human Connectedness. *International Journal of Human-Computer Interaction, 24*(2), 121–135.

Agamanolis, S., & Bove Jr., V. M. (2003). Viper: A Framework for Responsive Television. *IEEE MultiMedia, 10*(1), 88-98.

Alliez, D. (2008). Adapt TV Paradigms to UGC by Importing Social Networks. In A. Lugmayr, S. Kemper, M. Obrist, T. Mirlacher, & M. Tscheligi (Eds.), *Adjunct Proceedings of EuroITV 2008* (pp. 29–30). Tampere, Finland: Tampere University of Technology.

AmigoTV (2005). *AmigoTV: a social TV experience through triple-play convergence.* Whitepaper.http://www1.alcatellucent.com/publications/abstract.jhtml?repositoryItem=tcm%3A172-195491635

Anderson, C. (2006). *The Long Tail: Why the Future of Business Is Selling Less of More*. New York, Hyperion

Andrea, G., Arianna, I., & Sonia, M. (2007). Living@room: a Support for Direct Sociability through Interactive TV. In A. Lugmayr, & P. Gołębiowski (Eds.), *Interactive TV: A Shared Experience - TICSP Adjunct Proceedings of EuroITV 2007. Tampere International Center for Signal Processing. TICSP series #35* (pp 207-212). Amsterdam, the Netherlands.

Answer.com. (2008). Peer-to-Peer: Definition and Much More from Answer.com. Retrieved May 27, from http://www.answers.com/p2p&r=67

Arhippainen, L., & Tähti, M. (2003). Empirical Evaluation of User Experience in Two Adaptive Mobile Application Prototypes. *In Proceedings of the International*

Conference on Mobile and Ubiquitous Multimedia, Norrköping.

Arvola, M. (2003). *Good to Use!: Use Quality of Multi-User Applications in the Home*. Licentiate's Thesis. Linköping University, Sweden

Arvola, M., & Holmlid, S. (2000). IT-artefacts for Socializing: Qualities-in-use and Research Framework. *Proceedings of the 23rd Information Systems Research Seminar in Scandinavia*, Sweden

Associated Press. Average home has more TVs than people. Retrieved January 29, 2006, from USA Today: http://usatoday.com/life/television/news/2006-09-21-homes-tv x.htm, September 2006.

AT&T. (2008). *Technology Timeline: Picturephone*. Retrieved September 5, 2008, from http://www.corp.att.com/attlabs/reputation/timeline/70picture.html

B@Home. The Future of Broadband, Multimedia Services in the Home. From http://bathome.freeband.nl

Baca, M., & Holtzman, H. (2008). Television meets Facebook: Social Networks through Consumer Electronics. In A. Lugmayr, S. Kemper, M. Obrist, T. Mirlacher, & M. Tscheligi (Eds.), *Adjunct Proceedings of EuroITV 2008* (pp. 35–36). Tampere, Finland: Tampere University of Technology.

Baecker, R. M., Grudin, J., Buxton, W. A. S., & Greenberg, S. (1995). *Readings in Human-Computer Interaction: Toward the Year 2000*. San Francisco, CA: Morgan Kaufmann.

Bandura, A. (1971). *Social Learning Theory*. New York: General Learning Press.

Barker, C. (2003). *Cultural Studies: Theory and Practice*. London: Sage Publications.

Baron, R. A., Byrne, D., & Johnson, B. T. (1998). *Exploring Social Psychology* (4th ed.). Needham Heights, MA: Allyn and Bacon.

Bassoli, A., Moore, J., & Agamanolis, S. (2006). tunA: Socialising Music Sharing on the Move. In K. O'Hara, & B. Brown (Eds.), *Consuming Music Together: Social and Collaborative Aspects of Music Consumption Technologies* (pp. 151-172), Springer.

Baym, N. (1999). *Tune in, log on: Soaps, fandom, and online community*. Thousand Oaks, CA: Sage.

Beck, E., Obrist, M., Bernhaupt, R., & Tscheligi, M., (2008). Instant Card Technique: How and Why to apply in User-Centered Design. *In Proc. PDC2008*, Bloomington, Indiana, USA.

Beenen, G., Ling, K., Wang, X., Chang, K., Frankowski, D., Resnick, P., et al. (2004, November 6-11). *Using Social Psychology to Motivate Contributions to Online Communities*. Paper presented at the CSCW'04, Chicago, Illinois, USA.

Berg, M., Butterfield, S., Cosmas, J., Casagranda, P., Garrec, D., Guiraudou, M., Martinez, G., Launey, E., Mazieres, B., & Milenaso, D. (2003). *CISMUNDUS: Convergence of Digital Broadcasting and Mobile Telecommunications, International Broadcasting Conference 2003*, Amsterdam, the Netherlands.

Bernhaupt, R., Mihalic, K., & Obrist, M. (2007). Methods for Usability Evaluation of Mobile Applications. In J. Lumsden (Ed.), *Handbook of Research on User Interface Design and Evaluation for Mobile Technology*, IGI Global. (pp. 745-758).

Bernhaupt, R., Obrist, M., & Tscheligi, M. (2006). Usability and Usage of iTV Services: Lessons learned in an Austrian Field Trial. *In Proc. of EuroiTV 2006*, Athens, Greece.

Bernhaupt, R., Obrist, M., Weiss, A., Beck, E., & Tscheligi, M. (2008). Trends in the Living Room and Beyond: Results from Ethnographic Studies using Creative and Playful Probing. In *Computers in Entertainment*, 5(1) (March 2008).

Bernhaupt, R., Weiss, A., Obrist, M., & Tscheligi, M. (2007). Playful Probing: Making Probing more Fun. In *Proc. INTERACT 2007*, 10 -14 Sep 2007, Springer LNCS, (pp. 606-619).

Bernhaupt, R., Wilfinger, D., Weiss, A., & Tscheligi, M. (2008). An Ethnographic Study on Recommendations in the Living Room: Implications for Design of iTV Recommender Systems. In *Proc. of EuroiTV 2008*, (pp. 92-101).

Beyer, Y., Enli, G. S., Maasø, A. J., & Ytreberg, E. (2007). Small Talk Makes a Big Difference: Recent Developments in Interactive, SMS-Based Television. *Television & New Media, 8*(3), 213–234.

Bly, S. A., Harrison, S. R., & Irwin, S. (1993). Media Spaces: Bringing People Together in a Video, Audio, and Computing Environment. *Communications of the ACM, 36*(1), 28-47.

Boertjes, E. (2007). ConnecTV: Share the Experience. In A. Lugmayr & P. Gołębiowski (Eds.), *Interactive TV: A Shared Experience - TICSP Adjunct Proceedings of EuroITV 2007. Tampere International Center for Signal Processing. TICSP series # 35* (pp. 139-140). Amsterdam, The Netherlands.

Boertjes, E., Klok, J., & Schultz, S. (2008). ConnecTV: Results of the Field Trial. In A. Lugmayr, S. Kemper, M. Obrist, T. Mirlacher, & M. Tscheligi (Eds.), *Adjunct Proceedings of EuroITV 2008* (pp. 21–22). Tampere, Finland: Tampere University of Technology.

Bradner, E., & Mark, G. (2002, November 16-20). *Why Distance Matters: Effects on Cooperation, Persuasion and Deception.* Paper presented at the CSCW'02, New Orleans, Louisiana, USA.

Brembs, B. (2001). Hamilton's Theory. In S. Brenner & J. Miller (Eds.), *Encyclopedia of Genetics* (pp. 906-910). London, New York: Academic Press.

Brereton, P., & O'Connor, B. (2007). Pleasure and Pedagogy: The Consumption of DVD Add-Ons Among Irish Teenagers. *Convergence, 13*(2), 143-157

Brignull, H. & Rogers, Y. (2003). Enticing People to Interact with Large Public Displays in Public Spaces. In *Proceedings of INTERACT' 03*, Zurich.

Broekens, J. (2008). Want a Recommendation? Ask my Digital Personality! *EuroITV 2008 Workshop on Interactive Television Commerce and Advertising*, Salzburg, Austria.

Brown, B. & Barkhuus, L. (2006). The television will be revolutionized: Effects of PVRs and filesharing on television watching. In *Proceedings of the SIGCHI Conference on Human Factors in Computer Systems.* ACM, New York, NY, (pp. 663-666).

Brown, B. (2005). Play and Sociability in There. In R. Schroeder & A. Axelsson (Eds.), *Work and Play in Shared Virtual Environments.* Springer.

Bulterman, D. C. A. (2007). User-Centered Control within Multimedia Presentations. *Multimedia Systems Journal, 12*(4-5), 423-438.

Bulterman, D. C. A., & Hardman, L. (2005). Structured Multimedia Authoring. *ACM Transactions on Multimedia Computing Communications and Applications, 1*(1), 89-109.

Bulterman, D. C. A., & Rutledge, L. (2004). *SMIL 2.0: Interactive Multimedia for Web and Mobile Devices.* Springer-Verlag, Heidelberg.

Bulterman, D. C. A., Hardman, L., Jansen, J., Mullender, K. S., & Rutledge, L. (1998). GRiNS: A GRaphical INterface for Creating and Playing SMIL Documents. In *Proceedings of the WWW Conference.*

Bureau of Labor Statistics (2006). *American time use survey.* Technical report, U.S. Dept. of Labor, Washington, DC.

Buxton, B. (2007). *Sketching user experiences: getting the design right and the right design.* Morgan Kaufmann.

Cadiz, J. J., Balachandran, A., Sanocki, E., Grudin, J., & Jancke, G. (2000). Distance learning through distributed collaborative video viewing. In *Proceedings of the 2000 ACM Conference on Computer Supported Cooperative Work.* ACM, New York, NY, (pp. 135-144).

Cardoso, G., Espanha, R., Gomes, M. C., & Lapa, T. (2006). *Crianças e Jovens: A sua relação com as Tecnologias e os Meios de Comunicação*, report CIES-ISCTE. Retrieved August 25, 2007, from http://www.cies.iscte.pt/destaques/documents/CriancaseJovens_AsuarelacaocomasTecnologiaseosMeiosdeComunicacao.pdf

Cartwright, D., & Zander, A. F. (1953). *Group dynamics, research and theory*. Evanston, IL: Row, Peterson and Co.

Certeau, M. (1984). *The Practice of Everyday Life*. Berkley: University of California Press.

Cesar, P., Bulterman, D. C. A., & Jansen, J. (2006). An Architecture for End-User TV Content Enrichment. *Journal of Virtual Reality and Broadcasting (JVRB), 3*(9).

Cesar, P., Bulterman, D. C. A., Geerts, D., Jansen, J., Knoche, H., & Seager, W. (2008). Enhancing Social Sharing of Videos: Fragment, Annotate, Enrich, and Share. In *Proceedings of the 16th international Conference on Multimedia*, in press. New York: ACM Press.

Cesar, P., Bulterman, D. C. A., Obrenovic, Z., Ducret, J., Cruz-Lara, S. (2007). An Architecture for Non-intrusive User Interfaces for Interactive Digital Television. *EuroITV 2007*, (pp. 11-20), Springer LNCS.

Cesar, P., Bulterman, D., & Jansen, A. (2008). Usages of the Secondary Screen in an Interactive Television Environment: Control, Enrich, Share, and Transfer Television Content. In M. Tscheligi & M. Obrist (Eds.), *Changing Televion Environments - Proceedings of 6th European Conference, EuroITV2008, Salzburg, Austria, July 2008. LNCS, 5066*, 168-177. Heidelberg, Germany: Springer-Verlag.

Cesar, P., Chorianopoulos, K., & Jensen, J. F. (2008). Social Television and User Interaction. *ACM Computers in Entertainment, 6*(1).

Cheng, R., & Vassileva, J. (2005). *User Motivation and Persuasion Strategy for Peer-to-Peer Communities.* Paper presented at the 38th Hawaii International Conference on System Sciences, Hawaii.

Chorianopoulos, K, & Spinellis, D. (2006). User interface evaluation of interactive TV: a media studies perspective. *In Univ Access Inf. Soc.*, (2006) 5, 209-218.

Chorianopoulos, K. (2004). *Virtual Television Channels: Conceptual Model, User Interface Design and Affective Usability Evaluation.* Doctoral dissertation. University of Economics and Business, Athens. Retrieved April 9, 2005 from http://uitv.info/about/editors/chorianopoulos/thesis/phd.pdf

Chorianopoulos, K. (2006). Animated Character Likeability Revisited: The Case of Interactive TV. *Journal of Usability Studies, 1*(4), 171-184. UPA Press.

Chorianopoulos, K. (2007). Content-Enriched Communication: Supporting the Social Uses of TV. *The Journal of the Communications Network, 6*(1), 23–30.

Chorianopoulos, K. (2008). User Interface Design Principles for Interactive Television Applications. *International Journal of Human-Computer Interaction, 24*(6), 556-573.

Chorianopoulos, K., & Lekakos, G. (2008). Introduction to Social TV: Enhancing the Shared Experience with Interactive TV. *International Journal of Human Computer Interaction, 24*(2), 113-120.

Chorianopoulos, K., & Spinellis, D. (2006). User Interface Evaluation of Interactive TV: A Media Studies Perspective. *Universal Access in the Information Society*, 5(2), 209-218. Springer-Verlag (2006).

Chorianopoulos, K., & Spinellis, D. (2007). Coping with TiVo: Opportunities of the networked digital video recorder. *Telematics and Informatics, 24*(1), 48–58

Chuah, M. (2002). Reality Instant Messenger: The Promise of iTV Delivered Today. *Proceedings of TV'02: the 2nd Workshop on Personalization in Future TV - Second International Conference on Adaptive Hypermedia and Adaptive Web-based Systems*, Spain (pp. 65-74), Malaga

Chuah, M. (2003). Reality instant messaging: injecting a dose of reality into online chat. In *Proceedings of the SIGCHI Conference on Human Factors in Computer Systems*. ACM, New York, NY, (pp. 926-927).

Cisco. (2008). *The Exabyte Era* (White Paper): Cisco Systems, Inc.

Clark, M. S., & Mills, J. (1993). The difference between communal and exchange relationships: What it is and is not. *Personality and Social Psychology Bulletin, 19*, 684-691.

Coates, T. (2005, March 23). *Social Software for Set-Top Boxes*. Retrieved September 5, 2008, from http://www.plasticbag.org/archives/2005/03/social_software_for_settop_boxes/

Cohen, B. (2003). *Incentives Build Robustness in BitTorrent. 1st Workshop on Economics of Peer-to-Peer Systems*, Berkeley, CA.

Coppens T., Vanparijs F., Handekyn K. AmigoTV: A Social TV Experience Through Triple-Play Convergence. Alcatel, White paper, 2005.

Coppens, T., Trappeniers, L. & Godon, M., (2004). AmigoTV: towards a social TV experience. In Masthoff, J. Griffiths, R. & Pemberton, L. (Eds.), *Proceedings from the Second European Conference on Interactive Television "Enhancing the experience.* University of Brighton.

Coppens, T., Trappeniers, L., & Godon, M. (2004). AmigoTV: towards a social TV experience. In J. Masthoff, R. Griffiths, & L. Pemberton (Eds.), *Proceedings from the Second European Conference on Interactive Television "Enhancing the experience"*, UK, University of Brighton

Coppens, T., Trappeniers, L., & Godon, M. (2006). AmigoTV: towards a social TV experience. In *Proc. European Conf. on Interactive Television 2006.*

Coppens, T., Vanparijs, F., & Handekyen, K. (2005). *AmigoTV: A Social TV Experience Through Triple-Play Convergence* (Technology White Paper). Retrieved September 5, 2008, from http://www1.alcatel-lucent.com/com/en/appcontent/apl/T0205-Amigo_TV-EN_tcm172-195461635.pdf

Cornell, C., & Henry, B. (2002). The Almost But Still Not Quite Complete History of MST3K. *Satellite News.* Retrieved September 5, 2008, from http://www.mst3kinfo.com/history/

Crabtree, A., & Rodden, T. (2004). Domestic Routines and Design for the Home. *In CSCW: The Journal of Collaborative Computing, 13*(2), 191-220.

Craven, M., Benford, S., Greenhalgh, C., Wyver, J., Brazier, C., Oldroyd, A., and Regan, T. 2000. Ages of avatar: community building for inhabited television. In Proceedings of the Third international Conference on Collaborative Virtual Environments (CVE '00). ACM Press, 189-194. 2000.

Cui, Y., Chipchase, J., Jung, Y. Personal TV: A Qualitative Study of Mobile TV Users. EuroITV 2007: pages 195-204, Springer LNCS 2007.

Cullinan, C., & Agamanolis, S. (2002). Reflexion: a responsive virtual mirror. *Conference Companion, UIST 2002 Symposium on User Interface Software and Technology.*

Cullinan, C., Agamanolis, S., Roche, D., & Hegarty, F. (2004). Open Window: Reducing Isolation for Patients. *Adjunct Proceedings, UbiComp 2004.*

Daft, R. L, & Lengel, R. H. (1986). Organizational information requirements, media richness and structural design. *Management Science, 32*(5), (pp. 554-571).

Dahlbäck, N., Jönsson, A., & Ahrenberg, L. (1993). Wizard of Oz studies: why and how. *In Proce. of the 1st international Conference on intelligent User interfaces.* New York, NY: ACM.

Dahlgren, P. (Ed.). (1995). *Television and the Public Sphere: Citizenship, Democracy and the Media.* London: SAGE.

Daly-Jones, O., & Carey, R. (2000). *Interactive TV and electronic programming guides: usability guidelines.* Serco Usability Services.

Damásio, M. J. (2007). *Resumo do projecto MEDIA-CULT-PT/ Projectos de Investigação Científica e de Desenvolvimento Tecnológico em Todos os Domínios Científicos—Concurso 2006.* Fundação para a Ciência e Tecnologia. Retrieved January 11, 2008, from http://www.fct.mctes.pt/projectos/pub/2006/Painel_Result/vglobal_projecto.asp?idProjecto=70893&idElemConcurso=906

De Souza, C. S., & Preece, J. (2004). A framework for analyzing and understanding online communities. *Interacting with Computers, The Interdisciplinary Journal of Human-Computer Interaction, 16*(3)

De Valck, K. (2005). *Virtual Communities of Consumption: Networks of Consumer Knowledge and Companionship*. Erasmus University Rotterdam, Rotterdam.

Dees, W., & Shrubsole, P., (2007). Web4CE: Accessing Web-based Applications on Consumer Devices. *In proceedings of the 16th International World-Wide-Web Conference (WWW2007)* (pp. 1303-1304).

Dennis, A. R., & Kinney, S. T. (1998). Testing media richness theory in the new media: The effects of cues, feedback, and task equivocality. *Information Systems Research, 9*(3), 256-274.

Desmet, P. M. A., & Hekkert, P. (2007). Framework of product experience. *International Journal of Design, 1*(1), 57-66.

Diederiks, E. (2003). Buddies in a box: animated characters in consumer electronics. *In 8th international Conference on Intelligent User interfaces. IUI '03* (pp. 34–38). ACM Press.

Distance Lab web site: http://www.distancelab.org

Dix, A. J., Finlay, J., Abowd, G. D., & Beale, R. (1993). *Human-Computer Interaction*. Englewood Cliffs, NJP: Prentice-Hall.

Dix, A., Finlay, J., Abowd, G. D., & Beale, R. (2004). *Human-Computer Interaction* (3rd Edition), Prentice Hall Inc.

Dourish, P. (2006). Re-Space-ing Place: 'Place' and 'Space' Ten Years On. P*roceedings of the ACM 2006 Conference on Computer-Supported Cooperative Work - CSCW 2006*, Banff, Canada. pp. 299-308.

Drotner, K. (2008). Leisure Is Hard Work: Digital Practices and Future Competencies. In Buckingham, D. (Ed.), *Youth, Identity, and Digital Media* (pp. 167-184). Cambridge, MA: The MIT Press. Retrieved January 13, 2008, from http://www.mitpressjournals.org/toc/dmal/-/6

Ducheneaut, N., Moore, R. J., & Nickell, E. (2004). Designing for sociability in massively multiplayer games: an examination of the "third places" of SWG. *Proceedings of Other Players - Conference on Multiplayer Phenomena*, 2004 December 6-8, Copenhagen, Denmark

Ducheneaut, N., Moore, R. J., Oehlberg, L., Thornton, J. D., & Nickell, E. (2008). Social TV: Designing for Distributed, Sociable Television Viewing. *International Journal of Human-Computer Interaction, 24*(2), 136–154.

Dumas, J., & Redish, J. (1993). *A Practical Guide to Usability Testing*. Norwood, NJ: Ablex Publishing Corporation.

Durham, M. G., & Kellner, D. (Ed.) (2006). *Media and Cultural Studies: KeyWorks*. Malden (USA): Blackwell Publishing.

Ehn, P. (1992). Scandinavian Design: On Participation and Skill. *In* P. S. Adler, & T. A. Winograd (Eds.), *Usability: Turning technologies into tools*. New York: Oxford University Press, (pp. 96-132).

Ellis, C. A., Gibbs, S. J., & Rein, G. L. (1991). Groupware: some issues and experiences. *Communications of the ACM, 34*(1), 38-58.

Erickson, T., & Kellogg, W. A. (2000). Social Translucence: An Approach to Designing Systems that Support Social Processes. *ACM Transactions on Office Information Systems, 7*(1), 59-83.

Facer, K., & Williamson, B. (2004). *Designing technologies to support creativity and collaboration. A handbook from NESTA Futurelab* Bristol: NESTA Futurelab.

Fairbank, M. (1995). Sending signals: Kids speak out about values in the media. Maullin & Associates (USA). A Children Now Poll, Los Angeles, CA.

Fatah gen. Schieck, A. (2006). Towards an integrated architectural media space. In *First Monday, Special Issue #4: Urban Screens: Discovering the potential of outdoor screens for urban society*. http://www.firstmonday.org/issues/special11_2/

Federal Communications Commission (2006). *12th annual assessment of the status of competition in the market for the delivery of video programming*. Technical report, FCC, Washington, DC.

Festinger, L., Schachter, S., & Back, K. (1950). *Social pressures in informal groups: A study of human factors in housing*. New York: Harper & Brothers.

Fetter, M., & Gross, T. (2007). Time, Space, Connection: Scaling Ambient Intelligence. *UPGRADE - The European Journal for the Informatics Professional, VIII*(4), 44-49.

Fetter, M., & Gross, T. (2008). Contact Management on the Wall: A Card-Game Metaphor for Large Displays. *Proceedings of the Second International Conference on Tangible Embedded Interaction - TEI 2008*, (pp. 247-250). Bonn, Germany.

Fetter, M., Gross, T., & Zeller, B. ((accepted)). *Disclosure Templates: Vorlagen fuer die selektive Freigabe persoenlicher Informationen (Disclosure Templates: Templates for the Selective Disclosure of Personal Information; in German)*. Mensch & Computer - 8. Fachuebergreifende Konferenz fuer interaktive und kooperative Medien - M&C 2008, Luebeck, Germany.

Fields, B., Amaldi, P., Wong, W., & Gill, S. (2007). Editorial: In-use, In-situ: Extending Field Research Methods, In: *International Journal of Human Computer Interaction, 22*(1), 1-6.

Fink, M, Covell, M., & Baluja, S. (2006). Social – and Interactive – Television: Applications Based on Real-Time Ambient Audio Identification. In G. Doukidis, K. Chorianopoulos, & G. Lekakos (Eds.), *Proceedings of the 4th European Conference, EuroITV 2006* (pp. 138–146).

Fiske, J. (ed.). (1990). *Introduction to communication studies*. Routledge.

Fokker, J., Buntine, W., Pouwelse, J., De Ridder, H., & Westendorp, P. (2006, August 6-11). *Tagging in Peer-to-Peer Wikipedia: A method to induce cooperation.* Paper presented at the OSIR'06, Seattle, WA.

Fokker, J., De Ridder, H., Westendorp, P., & Pouwelse, J. (2007). *Inducing Cooperation in Peer-to-Peer Television Systems.* Paper presented at the EuroITV'07.

Fokker, J., De Ridder, H., Westendorp, P., & Pouwelse, J. (2008, June 4-6). *Exploring the Acceptability of Delayed Reciprocity in Peer-to-Peer Networks.* Paper presented at the Persuasive'08, Oulu, Finland.

Fokker, J., Pouwelse, J., & Buntine, W. (2007). Tag-Based Navigation for Peer-to-Peer Wikipedia. In R. K. J. Bandamutha (Ed.), *P2P Networks: An Introduction* (pp. 82-94). Hyderabad, India: Icfai University Press.

Fokker, J., Ridder, H., Westendorp, P., & Pouwelse, J. (2007). Psychological Backgrounds for Inducing Cooperation in Peer-to-Peer Television. In *Proceedings of EuroITV* (pp. 136-145).

Forlizzi, J., & Battarbee, K. (2004). Understanding experience in interactive systems. *In Proc. of the 2004 conference on Designing Interactive Systems*, ACM, New York, USA (p. 261).

Frey, B., & Jegen, R. (2001). Motivation crowding theory: A survey of empirical evidence. *Journal of Economic Surveys, 5*(5), 589-611.

Friedrich, O., Seeliger, R., & Arbanowski, S. (2008). Interactive and Personalized Services for an Open IMS-Based IPTV Infrastructure. *Seventh International Conference on Networking*, (pp. 302-307).

Gartner. (2007). Dataquest Insight. *A Vision of Telecom and Media Firms' Future Relationship with Consumers.*

Gauntlett, D., & Hill, A. (1999). *TV Living: Television, Culture and Everyday Life*. London: Routledge.

Gaver, W. H., Dunne, A., & Pacenti, E. (1999). Cultural Probes. *In Interactions, 6*(1), 21-29.

Gawlinski, M. (Ed.) (2003). *Interactive Television Production*. Oxford: Focal Press.

Geerts, D. (2006). Comparing voice chat and text chat in a communication tool for interactive television. In O. W. Bertelsen, S. Bødker, & K. Kuuti (Eds.), *Proceedings of the NordiCHI 2006*. (pp. 461-464). New York: ACM Press.

Geerts, D., Cesar, P., & Bulterman, D. (2008). The Implications of Program Genres for the Design of Social Television Systems. *Proceedings of uxTV2008*. Mountain View, California, October 22-24. ACM.

Geerts, D., Harboe, G., & Massey, N. (2007). *Social Interactive Television Workshop* [Presentation slides]. Retrieved September 5, 2008, from http://soc.kuleuven.be/com/mediac/socialitv/results.htm

Ghittino, A., Iatrino, A., Modeo, S., & Ricchiuti, F. (2007). Living@Room: A TV-Based Social Experience for Multimedia Content Streaming. In A. Lugmayr & P. Gołębiowski (Eds.), *Adjunct Proceedings of EuroITV 2007* (pp. 207–212). Tampere, Finland: Tampere International Center for Signal Processing.

Gibbons, J. F., Kincheloe, W. R., & Down, K. S. (1977). Tutored videotape instruction: a new use of electronics media in education. *Science, 195*(4283), 1139-1146.

Girgensohn, A., & Lee, A. (2002). Making web sites be places for social interaction. *In Proceedings of the ACM Conference on Computer Supported Cooperative Work*. ACM Press.

Goffman, E. (1966). *Behaviour in Public Places*. Free Press.

Goldenberg, S. (2007). Digital Video Recorders and Micro-Social Networking: Recreating the Shared Watching Experience of Television. In A. Lugmayr & P. Gołębiowski (Eds.), *Adjunct Proceedings of EuroITV 2007* (pp. 213–217). Tampere, Finland: Tampere International Center for Signal Processing.

Golder, S., Wilkinson, D. M., & Huberman, B. A. (2006). Rhythms of social interaction: messaging within a massive online network. *cs/0611137*. Retrieved May 28, 2008, from http://arxiv.org/abs/cs/0611137.

Goren, H., Kurzban, R., & Rapoport, A. (2003). Social loafing vs. social enhancement: Public goods provisioning in real-time with irrevocable commitments. *Organizational Behavior and Human Decision Processes, 90*, 277-290.

Grafen, A. (1990). Biological Signals as Handicaps. *Journal of Theoretical Biology, 144*(4), 517-546.

Greenberg, J., & Scott, K. S. (1996). *Why do workers bite the hands that feed them? Employee theft as a social exchange process.* (Vol. 18). Greenwich, CT: JAI Press.

Greenberg, S., Fitzpatrick, G., Gutwin, C., & Kaplan, S. (1999). Adapting the Locales framework for heuristic evaluation of groupware. *Proceedings of OZCHI'99*, (pp. 28-30).

Grigonis, R. (2008). Introducing P2P-Next, a European Union Internet Data and TV Standard. *TMCnet*. Retrieved May 21, 2008 from http://iptv.tmcnet.com/topics/iptv-technology/articles/21510-introducing-p2p-next-european-union-internet-data-tv.htm

Gross, T. (1997). Towards Flexible Support for Cooperation: Group Awareness in Shared Workspaces. *Proceedings of the Eighth International Workshop on Database an Expert Systems Applications - DEXA'97*, (pp. 406-411). Toulouse, France.

Gross, T., & Fetter, M. (2008). CoDaMine: Communication Data Mining for Feedback and Control in Ubiquitous Environments. *Proceedings of the Sixteenth Euromicro Conference on Parallel, Distributed, and Network-Based Processing - PDP 2008*, (pp. 539-546). Toulouse, France.

Gross, T., & Oemig, C. (2005). PRIMI: An Open Platform for the Rapid and Easy Development of Instant Messaging Infrastructures. *Proceedings of the 31st EUROMICRO Conference on Software Engineering and Advanced Applications - SEAA 2005*, Oporto, Portugal. (pp. 460-467).

Gross, T., & Oemig, C. (2006). From PRIMI to PRIMI-Faces: Technical Concepts for Selective Information Disclosure. *Proceedings of the 32nd EUROMICRO Conference on Software Engineering and Advanced Applications - SEAA 2006, Cavtat*, Dubrovnik, Croatia. (pp. 480-487).

Gross, T., Braun, S., & Krause, S. (2006). MatchBase: A Development Suite for Efficient Context-Aware Communication. *Proceedings of the Fourteenth Euromicro Conference on Parallel, Distributed, and Network-Based Processing - PDP 2006*, (pp. 308-315). Montbeliard-Sochaux, France.

Gross, T., Egla, T., & Marquardt, N. (2006). Sens-ation: A Service-Oriented Platform for Developing Sensor-

Based Infrastructures. *International Journal of Internet Protocol Technology (IJIPT), 1*(3), 159-167.

Gross, T., Fetter, M., & Paul-Stueve, T. (2008). Towards Advanced Social TV in a Cooperative Media Space. *International Journal of Human-Computer Interaction, 24*(2), 155-173.

Gross, T., Stary, C., & Totter, A. (2005). User-Centered Awareness in Computer-Supported Cooperative Work-Systems: Structured Embedding of Findings from Social Sciences. *International Journal of Human-Computer Interaction, 18*(3), =323-360.

Grudin, J. (1994). Computer-supported cooperative work: Its history and participation. *IEEE Computer, 27*(5), 19-26.

Guimarães, R. L., Costa, R. M. R., & Soares, L. F. G. (2008). Composer: Authoring Tool for iTV Programs. In *Proceedings of EuroITV* (pp. 61-71).

Gutwin, C., & Greenberg, S. (2002). A Descriptive Framwork of Workspace Awareness for Real-Time Groupware. *Computer Supported Cooperative Work: The Journal of Collaborative Computing, 11*(3-4), 411-446.

Gutwin, C., Greenberg, S., & Roseman, M. (1996). Short Paper: Supporting Awareness of Others in Groupware. Conference Companion of the *Conference on Human Factors in Computing Systems - CHI'96,* Vancouver, Canada. (p. 205).

Halskov, K., & Dalsgård, P. (2006). Inspiration card workshops. In Proceedings of the 6th ACM Conference on Designing interactive Systems. *In Proc. DIS '06.* ACM Press, New York, NY, (pp. 2-11).

Hamilton, W. D. (1964). The genetical evolution of social behaviour. I. *Journal of Theoretical Biology, 7*(1), 1-16.

Harboe, G., Massey, N., & Metcalf, C., (2007). Perceptions of Value: The Uses of Social Television. In Cesar, P., Chorianopoulos, K. & Jensen, J. (Eds.), *Interactive TV: A Shared Experience - Proceedings of EuroITV 2007. Lecture Notes in Computer Science, 4471*, 116-125. Amsterdam, The Netherlands: Springer.

Harboe, G., Massey, N., Metcalf, C., Wheatley, D., & Romano, G. (2008a, May). The Uses of Social Television. *Computers in Entertainment, 6*(1), 1–15.

Harboe, G., Massey, N., Metcalf, C., Wheatley, D., & Romano, G. (2007). Perceptions of Value: The Uses of Social Television. In *Proceedings of EuroITV* (pp. 116-125).

Harboe, G., Metcalf, C. J., Bentley, F., Tullio, J., Massey, N., & Romano, G. (2008). Ambient Social TV: Drawing People Into a Shared Experience. In *Proceeding of the Twenty-Sixth Annual SIGCHI Conference on Human Factors in Computing Systems* (pp. 1–10). New York: ACM Press.

Harboe, G., Metcalf, C. J., Bently, F., Tulio, J., Massey, N., & Romano, G. (2008). Ambient social TV: Drawing people into a shared experience. In *CHI'08: Proceedings of the twenty-sixth annual SIGGCHI conference on Human Factors in computing systems* (pp. 1–10), New York, NY, USA.

Hardy, C. L., & Van Vugt, M. (2006). Nice guys finish first: The competitive altruism hypothesis. *Personality and Social Psychology Bulletin, 32*(10), 1402-1413.

Harkins, S., & Szymanski, K. (1989). Social loafing and group evaluation. *Journal of Personality and Social Psychology, 56*, 934-941.

Harrison, C., & Amento, A. (2007). CollaboraTV: Using Asynchronous Communication to Make TV Social Again. In A. Lugmayr & P. Gołębiowski (Eds.), *Adjunct Proceedings of EuroITV 2007* (pp. 218–222). Tampere, Finland: Tampere International Center for Signal Processing.

Harrison, C., & Amento, B. (2007). Collaboratv: Using asynchronous communication to make TV social again. In A. Lugmayr & P.Gołębiowski (Eds.), *Interactive TV: A Shared Experience - TICSP Adjunct Proceedings of EuroITV 2007. Tampere International Center for Signal Processing. TICSP series # 35* (pp 137-138). Amsterdam, the Netherlands.

Harrison, S., & Dourish, P. (1996). Re-Place-ing Space: The Roles of Place and Space in Collaborative Systems. *Proceedings of the ACM 1996 Conference on Compu-*

ter-Supported Cooperative Work - CSCW'96, Boston, MA. (pp. 67-76).

Hassenzahl, M. (2003). The Thing and I: understanding the Relationship Between User and Product, In M. Blythe, C. Overbeeke, A. Monk & P. Wright (Eds.), *Funology: From Usability to Enjoyment* (pp. 31-42). Kluwer Academic Publishers, Dordrecht.

Hassenzahl, M. (2004). The interplay of beauty, goodness and usability in interactive products. *In Human Computer Interaction*, 19, 319-349.

Hassenzahl, M., & Tractinsky, N., (2006). User Experience - a research agenda. In *Behavior & Information Technology, 25*(2), 91-97.

Haythornthwaite, C. (1996). Social network analysis: An approach and technique for the study of information exchange. *Library & Information Science Research, 18*(4), 323-342. doi: 10.1016/S0740-8188(96)90003-1.

Heider, F. (1958). *The Psychology of Interpersonal Relations*: John Wiley & Sons.

Herlocker, J. L., Konstan, J. A., Terveen, L. G., & Riedl, J. T. (2004). Evaluating collaborative filtering recommender systems. *ACM Transactions on Information Systems, 22*(1), 5-53.

Heß, J. (2007). Supporting community building in iTV environments. In A. Lugmayr & P. Gołębiowski (Eds.), *Interactive TV: A Shared Experience - TICSP Adjunct Proceedings of EuroITV 2007. Tampere International Center for Signal Processing. TICSP series # 35* (pp. 129-130). Amsterdam, The Netherlands.

Hess, J. (2008). Community Awareness in Social TV Environments. In *Adjunct Proceedings of EuroITV*.

Hew, K., Gibbs, M., & Wadley, G. (2004). Usability and Sociability of the Xbox Live Voice Channel. *Proceedings of Australian Workshop on Interactive Entertainment 2004*.

Hill, B. (2001). *AOLTV for Dummies*. New York: John Wiley & Sons.

Hom, J. (1998). The Usability Methods Toolbox. *Jameshom.com*. Retrieved June 6, 2004, from http://jthom.best.vwh.net/usability/

Homans, G. C. (1958). Social Behavior as Exchange. *American Journal of Sociology, 63*(6), 597-606.

House, J. S., Landis, K. R., & Umberson, D. (1988). Social Relationships and Health. *Science, 241*(4865), 540-545.

House, N.V., Davis, M., Ames, M., Finn, M., & Viswanathan, V. (2005). The Uses of Personal Networked Digital Imaging: an Empirical Study of Cameraphone Photos and Sharing. In *Extended Abstracts on Human Factors in Computing Systems* (pp. 1853-1856).

Hu, J. (2003, February 18). America Online Confirms End of AOLTV. *CNET News*. Retrieved September 5, 2008, from http://news.cnet.com/2100-1023-984920.html

Hughes, J., King, V., Rodden, T., & Andersen, H. (1994). Moving out from the control room: Ethnography in system design. In *CSCW'94* (pp. 429-439). Chapel Hill: ACM Press. Retrieved September 13, 2007, from http://delivery.acm.org/10.1145/200000/193065/p429-hughes.pdf?key1=193065&key2=2263690711&coll=&dl=ACM&CFID=15151515&CFTOKEN=6184618

Human Connectedness group web site: http://www.medialabeurope.org/hc

Hutchinson, H., Mackay, W., Westerlund, B., Bederson, B. B., Druin, A., Plaisant, C., Beaudouin-Lafon, M., Conversy, S., Evans, H., Hansen, H., Roussel, N., Eiderbäck, B., Lindquist, S., & Sundblad, Y. (2003). Technology Probes: Inspiring Design for and with Families. *In Proc. of the CHI 2003*, ACM Press (2003), (pp. 17-24).

IMDb. (2008). *The Internet Movie Database (IMDb)*, http://www.imdb.com: IMDb.com Inc. (Accessed 11/7/2008).

Jabber. *Open instant messaging and presence*. From http://www.jabber.org

Jacko, J. A., & Sears, A. (Eds.) (2003). *The Human-Computer Interaction Handbook: fundamentals, evolv-*

ing technologies and emerging applications. Lawrence Erlbaum Associates

Jans, G., & Calvi, L. (2007). How to develop a mobile 2.0 application. *Proceedings of IADIS International Conference Web-based Communities* (pp. 217-224). Salamanca, Spain

Jenkins, H. (2003). Quentin Tarantino's Star Wars? Digital Cinema, Media Convergence and Participatory Culture. In D. Thorburn & H. Jenkins (Eds.), *Rethinking Media Change.* Cambridge, MA: The MIT Press.

Jenkins, H. (2006). *Convergence Culture: Where Old and New Media Collide.* New York and London: New York University Press.

Jenkins, H. (2006). *Confronting the Challenges of Participatory Culture: Media Education for the 21 st Century.* report to The John D. and Catherine T. MacArthur Foundation. Retrieved January 13, 2008, from http://digitallearning.macfound.org/atf/cf/{7E45C7E0-A3E0-4B89-AC9C-E807E1B0AE4E}/JENKINS_WHITE_PAPER.PDF

Jenkins, H. (1992). *Textual Poachers: Television Fans and Participatory Culture.* New York: Routledge, Chapman and Hall.

Jensen, J. F. (2005). Interactive Television: New Genres, New Format, New Content. In *Proceedings of the Australasian Conference on Interactive Entertainment* (pp. 89-96).

Joost N. V. (2008). *Joost-What's Joost?,* http://www.joost.com/whatsjoost.html: Joost N.V. (Accessed 1/5/2008).

Kameas, A., Bellis, S., Mavrommati, I., Delaney, K., Colley, M., & Pounds-Cornsih, A. (2003). An Architecture that Treats Everyday Objects as Communicating Tangible Components. *Proceedings of the 1st IEEE International Conference on Pervasive Computing and Communications - PerCom 2003,* Dallas-Fort Worth, TX. (pp. 115-122).

Kankainen, A. (2003). UCPCD: user-centered product concept design. In *Proc. of the 2003 conference on Designing for User Experiences.* ACM Press, San Francisco, California, USA.

Karapanos, E., Hassenzahl, M., & Martens, J-B., (2008). User Experience Over Time. Work in Prgress, In *Proc. CHI 2008,* ACM, New York, USA, (pp. 3561-3566).

Kawamoto, D., Hu, J., & Miles, S. (2000). AOLTV service to hit market Monday *CNET.* Retrieved May 6, 2004, from http://news.com.com/2100-1023_3-241984.html

Kharif, O. (2006). PARC to Make TV Watching More Social. *BuzinessWeek Online.* Retrieved March 22, 2008, from http://www.businessweek.com/the_thread/techbeat/archives/2005/03/parc_to_make_tv.html

Kindberg, T., Spasojevic, M., Fleck, R., & Sellen, A. (2005). I Saw this and Thought of You: Some Social Uses of Camera Phones, In *Extended Abstracts on Human Factors in Computing Systems* (pp. 1545-1548).

Kirk, D., Sellen, A., Harper, R., & Wood, K. (2007). Understanding Videowork. In *Proceeding of the SIGCHI Conference on Human Factors in Computing Systems* (pp. 61-70).

Klym, N., & Montpetit, M.J. (2008). *Innovation at the Edge: Social TV and Beyond* (VCDWG Working Paper). Cambridge, MA: MIT Communications Futures Program. Retrieved September 5, 2008, from http://cfp.mit.edu/publications/index.shtml

Knoche, H, McCarthy, J. D., & Sasse, M. A. (2008. How low can you go? The effect of low resolutions on shot types in mobile TV. *Multimedia Tools Appl., 36*(1-2), 145-166.

Knoche, H., & McCarthy, J. D. (2005). Design requirements for mobile TV. In *Proceedings of MobileHCI 2005.* ACM, New York, NY, (pp. 69-76).

Kraut, R., et al. (2002). Internet Paradox Revisited. *Journal of Social Issues, 58*(1), 49-74.

Kubey, R., & Csikszentmihalyi, M. (1990). Television and the Quality of Life: How Viewing Shapes Everyday Experiences. Lawrence Erlbaum.

Kujala, S. (2003). User involvement: a review of the benefits and challenges. *In Behaviour and Information Technology, 22*(1), 1-17.

Lafrance, J., (2005). Le phénomène télénaute ou la convergence télévision/ordinateur chez les jeunes, *Réseaux, Paris, nº 129-130, 23,* 311-321.

Lafrenière, D. (1996). CUTA: a simple, practical, low-cost approach to task analysis. *In Interactions,* 3(5) 35-39.

Lahanas, S. (2008). *Enable SOA Transformation and Cross-Domain Data Fusion.* DM Review Magazine.

Larson, R., & Csikszentmihalyi, M., (1983). The experience sampling method. In H. T. Reiss (Ed.), *Naturalistic approaches to studying social interaction, New directions for methodology of social and behavioral sciences.* Jossey-Bass, San Francisco, (pp. 41-56).

Latané, B. (1981). The psychology of social impact. *American Psychologist, 36*, 343-356.

Law, E., Thora, E., Hassenzahl, M., & Hvannberg, E. (Eds.) (2006). *UX'2006: 2nd International Workshop on User eXperience*, 14 October 2006, Olso, Norway, held in conjunction with Nordichi'2006, [online]. Available: http://cost294.org/upload/408.pdf (accessed: 30.05. 2008).

Law, E., Vermeeren, A., Hassenzahl, M., & Blythe, M., (Eds.) (2007). *Workshop "Towards a UX Manifesto" at the International Conference HCI 2007*, Lancaster, UK, September 3, 2007 [online]. Available: http://141.115.28.2/cost294/upload/506.pdf, (accessed: 30.05. 2008).

Lee, B., & Lee, R. S. (1995). How and why people watch TV: Implications for the future of interactive television. *Journal of Advertising Research, 35*(6), 9–18.

Leichtman Research Group, Inc. *Press release: DVDs now in over one of every five U.S. households.* Retrieved September 18, 2007, from: http://www.leichtmanresearch.com /press/082107release.html, August 2007.

Lemish, D. (2007). *Children and television: A global perspective.* Oxford UK: Blackwell Publishing.

Lenhart, A., & Madden, M. (2005).*Teen Content Creators and Consumers.* Washington, DC: Pew Internet & American Life Project. Retrieved September 13, 2007, from http://www.pewinternet.org/PPF/r/166/report_display.asp

Lenhart, A., Madden, M., Macgill, A., & Smith, A. (2007). Teens and Social Media. *Reports: Family, Friends & Community—Pew Internet.* Retrieved February 19, 2008, from http://www.pewinternet.org/PPF/r/230/report_display.asp

Lenhart, A., Rankin, M., & Smith, A., (2007). *Teens and Social Media: The use of social media gains a greater foothold in teen life as email continues to lose its luster.* Washington, DC: Pew Internet & American Life Project. Retrieved December 19, 2007, from http://www.pewinternet.org/pdfs/PIP_Teens_Social_Media_Final.pdf

Leurdijk, A. (2007). Will Broadcasters Survive in the Online and Digital Domain? In P. Cesar, K. Chorianopoulos, & J. Jensen, (Ed.), *Interactive TV: a Shared Experience 5th European Conference, EuroITV 2007, Amsterdam, The Netherlands, May 24-25, 2007. Proceedings* (pp. 86-95). *Lecture Notes in Computer Science, 4471.* Springer.

Lewis, T., Amini, F., & Landon, R. (2000). *A General Theory of Love*, Random House.

Liberty Alliance Project. (2008). Retrieved June 04, 2008, from http://www.projectliberty.org/

Light, A. (2004). A need to Commune. *ACM - Interactions, XI.2*, New York, March-April, 2004, (pp. 74-75).

Lipsman (2007). 3 Out of 4 U.S. Internet Users Streamed Video Online in May. *ComScore Press Releases.* Retrieved February 13, 2008, from http://www.comscore.com/press/release.asp?press=1529

Liu, Y., Shamma, D. A., Shafton, P., & Yang, J. (2007). Zync: the design of synchronized video sharing. In *Proceedings of Designing the User Experience.* Chicago, USA.

Livingstone, S. (2002). *Young People and New Media.* London: Sage Publications.

Livingstone, S. (2004).The Challenge of Changing Audiences Or, What is the Audience Researcher to do in the Age of the Internet? *European Journal of Communication, 19*(1), 75-78.

Livingstone, S. (2007). Interactivity and participation on the internet: Young people's response to the online invitation to engage in the civic sphere. In P. Dahlgren (Ed.), *Young Citizens and New Media: Learning for Democratic Participation.* London: Routledge.

Livingstone, S. (2008). Engaging with Media—a Matter of Literacy? *Communication, Culture & Critique.* 1 (2008), 51-62. Retrieved August 21, 2008, from http://www3.interscience.wiley.com/journal/120185561/abstract

Livingstone, S., Van Couvering, E., & Thumin, N. (2005). *Adult Media Literacy: A review of the research literature.* London: Ofcom. Retrieved January 9, 2008, from http://www.ofcom.org.uk/advice/media_literacy/medlitpub/medlitpubrss/aml.pdf

Logan, R., Augaitis, S., Miller, R., & Wehmeyer, K. (1995). Living room culture – an anthropological study of television usage behaviors. *In Human Factors and Ergonomics Society 39th Annual Meeting,* (pp. 326–330).

Lombard, M., & Ditton, T. (1997). At the Heart of It All: The Concept of Presence. *Journal of Computer-Mediated Communication, 3*(2).

Lowet, D., & Khmelinskaya, E. (2008). Social API for CE-HTML devices. *EuroITV 2008 Workshop on Sharing Content and Experiences with Social Interactive Television*, Salzburg, Austria.

Lowet, D., & Shrubsole, P. (2007). Content sharing and experience sharing with Web4CE. *In Adjunct Proceedings of EuroITV 2007: Interactive TV: A Shared Experience: TICSP*

Lozano-Hemmer, R. (2002). Alien Relationships with Public Space. In J. Brouwer & A. Mulder (Eds.), *Trans-Urbanism.* NAI Publishers.

Lu, K. Y. (2005). *Interaction Design Principles for Interactive Television.* Master's Thesis, Georgia Institute of Technology.

Lull, J. (1980). The social uses of television. *Human Communication Research, 6*(3)

Luyten, K., Thys, K., Huypensm S., & Coninx, K. (2006). Telebuddies: social stitching with interactive television. *In CHI '06: CHI '06 extended abstracts on Human factors in computing systems* (pp. 1049–1054). New York, NY, USA. ACM Press.

Madden, M. (2007). Online Video. *Reports: Technology & Media Use—Pew Internet.* Retrieved February 19, 2008, from http://www.pewinternet.org/PPF/r/219/report_display.asp

Magedanz, T., Witaszek, D., & Knuettel, K. (2005). The IMS Playground @ Fokus – An Open Testebed for Next Generation Network Multimedia Services. *In Proceedings of the 1st International Conference on Testbeds and Research Infrastructures for the DEvelopment of NeTworks and COMmunities (TRIDENTCOM'05)* (pp. 2-11).

Manovich, L. (2006). The Poetics of Urban Media Surfaces. In *First Monday, Special Issue #4: Urban Screens: Discovering the potential of outdoor screens for urban society.* http://www.firstmonday.org/issues/special11_2/

Mantzari, E., & Vrechopoulos, A. (2007). "My Social Tube": User Generated Content and Communication on Interactive Digital Television. In A. Lugmayr & P. Gołębiowski (Eds.), *Adjunct Proceedings of EuroITV 2007* (pp. 241–246). Tampere, Finland: Tampere International Center for Signal Processing.

Martins, D., & Quico, C. (2007). O desafio da Literacia dos Novos Media em Portugal: o contributo da iniciativa PT Escolas II. In 5º *Congresso SOPCOM* (pp. 2781-2731). Braga: Universidade do Minho—Centro de Estudos de Comunicação e Sociedade.

Masthoff, J. (2004). Group Modeling: Selecting a Sequence of Television Items to Suit a Group of Viewers. *User Modeling and User-Adapted Interaction, 14*(1), 37-85

Mauss, M. (2000). *The gift: the form and reason for exchange in Archaic societies* (W. D. Halls, Trans.). New York: Norton.

McAndrew, F. T. (2002). New Evolutionary Perspectives on Altruism: Multilevel-Selection and Costly-Signaling Theories. *Current Directions in Psychological Science, 11*(2), 79-82.

McKenna, K. Y. A., Green, A. S., & Gleason, M. E. J. (2002). Relationship formation on the Internet: What's the big attraction? *Journal of Social Issues, 58*(1), 9-31.

McQuail, D. (1998). *Mass Communication Theory: An introduction*. London: Sage Publications.

McQuail, D. (2000). *McQuail's Mass Communication Theory*. London: Sage Publications.

McQuail, D., & Windahl, S. (1993). *Modelos de Comunicação para o Estudo da Comunicação de Massas*. Lisboa: Editorial Notícias.

Mcquail, D., & Windahl, S. (Eds.). (1993). *Communication Models: For the Study of Mass Communications*. Prentice Hall

Merriam-Webster, I. (2008). *Merriam-Webster Online*, http://www.m-w.com/. (Accessed 21/3/2008).

Meschtscherjakov, A., Obrist, M., Mirlacher, T., Lankes, M., Tscheligi, M., (2008). *Evaluating the Emotional Attachment to Mobile Devices & Services*, HCI & Usability Unit: Technical Report 2008.

Miletich, M. (2008). Five years 'Buntes Fernsehen Engerwitzdorf': The Power of Homemade Television. *EuroITV 2008 Workshop on Sharing Content and Experiences with Social Interactive Television*, Salzburg, Austria.

Milgram, S. (1967). The small world problem. *Psychology Today, 2*(1), 60-67.

Milojicic, D. S., Kalogeraki, V., Lukose, R., Nagaraja, K., Pruyne, J., Richard, B., et al. (2002). *Peer-to-Peer Computing* (Technical report No. HPL-2002-57 (R.1)): HP Laboratories Palo Alto.

Moray, N. (1969). *Listening and Attention*. Baltimore, MD: Penguin.

Morley, D. (1986). *Family Television. Cultural Power and Domestic Leisure*. London: Comedia

Morley, D. (1993). *Television, Audiences and Cultural Studies*. London: Routledge.

Morley, D. (2007). *Media, Modernity and Technology: the Geography of the New*. Oxon: Routledge.

Mueller, F., Agamanolis, S., & Picard, R. (2003). Exertion Interfaces: Sports over a Distance for Social Bonding and Fun. *Proceedings of CHI 2003*, ACM Press.

Muller, M. J. (2001). Layered participatory analysis: new developments in the CARD technique. In *Proc. of the SIGCHI Conference on Human Factors in Computing Systems*.

Muller, M. J., & Kuhn, S. (1993). Participatory design. *Commun. ACM*, 36(6), 24-28.

Nardi, B. A., Whittaker, S., & Bradner, E. (2000). Interaction and Outeraction: Instant Messaging in Action. In *Proceedings of the ACM conference on Computer Supported Cooperative Work* (pp. 79-99). Philadelphia, PA: ACM Press.

Nardi, B., & Harris, J. (2006, November 4-8). *Strangers and Friends: Collaborative Play in World of Warcraft*. Paper presented at the CSCW'06, Banff, Alberta, Canada.

Nathan, M, Terveen, L., Harrison, C., Yarosh, S., Stead, L., & Amento, B. (2008). CollaboraTV: Making Television Viewing Social Again. In *Proceedings of UXTV 2008*, in press.

NetFlix, Inc. *Corporate facts sheet*. Retrieved September 18, 2007, from: http://ir.net ix.com.

Newcomb, H. (Ed.). (1997). *Encyclopedia of Television*. (1st ed.). New York: Taylor & Francis. Retrieved September 5, 2008, from http://www.museum.tv/archives/etv/

Nielsen, J. (1994). Heuristic evaluation. In J. Nielsen, & R. L. Mack (Eds.). *Usability Inspection methods*. New York, NY: John Wiley & Sons.

Nielsen, J. (2006). Participation Inequality: Encouraging More Users to Contribute. Retrieved October 10, 2006, from http://www.useit.com/alertbox/participation_inequality.html

Nielsen, J., & Molich, R. (1990). Heuristic evaluation of user interfaces. *Proceedings of the ACM CHI'90 Conference*, Seattle, Washington

Noll, A. M. (1992). Anatomy of a Failure: Picturephone Revisited. *Telecommunications Policy, 16*(4), 307–316.

Norman, D. (1988). *The Design of Everyday Things*. New York, NY: Basic Books

Norman, D. A. (1999). *The Invisible Computer*. Cambridge: The MIT Press.

NPD Group, Inc. *Npd group, inc. tv-dvds soften video sales decline*. Retrieved September 18, 2007, from: http://npd.com/press /releases/press061024.html, October 2006.

OASIS Security Assertion Markup Language (SAML). (2008). Retrieved June 04, 2008, from http://www.oasis-open.org/committees/tc_home.php?wg_abbrev=security#samlv20

Obrist, M., Beck, E., Kepplinger, S., Bernhaupt, R., & Tscheligi, M. (2008). Local Communities: Back to Life (Live) Through IPTV. In M. Tscheligi, M. Obrist, and A. Lugmayr (Eds.), *Proceedings of the 6th European Conference, EuroITV 2008* (pp. 148–157). Berlin: Springer Verlag.

Obrist, M., Bernhaupt, R., & Tscheligi, M. (2006). Interactive Television for the Home: An ethnographic study on users' requirements and experiences. *In Proc. of EuroiTV'06*, Athens, Greece, (pp. 349-358).

Obrist, M., Bernhaupt, R., Beck, E., & Tscheligi, M. (2007). Focusing on Elderly: an iTV Usability Evaluation Study with Eye-Tracking. In P. Cesar, K. Chorianopoulos & J. F. Jensen (Eds.), *Interactive TV: A shared Experience. 5th European Conference, EuroITV 2007* Amsterdam, The Netherlands, May 24-25, 2007, Proceedings, (pp. 66-75).

Obrist, M., Kepplinger, S., Beck, E., Tscheligi, M., & Muzak, P. (2008). Usability & User Experience: Preliminary Results from Evaluating an IPTV Community Plattform. In *Proc. of EuroiTV2008*, (pp. 303-307).

Oehlberg, L., Ducheneaut, N., Thomton, J. D., Moore, G. E., & Nickell, E. (2006). Social TV: Designing for Distributed, Sociable Television Viewing. *Proceedings of the European Conference on Interactive Television - EuroITV 2006*, Athens, Greece. (pp. 251-262).

Ofcom (2006). *Media Literacy Audit—Report on Media Literacy Amongst Children*. London: Ofcom. Retrieved January 11, 2008, from http://www.ofcom.org.uk/advice/media_literacy/medlitpub/medlitpubrss/children/

Okabe, D. (2004). Emergent Social Practices, Situations and Relations through Everyday Camera Phone Use. In *Proceedings of the Conference on Mobile Communication and Social Change*.

Oksman, V., Noppari, E., Tammela, A., Mäkinen, M., & Ollikainen, V. (2007). Mobile TV in Everyday Life Contexts - Individual Entertainment or Shared Experiences? *EuroITV 2007*, (pp. 215-225), Springer LNCS.

Open IPTV Forum (2008). *Open IPTV Forum Functional Architecture V1.1*. Retrieved from http://www.openiptvforum.org/downloads.html

Orwell, G. (1990). *Nineteen Eighty-Four*. London: Penguin Books.

Oumard, M., Mirza, D., Kroy, J., & Chorianopoulos, K. (2008). A Cultural Probes Study on Video Sharing and Social Communication on the Internet. *In Proc. of DIMEA 2008*. ACM press.

OWL (2008). *The OWL DL web ontology language*. W3C, http://www.w3.org/TR/owl-features

Pasquier, D., Buzzi, C., d'Haenens, L., & Sjöberg, U. (1998). Family Lifestyles and Media Use Patterns: An Analysis of Domestic Media Among Flemish, French, Italian and Swedish Children and Teenagers. *European Journal of Communication, 13*(4)

Patel, D., & Agamanolis, S. (2003). Habitat: awareness of life rhythms over a distance using networked furniture. *Adjunct Proceedings, UbiComp 2003*.

Pea, R., Mills, M., Rosen, J., Dauber, K., Effelsberg, W., & Hoffert, E. (2004). The DIVER Project: Interactive Digital Video Repurposing. *IEEE Multimedia, 11*(1), 54-61.

Peng, C. (2002). *Digital Television Applications.* Doctoral dissertation, University of Technology, Helsinki. Retrieved July 7, 2005 from http://lib.tkk.fi/Diss/2002/isbn9512261723/isbn9512261723.pdf

Perry, T. S. (1988). The Longest Survivor Loses Its Grip. *Spectrum, 25*(8), 16–20.

Persson, P. (2003). Exms: an animated and avatar-based messaging system for expressive peer communication. *In GROUP '03: Proceedings of the 2003 international ACM SIGGROUP conference on Supporting group work*, (pp. 31–39), New York, NY, USA: ACM Press.

Pimentel, M. G. C., Cattelan, R. G., Melo, E. L., & Texeira, C. A. C. (2008). End-User Editing of Interactive Multimedia Documents. In *Proceedings of ACM DocEng* (pp. 298-301).

Platow, M. J., Haslam, S. A., Both, A., Chew, I., Cuddon, M., Goharpey, N., et al. (2005). "It's not funny if they're laughing": Self-categorization, social influence, and responses to canned laughter. *Journal of Experimental Social Psychology, 41*(5), 542-550. doi: 10.1016/j.jesp.2004.09.005.

Ponte, C. (2007, May). *Web access of children: Possibilities of internet providers to promote media literacy of children.* Paper presented at the EU 2007 –Expert Conference on European Media Policy—Leipzig, Germany. Retrieved June 8, 2007, from http://www.fcsh.unl.pt/eukidsonline/docs/LeipzigCPONTE.pdf

Ponte, C., & Candeias, C. (2007). Crianças e Internet: que acesso e usos? Que potencialidades e que riscos dessa relação?. Anuário *da Comunicação 2005-2006*. Lisboa: Obercom. Retrieved August 25, 2007, from http://www.obercom.pt/client/?newsId=342&fileName=anuario_2005_2006.pdf

Postman, N. (1986). *Amusing ourselves to death: Public discourse in the age of show business.* New York, NY: Penguin.

Pouwelse, J. A., Garbacki, P., Epema, D. H. J., & Sips, H. J. (2005, February 24-25). *The BitTorrent p2p file-sharing system: Measurements and analysis.* Paper presented at the 4th International Workshop on Peer-to-Peer Systems (IPTPS'05), Ithaca, New York, USA.

Pouwelse, J. A., Garbacki, P., Wang, J., Bakker, A., Yang, J., Iosup, A., Epema, D. H. J., Reinders, M. J. T., van Steen, M. R., & Sips, H. J. (2007). Tribler: A Social-Based Peer-to-Peer System. *Concurrency and Computation: Practice and Experience, 20*(2), (pp. 127-138).

Powazek, D. (2001). *Design for Community: The Art of Connecting Real People in Virtual Places* (p. 307). New Riders Publishing. Retrieved May 28, 2008, from http://portal.acm.org/citation.cfm?id=516336.

Powazek, D. M. (2002). Design for community: The art of connecting real people in virtual places. Indianapolis, IN: New Riders Publishing.

Preece, J. (2000). *Online Communities: Designing Usability, Supporting Sociability.* Chichester, UK: John Wiley & Sons.

Preece, J., & Maloney-Krichmar, D. (2003). Online Communities: focusing on sociability and usability. In A. J. Jacko & A. Sears (Eds.), *Handbook of Human-Computer Interaction.* London: Lawrence Erlbaum Associates Inc.

Preece, J., Rogers, Y., & Sharp, H. (Ed.) (2002). *Interaction Design: beyond human-computer interaction.* New York: John Wiley & Sons, Inc.

Putnam, R. D. (1995). Tuning in, tuning out: The strange disappearance of social capital in America. *PS: Political Science and Politics, 28*(4), 664-683.

Putnam. R. (2000). *Bowling Alone: The Collapse and Revival of American Community.* Simon & Schuster.

Quico, C. (2008). *Audiências dos 12 aos 18 anos no contexto da convergência dos media em Portugal: emergência de uma cultura participativa?.* Unpublished doctoral dissertation. Universidade Nova de Lisboa—Faculdade de Ciências Sociais e Humanas.

Rasmussen, T. A. (2006). The Sociability of Interactive Television. *Proceedings of the 4th EuroITV Conference*, Athens.

Rauschenbach, U., Putz, W., Wolf, P., Mies, R., & Stoll, G. (2004). *A Scalable Interactive TV Service Supporting Synchronized Delivery over Broadcast and Broadband Networks, International Broadcasting Conference 2004*, Amsterdam, the Netherlands.

RDF (2008). *Resource Description Framework.* W3C, http://www.w3.org/RDF

Reeves, B., Lang, A., Kim, E., & Tartar, D. (1999). The effects of screen size and message content on attention and arousal. *Media Psychology, 1*, 49-68.

Regan, T., & Todd, I. (2004). Media center buddies: instant messaging around a media center. In A. Hyrskykari (Ed.), *Proceedings of the Third Nordic Conference on Human-Computer interaction* (Tampere, Finland, October 23 - 27, 2004). NordiCHI '04, vol. 82 (pp. 141-144). New York: ACM Press.

Regan, T., & Todd, I. (2004). *Media Center Buddies: Instant Messaging Around a Media Center* (Technical Report MSR-TR-2004-47). Redmond, WA: Microsoft Research. Retrieved September 5, 2008, from ftp://ftp.research.microsoft.com/pub/tr/TR-2004-47.pdf

Rehn, A. (2001). *Electronic Potlatch: A study on new technologies and primitive economic behaviors.* Stockholm.

Reimers, U. (2006). DVB – The Family of International Standards for Digital Video Broadcasting. *Proceedings of the IEEE, 94*(1), 173-182

Rein, L., & Winningham, T. (2008). *Video metadata model.* The Microformats Project. Retrieved May, 2008, from http://microformats.org/wiki/video-metadata-model.

Resnick, P. (2001). Beyond Bowling Together: SocioTechnical Capital. In J. M. Carroll (Ed.), *Human-Computer Interaction in the New Millennium* (pp. 647-672). New York: Addison-Wesley.

Resnick, P., & Varian, H. R. (1997). Recommender Systems. *Communications of the ACM, 40*(3), 56-58.

Ribeiro, S. (2007). Current TV: a televisão feita por todos está na berra. *Jornal Público.* Retrieved February 29, 2008, from http://ultimahora.publico.clix.pt/noticia.aspx?id=1307974

Ripeanu, M., Mowbray, M., Andrade, N., & Lima, A. (2006). Gifting technologies: A BitTorrent case study. *First Monday, 11.*

Robida, A. (2004). *The Twentieth Century.* In P. Willems (Trans.), A. B. Evans, (Ed.). Middletown, CT: Wesleyan University Press. (Original work published 1882).

Roman, M., Hess, C., Cerqueira, R., Ranganathan, A., Campbell, R. H., & Nahrstedt, K. (2002). A Middleware Infrastructure for Active Spaces. *IEEE Pervasive Computing, 1*(4), 74-83.

Rosenberg, J., Schulzrinne, H., Camarillo, G., Johnston, A., Peterson, J., Sparks, R., Handley, M., & Schooler, E. (2002). *SIP: Session Initiation Protocol - RFC 3261*, http://www.ietf.org/rfc/rfc3261.txt: The Internet Society. (Accessed 11/7/2008).

Rosson, M. B., & Carroll, J. M., (2002). *Usability Engineering. Scenario-Based Development of Human-Computer Interaction.* San Francisco: Morgan Kaufmann Publishers.

Rubin, A. (1983). Television uses and gratifications: The interaction of viewing patterns and motivations. *Journal of Broadcasting & Electronic Media, 27*(1), 37–51.

Ryan, R. M., & Deci, E. L. (2000). Self-Determination Theory and the Facilitation of Intrinsic Motivation, Social Development, and Well-Being. *American Psychologist, 55*(1), 68-78.

Saint-Andre, P. (2004). *Extensible Messaging and Presence Protocol (XMPP): Core - RFC 3920*, http://www.ietf.org/rfc/rfc3920.txt: The Internet Society. (Accessed 11/7/2008).

Salem, B., & Earle, N. (2000). Designing a non-verbal language for expressive avatars. *In CVE '00: Proceedings of the third international conference on Collaborative virtual environments* (pp. 93–101), New York, NY, USA: ACM Press.

Sanders, E. (2002). *Ethnography in NPD Research. How Applied Ethnography can improve your NPD Research Process*. Retrieved October 9, 2005, from http://www.pdma.org/ visions/apr02/applied.html.

Sassenberg, K. (2002). Common bond and common identity groups on the Internet: Attachment and normative behavior in on-topic and off-topic chats. *Group Dynamics: Theory, Research, and Practice, 6*(1), 27-37.

Schatz, R., & Egger, S. (2008). Social interaction features for mobile TV services. In *Proceedings of IEEE Broadband Multimedia Systems and Broadcast Symposium 2008*, (pp. 1-6).

Schatz, R., Baillie, L., Fröhlich, P., & Egger, S. (2008). Getting the Couch Potato to Engage in Conversation: Social TV in a Converging Media Environment. *EuroITV 2008 Workshop on Sharing Content and Experiences with Social Interactive Television*, Salzburg, Austria.

Schatz, R., Wagner, S., Egger, S., & Jordan, N. (2007). Mobile TV Becomes Social: Integrating Content with Communications. In V. Luzar-Stiffler & V .H. Dobric (Eds.), *Proceedings of the ITI 2007 29th International Conference on Information Technology Interfaces* (pp. 263–270). Zagreb, Croatia: SRCE University of Zagreb.

Schmidt, A., & Van Laerhoven, K. (2001). How to Build Smart Appliances? *IEEE Personal Communications, 8*(4), 66-71.

Schroder, K., Drotner, K., Kilne, S., & Murray, C. (2003). *Researching Audiences*. London: Arnold.

Schulzrinne, H., Casner, S., Frederick, R., & Jacobson, V. (1993). *RTP: A Transport Protocol for Real-Time Applications - RFC 1889*, http://www.ietf.org/rfc/rfc1889.txt: The Internet Engineering Task Force. (Accessed 2/10/2006).

Seiter, E. (2004). *Qualitative Audience* Research. In Allen, R. C. & Hill, A. (Ed.) The Television Studies Reader. London: Rouledge.

Senot, C., Toms, Y., Aghasaryan, A., & Betgé-Brezetz, S. (2007). Multi-Platform User and Usage Profiling Demonstration: Video-on-Demand Services in IPTV and Mobile Video environments. *Interactive TV: A Shared Experience: TICSP Adjunct Proceedings of EuroITV 2007*, Amsterdam, Netherlands.

Shamma, D. A., Bastea-Forte, M., Joubert, N., & Liu, Y. (2008). Enhancing online personal connections through the synchronized sharing of online video. In *Proceedings of ACM CHI 2008 Conference on Human Factors in Computing Systems*, April 5-10, 2008 (pp. 2931-2936). New York: ACM Press.

Shamma, D. A., Shaw, R., Shafton, P. L., & Liu, Y. (2007). Watch What I Watch: Using Community Activity to Understand Content, In *Proceedings of the International Workshop on Multimedia Information Retrieval* (pp. 275-284).

Shneiderman, B., & Plaisant, C. (2005). *Designing the User Interface: Strategies for Effective Human-Computer Interaction. (4th ed.)*. University of Maryland, College Park: Pearson/Addison-Wesley

Simmel, G. (1949). The sociology of sociability. *American Journal of Sociology, 55*. 254-261

Skinner, B. F. (1953). *Science and Human Behaviour* (1st ed.). New York: The Macmillan Company.

Smith, B., & McClave, P. (2001). Similarity vs. Diversity. In D. W. Aha & I. Watson (Eds.), *Proc. of the ICCBR 2001, LNAI 2080*, (pp. 347-361).

Smith, M. A., Farnham, S. D., & Drucker, S. M. (2000). The social life of small graphical chat spaces. *In CHI '00: Proceedings of the SIGCHI conference on Human factors in computing systems*, (pp. 462–469), New York, NY, USA: ACM Press.

Södergård, C. (2003). *Mobile television - technology and user experiences Report on the Mobile-TV project (Rep. No. P506) VTT Information Technology.*

Spinuzzi, C. (2005). The methodology of participatory design. *In Technical Communication, 52*(2).

Stappers, P. J., & Sanders, E. B.-N. (2003). Generative tools for contextmapping: tuning the tools. In *Proc. of the Third International Conference on Design & Emotion*, 2003 in Loughborough.

Steele, C. M. (1988). The psychology of self-affirmation: Sustaining the integrity of the self. In L. Berkowitz (Ed.), *Advances in experimental social psychology, 21*, 261-301. New York: Academic Press.

Steen, M., Kuijt-Evers, L., & Klok, J., (2007). *Early user involvement in research and design projects—A review of methods and practices.*

Stone, D., Jarrett, C., Woodroffe, M., & Minocha, S. (2005). *User Interface Design and Evaluation.* Amsterdam et al.: Morgan Kaufmann Publishers.

Strauss, R. (Ed.) (1997). *Managing Multimedia Projects.* Boston: Focal Press.

Sun Microsystems, I. (2007). *J2SE 5.0.* http://java.sun.com/j2se/1.5.0/. (Accessed 31/7/2008).

Syeda-Mahmood, T., & Ponceleon, D. (2001). Learning video browsing behavior and its application in the generation of video previews. In *Proceedings of the ninth ACM international conference on Multimedia* (pp. 119-128). Ottawa, Canada: ACM. doi: 10.1145/500141.500161.

Tadayoni, R., & Sigurdsson, H. (2006). IPTV Market Development and Regulatory Aspects. *16th Biennal Conference of the International Telecommunications Society*, Beijing, China.

Tajfel, H., & Turner, J. C. (1986). The social identity theory of intergroup behavior. In S. Worchel & W. Austin (Eds.), *Psychology of Intergroup Relations* (pp. 7-24). Chicago: Nelson Hall.

Taylor, A. S., & Harper, R. (2002). Age-Old Practices in the 'New World': a Study of Gift-Giving between Teenage Mobile Phone Users, In *Proceeding of the SIGCHI Conference on Human Factors in Computing Systems* (pp. 439-446).

Taylor, A., & Harper, R. (2003). Inside the Smart Home, chapter Switching on to Switch O (pp. 115–126). London: Springer-Verlag.

Taylor, D. (2007, April). *The digital water cooler.* Retrieved June 9, 2008, from http://www.fabricoffolly.com/2007/04/digital-water-cooler.html, April 2007.

Taylor, K. (2006). Programming video art for urban screens in public space. In *First Monday, Special Issue #4: Urban Screens: Discovering the potential of outdoor screens for urban society.* http://www.firstmonday.org/issues/special11_2/

Television Bureau of Advertising (2006). *Media comparisons study.* Technical report, TBA, New York, NY.

Thomas, P., & Macredie, R. D. (2002). Introduction to the new usability. *ACM Transactions on Computer-Human Interaction, 9*(2)

Tichi, C. (1991). Electronic Hearth: Creating an American Television Culture. New York: Oxford University Press.

Trivers, R. (1971). The Evolution of Reciprocal Altruism. *Quarterly Review of Biology, 46*, 35-57.

Tudor, L. G., Muller, M. J., Dayton, T., & Root, R. W. (1993). A participatory design technique for high-level task analysis, critique, and redesign: The CARD method. In *Proc. HFES'93.*

Tullio, J., & Harboe, G. (2008) Investigating the Use of Voice and Text Chat in a Social Television System. *Changing Television Environments, Proceedings of the 6th European Interactive TV Conference (EuroITV 2008)*, Springer

TV-Anytime (2008). *ETSI TV-Anytime XML Schema. ETSI Technical Specification 102 822.* Retreived May 2008, from http://www.etsi.org/WebSite/Technologies/TVAnytime.aspx

Ursu, M. F., Kegel, I. A., Williams, D., Thomas, M., Mayer, H., Zsombori, V., Tuomola, M. L., Larsson, H., & Wyver, J. (2008). ShapeShifting TV: Interactive Screen Media Narratives. *Multimedia Systems Journal, 14*(2), 115-132.

Utz, S. (2003). Social identification and interpersonal attraction in MUDs. *Swiss Journal of Psychology, 62*(2), 91-101.

van der Heijden, K. (1996). *Scenarios*. The art of strategic conversation. John Wiley & Sons.

Van Rompaey, V., Hemmeryckx-Deleersnijder, B., Van Der Meerssche, B., De Mondt, H., & Godon, M. (2005). Beyond Marketing. Applying qualitative user experience research techniques on social media applications. In *Journal of The Communications Network, 4*(3), 26-30.

Van Rompaey, V., Van Der Meerssche, B., Godon, M., Vanden Abeele, M., & Charliers, K., (2005). Connecting the family home: Co-designing new technologies for Community Communication. In *Proc ECCR/ECA* Conference.

van Setten, M. (2005). Supporting People in Finding Information. *Hybrid recommender systems and goal-based structuring*. Doctoral dissertation, University of Twente, The Netherlands.

Vasconcelos, P., Fava, F., Kampf, T., Schilling, A., & Furtado, E. (2007). Ethnographic Investigational Methodology and Evaluation on Local Television Channel Creation that Allows Interaction with the Community. In A. Lugmayr & P. Gołębiowski (Eds.), *Adjunct Proceedings of EuroITV 2007* (pp. 282–286). Tampere, Finland: Tampere International Center for Signal Processing.

Veinott, E. S., Olson, J., Olson, G. M., & Fu, X. (1999). Video helps remote work: speakers who need to negotiate common ground benefit from seeing each other. *In Proceedings of the SIGCHI Conference on Human Factors in Computing Systems (CHI '99)* (pp. 302-309). ACM Press.

VideoLAN Team. (2008). *VideoLAN - Free and Open Source Software and Video Streaming Solutions for Every OS!*, http://www.videolan.org/: VideoLAN Team. (Accessed 10/7/2008).

Virilio, P. (1997). *Open Sky*. NY: Verso.

Voida, A., Newstetter, W. C., & Mynatt, E. D. (2002). When Conventions Collide: The Tensions of Instant Messaging Attributed. In *Proceedings of the SIGCHI Conference on Human Factors in Computing Systems: Changing Our World, Changing Ourselves* (pp. 187–194). New York: ACM Press.

Vosgerau, J., Wertenbroch, K., & Carmon, Z. (2006). Indeterminacy and live television. *Journal of Consumer Research, 32*, 487-495.

Walster, E., Wlaster, G. W., & Berscheid, E. (1978). *Equity: Theory and research*. Boston: Allyn and Bacon.

Wang, J., Pouwelse, J., Fokker, J., De Vries, A. P., & Reinders, M. J. T. (2008). Personalization on a peer-to-peer television system. *Multimedia Tools and Applications, 36*(1-2), 89-113.

Wasserman, S., & Faust, K. (1994). *Social Network Analysis: Methods and Applications* (p. 825). Cambridge UniversityPress.

Watts, D. J. (2003). *Six Degrees: The Science of a Connected Age* (p. 368). W. W. Norton & Company.

Weiser, M. (1991). The Computer of the 21st Century. *Scientific American, 265*(9), 94-104.

Weiser, M. (*1993). Some Computer Science Issues in Ubiquitous Computing*. Communications of the ACM, 36(7), 75-84.

Weisz, J. D. & Kiesler, S. (2008). How text and audio chat change the online video experience. To appear in the *First International Conference on Designing Interactive User Experiences for TV and Video* (uxTV'08). ACM, New York, NY.

Weisz, J., Kiesler, S., Zhang, H, Ren, Y., Kraut, R., & Konstan, J. (2007). Watching together: integrating text chat with video. In M. Rosson, & D. Gilmore, (Eds.), *Proceedings of the SIGCHI conference on Human factors in computing systems* (pp. 877-886). New York: ACM Press.

Welford, A. T. (1967). Single-channel operation in the brain. *Acta psychologica, 27,* (pp. 5-22). doi: 4863177.

Wells, M. (2004). The TV Revolution Starts Here. *The Guardian*, December 4, 2004, UK.

Westerlund, B., Lindquist, S., Mackay, W., & Sundblad, Y., (2003). Co-designing methods for designing with and for families, In *Proc. for 5th European Academy of Design Conference,* Barcelona.

Wickens, C. D. (2002). Multiple resources and performance prediction. *Theoretical Issues in Ergonomics Science, 3*(2), 159-177.

Wilson, D. S., & Sober, E. (1994). Reintroducing group selection to the human behavioral sciences. *Behavioral and Brain Sciences, 17*(4), 585-654.

Winer, D. (1999). *XML-RPC Specification.* http://www.xmlrpc.com/spec. (Accessed 12/12/2007).

Wolton, D. (Ed.). (1997). *Penser la communication.* Paris: Flammarion.

Zahariadis, P., & Biddle, S. (2000). Goal Orientations and Participation Motives in Physical Education and Sport: Their relationships in English schoolchildren. *Athletic Insight—The Online Journal of Sport Psychology, 2*(1).

Zaslow, J. (2002, November). *If tivo thinks you are gay,here's how to set it straight.* Retrieved June 5, 2008, from WSJ Online: http://online.wsj.com /article email/ SB1038261936872356908.html, November 2002.

Zelenkauskaite, A., & Herring, S.C. (2008). Television-Mediated Conversation: Coherence in Italian iTV SMS Chat. In *Proceedings of the 41st Hawaii International Conference on System Sciences* (pp. 145–155). Los Alamitos, CA: IEEE Computer Society.

Zenith Electronics. (2008). In *Wikipedia, the free encyclopedia.* Retrieved September 5, 2008, from http://en.wikipedia.org/wiki/Zenith_Electronics

About the Contributors

Pablo Cesar is a postdoctoral researcher at CWI (The National Research Institute for Mathematics and Computer Science in the Netherlands) in Amsterdam. He received a Dr. Tech. degree (December 2005) from Helsinki University of Technology (Finland) and a MSc degree (February 2002) from Universidad Politecnica de Madrid (Spain). Pablo Cesar has (co)authored over 40 articles (conference papers and journal articles) about multimedia systems and infrastructures, media sharing, interactive media, multimedia content modelling, and user interaction. He is, as well, involved in standardization activities (e.g., SMIL from W3C). He is co-editor of the TOMCCAP special issue "Human-Centred Television: Directions in Interactive Digital Television Research". In addition, he was the general chair of the 5th European Interactive conference (EuroITV2007) and has given tutorials about Interactive Digital Television in prestigious conferences such as ACM Multimedia and WWW Conferences.

David Geerts has a master in Communication Science at the K.U.Leuven and a master in Culture and Communication at the K.U.Brussel. He leads the Centre for User Experience Research (CUO), part of the Katholieke Universiteit Leuven (Belgium) as well as the Interdisciplinary Institute for Broadband Technology (IBBT), is involved in several research projects on user-centered design and evaluation, and is taking his doctor's degree on Sociability of Interactive Television, with the purpose of developing guidelines and heuristics for designing and evaluating social television interfaces. He organized several workshops and SIGs at CHI2006, CHI2007 and CHI2008 and two workshops on Social Interactive Television at EuroITV2007 and EuroITV2008. David Geerts is co-founder of the Belgian SIGCHI.be chapter and program chair of EuroITV2009, the 7th European Interactive TV Conference.

Konstantinos Chorianopoulos is a Lecturer at the Department of Informatics at the Ionian University, Corfu, Greece. He holds an MEng (Electronics and Computer Engineering, 1999) an MSc. (Marketing and Communication, 2001), and a PhD (Human-Computer Interaction, 2004). During his studies and research, he has been affiliated with engineering, business, and applied arts universities. Since 1997, he has worked in four academic research labs (Greece, UK, Germany), which specialize in the areas of multimedia, e-commerce, intelligent systems and interaction design. He has participated in many EC-funded research projects in the field of human-computer interaction for information, communication and entertainment applications in TV, mobile, and situated computing devices. In 2002, he founded UITV.INFO, which is a newsletter and web portal for interactive television research resources (papers, theses), news and events. He is the main author of more than ten journal papers and he has lectured internationally (conferences, tutorials, seminars, guest lectures) on several aspects (design, engineering, science, art) of interactive TV. He is serving on the steering committee of the European Interactive

TV organization and on the editorial boards of ACM Computers in Entertainment and of the Journal of Virtual Reality and Broadcasting.

* * *

Daniele Abbadessa is an experienced consultant with more than 13 years of experience in the Telecommunication Industry. He received his PhD from the University of Bristol (UK) in 1998. During his career he worked for leading network equipment vendors, Telcos and Consulting companies on a variety of projects on mobile networks, VoIP, service platforms, IP Multimedia Subsystem (IMS) and IPTV. He is also author of several patents. He is working at NEC Europe since April 2003. He is currently leading a few projects in the area of advanced services for IMS-based IPTV and multi-domain IPTV.

Jorge Trinidad Ferraz de Abreu got his graduation and Master's degree in Electronic and Telecommunication by the University of Aveiro, Portugal. After his participation in several European projects he joined the Department of Communication and Arts and concluded is PhD in Sciences and Communication Technologies. Actually he is a lecturer in the undergraduate course of New Communication Technologies, in the Master in Multimedia Communication (where is member of its Scientific Committee) and in the PhD program in Information and Communication on Digital platforms. As a member of the research unit CETAC.MEDIA and the Lab SAPO/UA, he develops his research activities in new media, cross-platform content and Interactive Television with particular interest in the development and evaluation of Social iTV applications. He is also interested in awareness, communication and sharing content applications in sailing scenarios.

Stefan Agamanolis (http://www.agamanolis.com) is the Chief Executive and Research Director of Distance Lab, a creative research institute bringing together technology, design and the arts to overcome the disadvantages of distance. Stefan has over 10 years of experience inside the MIT Media Lab, first as a student and later as one of the founding research directors at Media Lab Europe, its former sister lab in Dublin, Ireland. There he led the Human Connectedness group, an interdisciplinary team that explored the future of human relationships as mediated by technology. His work and that of his researchers has won a number of awards and has been exhibited in diverse venues including Ars Electronica (Austria), the Osaka National Museum of Art (Japan), the Victoria and Albert Museum (UK), and the Wired NextFest (USA). He speaks and consults internationally on a variety of digital media topics. He holds Masters and PhD degrees in Media Arts and Sciences from the Massachusetts Institute of Technology, and a Bachelor of Arts degree in Computer Science from Oberlin College. He was born and raised in the state of Ohio in the United States.

Pedro Alexandre Ferreira dos Santos Almeida got is graduation in New Communication Technologies by the University of Aveiro and is PhD in the same university in Sciences and Communication Technologies. Actually he is a lecturer in the Communication and Arts Department in the undergraduate course of New Communication Technologies, in the Master in Multimedia Communication (which he coordinates) and in the PhD programs in "Information and Communication on Digital Platforms" and "Multimedia in Education". As a member of the research unit CETAC.MEDIA and the Lab SAPO/UA, he develops his research activities in new media, cross-platform content and in communication technologies and social media applied to education and organizational scenarios. He is a member, among others, of the research team responsible for the project Second.UA – the Island of Aveiro University in Second Life.

Brian Amento has been a Principal Technical Staff Member at AT&T Labs – Research for 10 years, working in the Human Computer Interface Research group. Brian received his PhD in Computer Science from Virginia Polytechnic Institute and State University. His research interests include novel interaction techniques, mining implicit social data, and enabling ubiquitous collaboration. Brian's current research on user interfaces for speech technologies is funded by the Department of Defense. He is currently an adjunct faculty member at New York University's Courant Institute of Mathematical Sciences and a research professor at the New Jersey Institute of Technology exploring next generation social mobile computing. He has published over 30 scientific papers.

Regina Bernhaupt is currently on a sabbatical leave at the University of Toulouse III, IRIT to accomplish her so called habilitation. Since 2002 she is working as an assistant professor at the University of Salzburg on usability and user experience evaluation in non-traditional environments. Some of her recent publications include the usage of usability evaluation and user experience evaluation methods for new forms of personalized and individualized interactive TV. Regina Bernhaupt was involved in several (industrial) projects focusing especially on usability evaluation and user experience evaluation in such non-traditional environments like the home, public places, games or even human-robot interaction in industrial settings. She is involved in several scientific conferences in various roles (EuroiTV 2008, ACE 2007, MobileHCI 2007, 2006, Interact 2007) and acts as vice-chair of the IFIP 13.2. working group on user-centered methods.

Erik Boertjes is a scientist at TNO Information and Communication Technology in the Multimedia Technology group in Delft, The Netherlands. He obtained his MSc at the Technical University in Enschede. Before joining TNO in 2003 he worked as a research assistant at the EPFL in Lausanne and at the Free University in Amsterdam. His research interests include Interactive TV Services, and the use of personalization and visualization techniques in future Electronic Program Guides.

Joost Broekens has a PhD from Leiden University, The Netherlands, in artificial intelligence in the area of affective computing. His interests include cognitive modeling, human computer interaction, entertainment computing and artificial intelligence. He has worked at the Telematica Instituut as an AI researcher and currently holds a post-doc position at Delft University of Technology, The Netherlands.

Dick C.A. Bulterman is a senior researcher at CWI in Amsterdam, where since 2004 he heads the Distributed Multimedia Languages and Interfaces theme. From 1988-1994 (and briefly in 2002), he led CWI's Department of Computer Systems and Telematics and from 1994 to 1998, he was head of the Multimedia and Human Computer theme. In 1999, he and two other brave souls started Oratrix Development BV, a CWI spin-off company that transferred the group's SMIL-based GRiNS software to many parts of the civilized world. In 2002, after handing the responsibilities of CEO over to Mario Wildvanck, he returned to CWI and started up a new research activity at CWI on distributed multimedia systems. Prior to joining CWI in 1988, he was on the faculty of the Division of Engineering at Brown, where he was part of the Laboratory for Engineering Man/Machine Systems. Other academic appointments include visiting professorships in computer science at Brown (1993-94) and in the information theory group at TU Delft (1985) and a part-time appointment in computer science at the University of Utrecht (1989-1991). Dr. Bulterman received a PhD in computer science from Brown University (USA) in 1982. He also holds a Sc.M. in computer science from Brown (1977) and a B.A. in economics from Hope College (1973).

Wijnand Derks is Research Scientist at Telematica Instituut, Enschede. He has over ten years of research experience in the areas of telecommunication services and related technologies, including data and process management of telecommunication services at KPN Telecom. Derks holds a PhD in Computer Science from the University of Twente. At Telematica Instituut he has specialized in novel interaction concepts and business models for multi-party services in the entertainment and administrative domain.

Dirk-Jan van Dijk completed his master project on distributed multi agent platforms at TNO Defense and Safety in 2006. He received his Master in Computer Science with Software Engineering as specialization from Delft University of Technology in that same year. From early 2007 until now he has been working as application engineer at Telematica Instituut. He is involved in multiple projects dealing with identity management, mobile technologies and context management.

Hugo ter Doest is Research Scientist at the Telematica Instituut. As a project manager he has been responsible for several innovation projects in the public sector, finance and manufacturing industry. His main focus is on enterprise and business architectures, business modeling and the alignment between these areas. Furthermore, he is head of software development and as such responsible for the development of software prototypes for research projects. He holds a PhD in computer science from the University of Twente, The Netherlands.

Nicolas Ducheneaut is a member of the Research Staff in the Computer Science Laboratory at the Palo Alto Research Center. He joined PARC after completing his PhD at the University of California, Berkeley. His research focuses on software and technologies that support and enhance social interactions. He observes interactions in online environments; builds tools to analyze these interactions further; and designs, deploys and evaluates new social technologies to help people communicate and collaborate with each other. He uses a broad combination of methods in research: "virtual" ethnography (participant observation and in-depth interviews in online spaces), data mining and statistical analysis, and information visualization. His publications have appeared in areas including Human-Computer Interaction, Computer-Supported Cooperative Work, and sociology.
http://www2.parc.com/csl/members/nicolas/

Mirko Fetter is a Ph.D candidate and research and teaching assistant at the CSCW group of the Bauhaus-University Weimar, Faculty of Media. He received a diploma in Computer Science from the University of Applied Science Wuerzburg and is a graduate of the Academy for Communication Kassel (AKK) and the Media Design Academy, Wuerzburg. He worked for Hewlett Packard in Boeblingen, Germany and did research work in the fields of pervasive and location based gaming at the Fraunhofer-Institute for Computer Graphics IGD in Darmstadt, Germany.

Jenneke Fokker is a PhD student at Delft University of Technology within the department of Human Information Communication Design. She holds a Master's Degree in Industrial Design Engineering, specializing in Design for Interaction (2003). During her master's she worked at the design agency Bureau Mijksenaar that specializes in consultancy and the creation of visually oriented information systems. Her work was published in *Else/Where: Mapping. New Cartographies of Networks and Territories*, University of Minnesota Design Institute (2006). In September 2004 she started on a PhD project called *Inducing Human Cooperation in Peer-to-Peer Systems*. Believing that users are the decisive factor for the success of any Peer-to-Peer system, she investigates what social psychological phenomena could stimulate users to engage voluntarily and massively in specific forms of cooperative behavior. She is part of the Delft based Tribler team.

Maxine Glancy is a Behavioural Scientist and User Experience Designer at the BBC (Research and Innovation) in London, England. Her research explores behavioural and design factors that shape the design and use of broadcast content and technologies. This includes subjects such as location-based computing, targeted help & accessibility, public displays, mobile collaborative 'learning-games', user generated content, pervasive media, broadcast entertainment services, and broadcast system design. She has degrees in Psychology, Computing, Design, and Environmental Science.

Daniel Görgen is a Senior Scientist at Philips Research in Eindhoven Netherlands. He joined the Connected Consumer Solutions group of Philips Research December 2006. He holds a PhD in Computer Science from the University of Trier, Germany in the area of routing protocols and middleware support for mobile ad-hoc networks. His current research interests are in the area of ubiquitous computing and networked multimedia.

Robert de Groote studied Computer Science at the University of Twente, the Netherlands. He has worked in both industry and research and has built up experience in various areas including security and artificial intelligence. In 2007 he joined Telematica Instituut, where he works as an Application Engineer with a strong focus on Identity Management.

Tom Gross is professor and chair of Computer-Supported Cooperative Work, and Vicerector (i.e., Vice-President) of the Bauhaus-University Weimar, Germany. His research interests are particularly in the fields of Computer-Supported Cooperative Work, Human-Computer Interaction, and Ubiquitous Computing. In these areas he has published numerous articles in journals, conference proceedings, books and book chapters. And he has been teaching at various universities across Europe. He has participated in and coordinated activities in various national and international research projects. He is an expert member of the IFIP Technical Committee on 'Human Computer Interaction' (TC.13). He has been conference co-chair and organiser of many international conferences (e.g., most recently program co-chair of the INTERACT 2009 conference). Further information can be found at: http://www.tomgross.net.

Rodrigo Laiola Guimarães is Computer Engineer at the Federal University of Espírito Santo, Brazil (2004) and Master in Computer Science at the Pontifical Catholic University of Rio de Janeiro, Brazil (2007). He has been researching multimedia/hypermedia authoring since 2005 and he wrote his master's dissertation (Composer: an Authoring Tool of NCL Documents for Interactive Digital TV) on this topic. He worked in the TeleMídia Laboratory (Pontifical Catholic University of Rio de Janeiro) from February 2005 until August 2007. Currently, he is a PhD student at the National Research Institute for Mathematics and Computer Science in the Netherlands (CWI).

Mark Gülbahar received his Diploma degree in Computer Science from the University of Applied Sciences in Landshut. He is currently working as Principal R&D Engineer within IRT, leading research in the application of metadata aggregation, integration, conversion and distribution. He also undertakes research in the field of interactive television and P2P technology for the broadcasters and the industry market. In the last 6 years, he has worked in a variety of roles within broadcasting and new digital media, including DVB MHP and TV-Anytime standardisation, implementation and software design efforts at IRT. Mark is active within DVB and EBU focus groups relating to Metadata, IPTV, iTV, P2P and IP infrastructures.

Gunnar Harboe is a Senior Research Engineer at Motorola. He received his B.A. from Cambridge University in 2003, where he studied Computer Science, and a Master of Human-Computer Interaction degree from Carnegie Mellon University in 2004. He subsequently joined Motorola Labs as a user researcher. He has been doing research in the field of social television for several years, co-chairing a EuroITV workshop on the subject in 2007 and 2008. He has also explored social uses of music and other media. Currently he is focusing on creating new TV user experiences.

Chris Harrison received his bachelors and masters degrees in computer science from New York University in 2005. He worked at IBM Research and AT&T Labs before starting his PhD in Human-Computer Interaction at Carnegie Mellon University in 2007. Chris' research interests include input methods, display technologies, perceptually augmented user interfaces, and inexact and inattentive interaction techniques.

Cristian Hesselman is a research scientist at the Telematica Instituut, the Netherlands. His research interests include networked multimedia, ubiquitous computing, and the combination thereof. Cristian is currently the coordinator of iNEM4U, an EU-funded research project on enabling technologies for integrated multimedia experiences. Cristian holds a PhD in Computer Science from the University of Twente, the Netherlands. He was with Lucent Technologies from 1996 until 1998.

Elaine M. Huang is a Senior Research Scientist at Motorola Applied Research and Technology Center in the Applications Experiences group. She received a PhD in Computer Science in 2006 from the Georgia Institute of Technology with a focus area of Human-Computer Interaction (HCI). She was the recipient of an Alexander von Humboldt Fellowship and subsequently spent a year as a visiting research fellow at the RWTH Aachen University in the Media Computing Group. Her research interests lie at the intersection of HCI, Ubiquitous Computing, and Computer-Supported Cooperative Work. Recent topics of research include ecological sustainability and technology, groupware applications for large displays, social networking and social exchange, and enriching communication experiences. She has published extensively and participates actively within the international HCI and Ubiquitous Computing research communities.

Jack Jansen is a scientific programmer at CWI, with over 25 years of experience in multimedia and distributed systems. Empowering people to put available technology to a use they themselves envision is his driving principle. This results in activities ranging from languages, such as Python, via web standardization work (SMIL, Rich Web Application Backplane) to implementing systems for accessible and reusable multimedia (Ambulant). His short term plan is to finally obtain his masters degree from the Vrije Universiteit Amsterdam, some time in the coming year.

Ferry de Jong studied Computer Science at the Technical University of Eindhoven, The Netherlands, where he received his Master degree in 1997. He graduated on the possibilities of implementing interactive documents/content. After his study, Ferry joined Logica as a software engineer, mainly in the area of application interfaces. From that, he has grown into the role of technical team leader of several projects. After working hours, Ferry spends most of his free time as composer, orchestrator and performing musician.

Jente Klok works as a consultant/scientist at the User Centered Innovation group of TNO Information and Communication Technology in Groningen, the Netherlands. Before joining TNO in 2003, she was employed in various functions at KPN Research. Her main focus lies on the field of user centered engineering, spanning applications that vary from gaming to interactive tv.

Rogier van Laar is Senior Management Consultant at Logica Management Consulting. He has a broad track record from strategic marketing to business development and real estate. His current expertise is strategy consulting where he identifies relevant economical and social trends, organization specific elements and links them to a solid business model. Rogier is an active blogger on Yellowlab.

Yiming Liu received a Master's degree in Information Management and Systems from the University of California, Berkeley, and a BS in Computer Science from Northwestern University. His research work has focused on the design and analysis of collaborative and social media systems. He is a co-founder of Popcuts, an online music startup that uses monetary and social incentives to encourage digital music purchasing and discovery.

Noel Massey joined Motorola after completing a Master's degree in EE/CS from MIT. Noel spent 11 years in Motorola R&D (which existed under various names prior to becoming Motorola Applied Research and Technology Center) working with text-to-speech systems, avatars, intelligent agents, machine learning systems, and most recently the Driver Advocate project. In 2005 Noel took a sabbatical to launch a local startup company which was being formed to provide technical consulting services to small businesses. Noel spent a year working with business owners to increase profitability primarily through the use of web-centric tools. In July 2006 Noel rejoined Motorola Labs as part of the UCSL where he is creating ideas and prototypes of future enhanced communication experiences in order to help people feel more connected. In addition to his work, Noel also enjoys woodworking, reading, creating advanced motion control systems, and is very happy to be playing with LEGOs again as he tries to build strong connections with his children.

Crysta Metcalf joined Motorola in 2000, after completing her PhD in Business and Organizational Anthropology at Wayne State University under Dr. Marietta Baba and Dr. Elizabeth Briody. Her dissertation, entitled "Cultural influences on the adoption and use of information and communication technologies in organizations," involved field research in 8 different U.S. Air Force divisions. She has a Master's in Cultural Anthropology from Wayne State, and her Bachelor's in Anthropology from the University of South Florida. Her specialty areas include economic anthropology, business and organizational anthropology, and design anthropology. Since joining Motorola, she has worked in applied research on a variety of projects utilizing team-based, trans-disciplinary methods for innovation. The majority of her work involves application design in sociable media and communication technologies. She regularly publishes and speaks about the use of rigorous research techniques as part of the invention process.

Mukesh Nathan is a PhD student in the Department of Computer Science at the University of Minnesota. Before attending the University of Minnesota, Mukesh received his BS and MS in Computer Technology from Bharathiar University, India, in 2000 and 2002 respectively. Following this, he spent 2 years as a software engineer at Novell Inc, working on web services products including Novell's UDDI server. Mukesh's research interests in human computer interaction primarily focus on collaborative systems and location-aware systems. Currently he is working on projects related to social television in collaboration with AT&T Labs, NJ.

Omar Niamut is a researcher at TNO Information and Communication Technology, the largest independent ICT research organisation in The Netherlands. He works on projects related to IPTV networks, services and standardisation. His current focus is on the combination of IMS and IPTV to leverage social TV and service blending, the perceived quality of experience of HD and, as a recording artist, user-generated content. He has masters and PhD degrees in electrical engineering from Delft University of Technology. From 2001 to 2005 he worked in the national and European research projects SiCAS and ARDOR on algorithms for universal audio coding.

Ashley Novak is a Senior Research Engineer within the Applications Experiences group at Motorola's Applied Research and Technology Center. He received his BS in Computer Engineering from Syracuse University in 2006. His expertise is in Interaction/Interface Design and Human-Computer Interaction with specific interests in software & hardware prototyping, building user interfaces, and designing/creating new experiences. Ashley has focused on several areas from context aware mobile applications to improving retail experiences. Recently, his work has centered on creating future experiences for television.

Marianna Obrist is Assistant Professor in the HCI & Usability Unit of the ICT&S Center at the University of Salzburg. She holds a doctoral degree in Communication Science from the University of Salzburg. The focal point of her research lies in user-centered design methods, and in particular the user involvement into the development of new products/systems. She is involved in several research projects concerned with the study of user requirements and the evaluation of user experience and acceptance for iTV, IPTV (i.e. iiTV@home, iTV 4 all, CITIZEN MEDIA). She was part of the organization team for the MobileHCI 2005, ACE 2007, and was conference co-chair for EuroiTV 2008.

Kenton O'Hara is a Principal Scientist at the Commonwealth and Scientific Industrial Research Organisation, in Sydney Australia, where he is Director of the HxI initiative. His research explores the social and behavioural factors that shape the design and use of emerging technologies. Over the years he has looked at a wide variety of topics including social and collaborative aspects of mobile video cosnumption, collaborative music consumption, location-based computing, public and situated displays, pervasive media, local and remote collaboration, context aware computing, and mobile communication. He has worked at key research labs such as HP Labs and Rank Xerox EuroPARC. He has published many articles on these areas of interest as well as the books, "Social and collaborative aspects of new music technology" and "Public and Situated Displays". He has a degree in Psychology and a PhD in Human-Computer Interaction.

Thilo Paul-Stueve is a Ph.D candidate and research and teaching assistant at the CSCW group of the Bauhaus-University Weimar, Faculty of Media. He received a diploma in Computer Science from the University of Bielefeld. Before joining the Bauhaus-University he worked at the Networks and Distributed Systems group and at the Neurobiology group of the University of Bielefeld on multimedia e-learning platforms.

Remco Poortinga studied Electrical Engineering at the University of Twente, The Netherlands, where he received his Master degree in 1997. After working at Ericsson Enschede on DECT systems, he joined the Telematica Instituut in 1999. He has worked in numerous national and European projects in the area of networking and middleware, accounting, mobility management, IPv6 and multimedia IP-based streaming, as well as context management.

Johan Pouwelse is an assistant professor at Delft University of Technology, at the PDS group. He is the technical coordinator of the Tribler project. Tribler is being developed by a team of 18 full-time people. Some time ago Dr. Pouwelse conducted one of the largest measurements of the Bittorrent P2P network. This detailed measurement study ran over a period of two years and discovered many unique properties of this P2P market leader. He delivered a statement for the Federal Trade Commission in Washington, was a visiting scientist at Massachusetts Institute of Technology (MIT), and recently spent a few months at Harvard Business School to study the economic impact of movie downloads on Hollywood.

Célia Quico is a project manager of digital interactive television and multimedia at ZON TV Cabo, researcher and teacher at Universidade Lusófona. During 2006 and 2007, Celia also worked as consultant for the social responsibility project PT Escola for Portugal Telecom, which promoted digital literacy among Portuguese young people. Currently, Celia is a member of the research team "MEDIACULT-PT - Participatory culture and media usage" (2008-2009) from Universidade Lusófona. In September 2008, she obtained her PhD in Communication Sciences by Universidade Nova de Lisboa, with a thesis about young people and their usage of media and information and communication technologies in the current media convergence context in Portugal.

Huib de Ridder is full professor of Informational Ergonomics in the Faculty of Industrial Design Engineering (IDE) at Delft University of Technology, The Netherlands. He holds a MSc degree in Psychology from the University of Amsterdam and a PhD degree from Eindhoven University of Technology. From 1982 till 1998 he has been affiliated with the Institute for Perception Research (IPO), Eindhoven, The Netherlands, where his research focused on both fundamental and applied visual psychophysics. From 1987 till 1992 his research on the fundamentals of image quality metrics was supported by a personal fellowship from the Royal Netherlands Academy of Arts and Sciences (KNAW). In November

1998 he moved to Delft University of Technology where he initiated the research program on Product Intelligence, in particular the subprogram Transparent Interfaces covering topics like user understanding (e.g. intention tracking in user-product interaction, interaction with embedded intelligence) and information presentation (e.g. picture perception, form perception, image quality). He is vice-director of the Delft ICT Research Center and involved in research on ambient intelligence and wireless communication. Current research topics include trust in Peer-to-Peer networks, inducing cooperation in a P2P television system called Tribler, Willingness-to-Share in virtual communities and vision-based gesture recognition.

Guy Romano is a Distinguished Member of the Technical Staff at Motorola's Applied Research and Technology Center (formerly Motorola Labs). Guy has spent over 14 years working in corporate research on a variety of projects that have both technical and user centered aspects. Over the past seven years Guy has led a number of projects, such as Social TV, that enable new experiences by blending content and communication together. Guy holds a Masters degree in Computer Science from the Illinois Institute of Technology, a Bachelors degree in Electrical Engineering from the Illinois Institute of Technology, and an MBA from the Northwestern University Kellogg School of Management.

David Ayman Shamma is a research scientist in the Internet Experiences Group at Yahoo! Research. He is interested in personal expression and how communities disseminate media. His research focuses on media experience and understanding creativity as well as building new creative models and tools. Using models of creativity from his research, Ayman creates media art installations that have been reviewed by The New York Times, International Herald Tribune, and Chicago Magazine and exhibited internationally, at Second City Chicago, SIGGRAPH ETECH, Chicago Improv Festival, Wired NextFest and NextMusic. Ayman holds a BS and a MS in Computer Science from the Institute for Human and Machine Cognition at The University of West Florida and a PhD in Computer Science from the Intelligent Information Laboratory at Northwestern University. Prior to receiving his PhD, he was a visiting research scientist for the Center for Mars Exploration at NASA Ames Research Center.

João da Silva graduated at the University of Aveiro, Portugal, receiving a Diploma degree on Computer and Telematics Engineering in 2007. He joined NEC Europe Ltd. for an internship during his studies and later continued his work as a Software Engineer. J. da Silva is currently working on Identity Management and IPTV.

Sander Smit is Management Consultant at Logica Management Consulting, specialized in Business Innovation. During his career, he has built up extensive experience with the first of a kind development of various CE-devices like interactive set-top boxes and robotic systems and research on new broadband services and concepts. Sander is currently performing research in the area of future business models for the networked multimedia domain.

Martin Snijders is software engineer holding a Bachelor's degree in applied physics as well as computer engineering. After working in the IT sector for some years he joined the Telematica Instituut where he has been involved in numerous projects with regard to computer engineering. His interests are amongst others mobile technologies, middleware, computer aided collaboration and recommender systems.

Martijn Staal (1983) works as innovator and consultant at TNO Information and Communication Technology. He has a deep expertise on the field of market analyses, business cases and marketing on the subject of old and new media, such as digital TV, advertising and User Generated Content. Martijn works for clients such as Dutch Telecom operators, advertisers, publishers, and the European Commission. In 2005 Martijn Staal accomplished his Master degree in Business Informatics at the Free University in Amsterdam.

Loren Terveen is an Associate Professor of Computer Science at the University of Minnesota. Before joining the University of Minnesota, he received his PhD in Computer Sciences from the University of Texas at Austin, then spent 11 years at AT&T Labs / Bell Labs. Terveen's research interests are human-computer interaction and computer-mediated communication. Currently, he is working on projects in two areas. In the area of online communities, he works to apply theories from the social sciences to create new interaction techniques and algorithms that elicit more positive participation from community members. In the area of location-based systems, he is working to create mobile and web-based systems that let individuals and groups organize and access information about geographical places and areas. He has published over 60 scientific papers, holds 10 patents, and has held many leadership positions in his professional community.

Andrew Tokmakoff is a Senior Research Scientist at Philips Research Laboratories Europe, located at the High Tech Campus in Eindhoven. He has extensive experience in the areas of Mobile Services Platforms, MPEG-21 technologies and Systems Architecture, Design and Implementation. He graduated from the University of South Australia with a Bachelor of Engineering in Computer Systems Engineering (with honours) in 1992. He holds a PhD from the University of South Australia, in the area of Formal Methods and Service Brokering in Distributed Systems, which he received in 1998. Andrew currently leads the Smart Networked Environments cluster within the Connected Consumer Solutions Group in the Philips Research Balanced Lifestyle Sector. His current research interests involve the application of emerging networking and sensing technologies towards context-aware ubicomp applications, and next-generation IPTV applications and standards.

Manfred Tscheligi is one of the key persons involved in developing the field of human-computer interaction in Europe, distinguished speaker at conferences, workshops, tutorials and seminars, author of several publications. He has been initiating and managing a broad variety of research and industrial projects. Being a member of various national and international expert, advisory and conference committees (e.g. CHI conference series), his work is based mainly on the interdisciplinary synergy of different fields to enrich the interaction between humans and systems. Since March 1st 2004, Manfred Tscheligi is the first full professor assigned to the new ICT&S Center. Manfred Tscheligi holds a master's degree in business informatics and PhD in social and economic science (specialisation in: applied computer science). Being an associate professor for applied computer science at the University of Vienna, he is very active in the field for more than ten years.

Joe Tullio is a Research Scientist at Motorola Labs. He received his PhD from the Georgia Institute of Technology in 2005. His past research projects range from group support tools to intelligent user interfaces to ambient displays, but share a common theme of promoting communication within groups of friends or colleagues. His work has been published in CHI, UIST, IUI, and Interact. He was recently a co-organizer of the EuroITV 2008 Workshop on Sharing Content and Experiences with Social Interactive Television.

John Vester is a Principal Management Consultant with over 12 years of experience in the telecom industry. He joined Logica in 1995 as a consultant and has worked for various customers in the Telecom domain in a variety of roles (business consultant, test management, project management). Besides his main expertise, the business and process side of CRM & Billing in telecommunications, he is active in research. Especially in Business Modeling for the (cross) media domain.

Justin D. Weisz is a PhD candidate at Carnegie Mellon University, Pittsburgh, PA. He received the BS degree in computer science and the MS degree in computer science from Carnegie Mellon University in 2003 and 2007. His research is focused on designing social online video experiences for audiences of millions of viewers by creating novel summarization, visualization and recommendation systems.

Piet Westendorp holds a PhD in Industrial Design Engineering from Delft University of Technology (2002). Major fields of study are user assistance design and visual communication design. He is an associate professor of Visual Communication Design at Delft University of Technology and an assistant professor of Communication Design at Eindhoven University of Technology. He has been a consultant and trainer for major international companies, such as Philips (Consumer Electronics, Lighting, Medical Systems, Small Domestic Appliances, Telecommunication and Research; Oce (Copiers), Lucent (telecommunication) & the Dutch Defense Department. Dr. Westendorp published various books (such as: Open here: the art of instructional design, New York (NY): Elffers Books (2000) (including translations into German, Spanish, French, Dutch and Japanese); Interface Design & Document Design, Amsterdam: Rodopi (2001), and many articles in the field of visual communication design. He was a general editor of the Information Design Journal (2001-2005) (Amsterdam / Atlanta: John Benjamins Publishing Company, and a guest editor of Visible Language, Rhode Island: Rhode Island School of Design (2007).

Florian Winkler studied Computer Science and Engineering at the University of Mannheim, Germany and received his Diploma degree in 2003. He graduated on disparity estimation in stereo images, a topic of computer vision. During his studies, he worked as an intern at IBM's Network Laboratories in Heidelberg, Germany. In April 2004 F. Winkler joined the Internet Services Group of NEC Europe Ltd. as a Software Engineer working on the development of a Service Integration Environment. Since 2007 F. Winkler is working as a Research Scientist in the area of IPTV. He is an author of several publications, both for scientific conferences and print media.

Xiaoming Zhou is a Senior Scientist at Philips Research in Philips Research Laboratories, Eindhoven, the Netherlands. He is working in the Connected Consumer Solutions group of Philips Research since January 2007. He holds a PhD in Telecommunications from Delft University of Technology, the Netherlands, in the area of Internet performance and analysis. His current research interests are in the area of next-generation IPTV applications and standards.

Index

F

G

H

I

J

K

M

N

O

P

R

S